inelastic
ion–surface
collisions

Academic Press Rapid Manuscript Reproduction

inelastic ion–surface collisions

EDITED BY

N. H. TOLK
J. C. TULLY

Bell Telephone Laboratories
Murray Hill, New Jersey

W. HEILAND

Max Planck Institut für Plasmaphysik
Garching, Germany

C. W. WHITE

Oak Ridge National Laboratory
Oak Ridge, Tennessee

ACADEMIC PRESS, INC. New York San Francisco London 1977
A Subsidiary of Harcourt Brace Jovanovich, Publishers

COPYRIGHT © 1977, BY ACADEMIC PRESS, INC.
ALL RIGHTS RESERVED.
NO PART OF THIS PUBLICATION MAY BE REPRODUCED OR
TRANSMITTED IN ANY FORM OR BY ANY MEANS, ELECTRONIC
OR MECHANICAL, INCLUDING PHOTOCOPY, RECORDING, OR ANY
INFORMATION STORAGE AND RETRIEVAL SYSTEM, WITHOUT
PERMISSION IN WRITING FROM THE PUBLISHER.

ACADEMIC PRESS, INC.
111 Fifth Avenue, New York, New York 10003

United Kingdom Edition published by
ACADEMIC PRESS, INC. (LONDON) LTD.
24/28 Oval Road, London NW1

Library of Congress Cataloging in Publication Data

International Workshop on Inelastic Ion-Surface
    Collisions, Murray Hill, N.J., 1976.
    Inelastic ion-surface collisions.

      Workshop held at the Bell Laboratories in July,
1976.
    Includes index.
      1.   Collisions (Nuclear physics)—Congresses.
2. Ions—Scattering—Congresses. 3. Surfaces
(Physics)—Congresses. I. Tully, John C.
II. Bell Telephone Laboratories, inc. III. Title.
QC794.6.C6I6   1976   539.7'54   77-12122
ISBN 0-12-703550-8

PRINTED IN THE UNITED STATES OF AMERICA

contents

LIST OF CONTRIBUTORS — vii
PREFACE — ix

Low Energy De-excitation and Neutralization
Processes Near Surfaces — 1
   H. D. Hagstrum

Neutralization and Inelastic Energy Losses
in Low-Energy Ion Scattering — 27
   W. Heiland and E. Taglauer

Neutralization Behavior in Medium Energy Ion Scattering — 47
   T. M. Buck

Oscillatory Scattered Ion Yields in Low Energy
Ion-Surface Scattering — 73
   T. W. Rusch and R. L. Erickson

Nonadiabatic Neutralization at Surfaces:
Oscillatory Ion Scattering Intensities — 105
   J. C. Tully and N. H. Tolk

Sputtering Processes: Collision Cascades and Spikes — 121
   P. Sigmund

Secondary Ion Production due to Ion–Surface Bombardment — 153
   K. Wittmaack

Optical Emission from Low-Energy Ion–Surface Collisions — 201
   C. W. White, E. W. Thomas, W. F. van der Weg,
   and N. H. Tolk

Electron Pickup by Fast Ions in Solids — 253
   M. C. Cross

Processes of Charge Exchange into the Continuum of Ionic
Projectiles Interacting with Gases and Solids — 283
   W. Meckbach and R. A. Baragiola

Orientation and Alignment in Beam Tilted-Foil Spectroscopy 309
  H. G. Berry

Optical Polarization in High Energy Ion Surface Scattering
at Grazing Incidence 329
  H. J. Andrä, R. Fröhling, and H. J. Plöhn

INDEX 349

list of contributors

Numbers in parentheses indicate the pages on which authors' contributions begin.

H. J. ANDRÄ (329), Institut für Atom und Festkörperphysik, Freie Universität Berlin, D-1 Berlin, Germany

R. A. BARAGIOLA (283), Comisión Nacional de Energia Atómica, Centro Atómico Bàriloche, 8400 San Carlos de Bàriloche, Argentina

H. G. BERRY (309), Department of Physics, The University of Chicago, Chicago, Illinois 60637

T. M. BUCK (47), Bell Laboratories, Room 1E-448, Murray Hill, New Jersey 07974

M. C. CROSS (253), Bell Laboratories, Room 1D-438, Murray Hill, New Jersey 07974

R. L. ERICKSON (73), Central Research Laboratories, 3M Company, St. Paul, Minnesota 55133

R. FRÖHLING (329), Institut für Atom und Festkörperphysik, Freie Universität Berlin, D-1 Berlin, Germany

H. D. HAGSTRUM (1), Bell Laboratories, Room 1C-326, Murray Hill, New Jersey 07974

W. HEILAND (27), Max Planck Institut für Plasmaphysik, 8046 Garching bei München, Germany

W. MECKBACH (283), Comisión Nacional de Energia Atómica, Centro Atómico Bàriloche, 8400 San Carlos de Bàriloche, Argentina

H. J. PLÖHN (329), Institut für Atom und Festkörperphysik, Freie Universität Berlin, D-1 Berlin, Germany

T. W. RUSCH (73), Central Research Laboratories, 3M Company, St. Paul, Minnesota 55133

P. SIGMUND (121), Physical Laboratory II, 4.C. Orsted Institute, DK-2100 Copenhagen, Denmark

E. TAGLAUER (27), Max Planck Institut für Plasmaphysik, 8046 Garching bei München, Germany

E. W. THOMAS (201), School of Physics, Georgia Institute of Technology, Atlanta, Georgia 30332

N. H. TOLK (105, 201) Bell Laboratories, Room 1E-444, Murray Hill, New Jersey 07974

J. C. TULLY (105), Bell Laboratories, Room 1A-357, Murray Hill, New Jersey 07974

W. F. VAN DER WEG (201), Philips Research Laboratories W.P.B., Eindhoven, The Netherlands

C. W. WHITE (201), Solid State Division, Oak Ridge National Laboratory, Oak Ridge, Tennessee 37830

K. WITTMAACK (153), Gesellschaft für Strahlenforschung, D-8042 Neuherberg, Germany

preface

Inelastic outershell processes occurring in collisions of fast atomic and molecular particles with surfaces have received an extraordinary amount of recent attention. This is due in part, of course, to a need to understand the energy exchange interactions of a confined plasma with the first wall surface, which constitutes one of the most important problems in the effort to achieve controlled thermonuclear fusion, and to the fact that these processes form the basis of a number of surface analysis techniques, for example, ISS, SCANIIR, and SIMS. In addition, this subject has proven to be a fruitful area of fundamental research with substantial overlapping interests in atomic physics, solid state physics, nuclear physics, surface physics, chemical physics, and even space physics.

With the exception of Chapter 6 on sputtering processes by Peter Sigmund, this book is a compilation of major presentations given at the International Workshop on Inelastic Ion–Surface Collisions held July 1976 at Bell Laboratories, Murray Hill, New Jersey. The major emphasis throughout the book is on identifying and studying the underlying physical mechanisms associated with ionization, neutralization, and excitation responsible for the observed sputtering and particle backscattering outershell inelastic collision phenomena. It has been our intent to examine and synthesize information and concepts from a number of disciplines that in the past have been widely separated but are suddenly converging on identical theoretical and in some cases experimental challenges.

In addition to the excellent treatments on the role of the bulk, the surface and the near surface interaction regions on the degree of ionization, neutralization, and excitation, there are a number of contributions on newly observed and in some instances controversial phenomena. These new topics include oscillations in the energy dependence of backscattered ions, wake riding states, and optical polarization effects from beam transmission through tilted foils and from beam particle bombardment of surfaces at grazing incidence. Clearly this area is in its dynamic infancy and holds great promise for equally dynamic and exciting growth in the future.

The editors would like to thank Academic Press for patient encouragement, Walter Brown for conversations held on an airplane returning from Amsterdam, which initiated this project, and the participants in the workshop for their stimulating contributions.

low energy de-excitation and neutralization processes near surfaces

Homer D. Hagstrum
Bell Laboratories

The electronic transitions near solid surfaces that de-excite and neutralize atomic particles carrying potential energy, such as Auger neutralization and the two-stage process of resonance neutralization followed by Auger de-excitation, are discussed. Experimental evidence for atomic energy level variation near surfaces is reviewed and the critical distance separating regions outside the surface where different processes occur is defined. Further evidence for the general validity of the theoretical picture is that provided in the investigation of the processes occurring for the metastably excited $He^+(2s)$ ion.

I. INTRODUCTION

An important facet of particle-solid interaction is the electronic transition process that occurs outside the surface of a solid for an atomic particle moving either toward or away from the solid. Such a process is driven by the circumstance that the atomic particle carries potential energy by virtue of the fact that it is excited or ionized or both. The excitation or ionization of an atom or molecule requires energy that is, in a sense, stored in the particle and can be at least partially reclaimed when the particle is de-excited or neutralized near a solid surface. We shall discuss these processes here in some detail and review experimental evidence that the theoretical picture developed in earlier work is a valid one. Of particular interest is the evidence for energy level variation in the atomic particle near the surface and for a critical distance separating spatial regions where different transition processes occur.

II. TRANSITION PROCESSES NEAR SURFACES

If we consider how an ion, for example, could be neutralized at a solid surface we come up with three possibilities: radiative, resonance and Auger neutralization. The neutralization process in which the neutralization energy is emitted as a photon is a possibility of low probability because its initial state lifetime of 10^{-8} sec is about 10^6 times longer than that for either the resonance or Auger processes.

Resonance tunneling processes are illustrated in the electron energy diagram of Fig. 1.(1) Here the filled portion of the conduction band in the metal is shown shaded. Resonance neutralization of an ion to an excited level of the neutralized atom that lies below the Fermi level is indicated by the electronic transition 1 into the atomic well of an incoming particle outside the surface. Here an electron moves from a filled level in the solid below the Fermi level into the excited level at the same energy. Such energy level equality characterizes the resonance processes. The arrow labelled 2 illustrates the resonance transition of an electron from a filled excited state of a neutral atom into an unfilled level in the metal above the Fermi level.

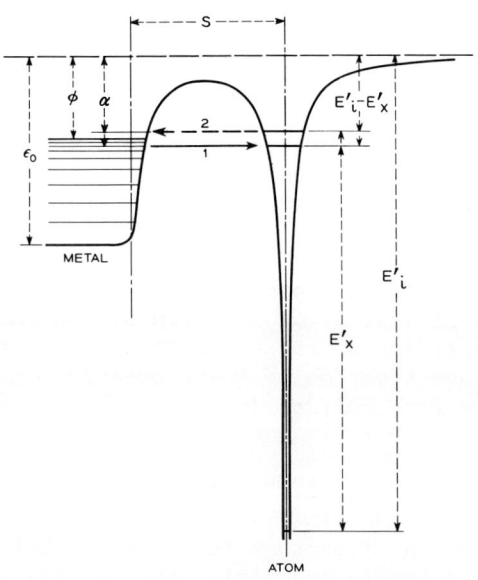

Fig. 1

Figure 1 serves also to define several energy quantities such as the work function of the solid ϕ, which is the energy separation of the Fermi and vacuum levels, and the effective excitation and ionization energies, E'_x and E'_i respectively, of the atom outside the solid. s is the distance of the atom from the solid surface.

The Auger neutralization and de-excitation processes are illustrated, respectively, in Figs. 2 and 3.(1) In the two-electron Auger neutralization process (Fig. 2) one electron tunnels into the ion ground state from the level labelled 1 while a second electron is simultaneously excited from another level labelled 2. The "down" and "up" electrons lose and gain,

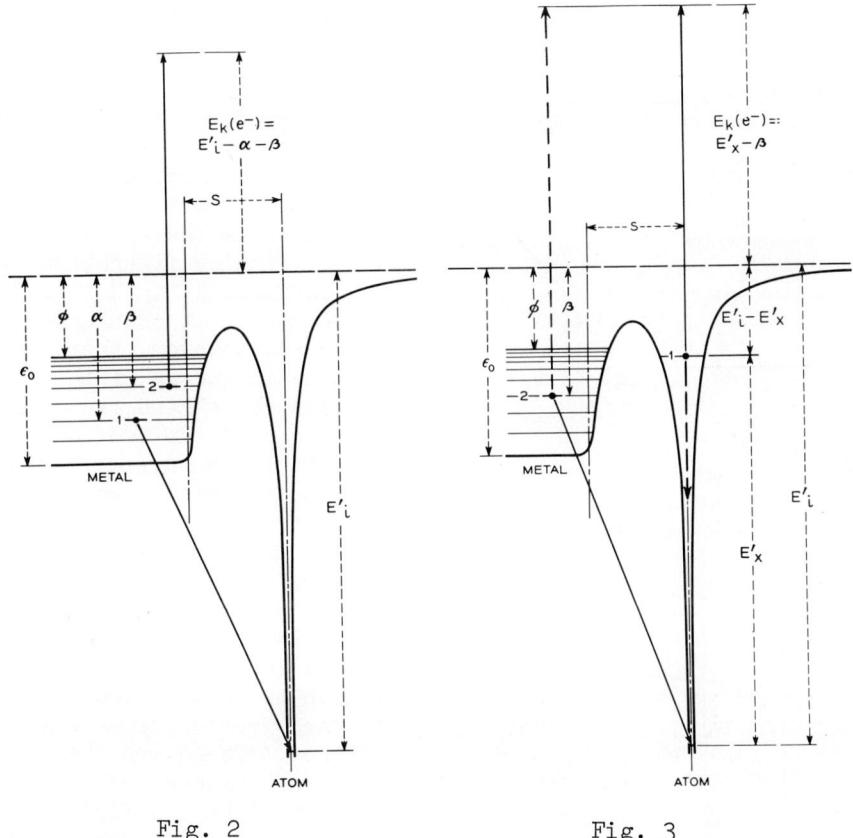

Fig. 2   Fig. 3

respectively, the same amount of energy in this radiationless process. The excited electron may leave the solid and be collected outside if moving with sufficient momentum normal to the surface. In the Auger de-excitation process of Fig. 3 an electron in an excited state 1 is either ejected or drops to the ground state while a metal electron 2 performs the other transition.

III. INTERRELATION OF TRANSITION PROCESSES

The two resonance and two Auger processes just discussed form a "family" of processes in that they are interrelated. This interrelation is conveniently described by the "triangle" diagram of Fig. 4. Here the atomic particle and the electrons, in the solid or free, are indicated in a chemical-like notation. If the system of the X^+ ion and the n electrons resident in the metal, $X^+ + ne_M^-$, is resonance neutralized to

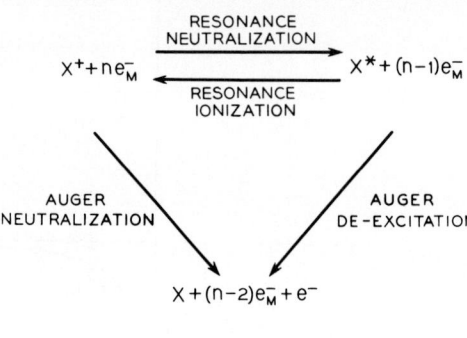

Fig. 4

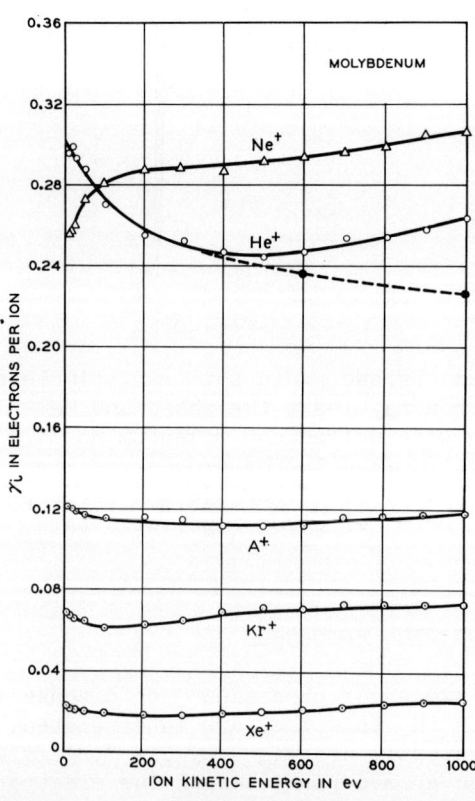

Fig. 5

an excited state, the system $X^* + (n-1)e_M^-$ results. Similarly this latter system can be returned, if energies are right, to the former via resonance ionization. The two Auger processes transform their respective starting configurations to the final state $X + (n-2) e_M^- + e^-$. It should be noted, however, that the kinetic energy of the free electron e^- and the energy levels of the two holes left in the system differ between these Auger processes.

The triangle of Fig. 4 indicates that there are two distinguishable processes by which an ion incident on a solid can be neutralized with the ejection of an electron. These are the direct process of Auger neutralization first discussed by Shekhter[2] and the two-stage process of resonance neutralization followed by Auger de-excitation discussed by Massey[3], Shekhter[2], and Cobas and Lamb[4]. Oliphant and Moon[5] had earlier suggested the possibility of resonance neutralization of an ion to an excited state.

IV. EXPERIMENTAL MEASUREMENTS

Our means of observing the occurrence of an Auger process is the

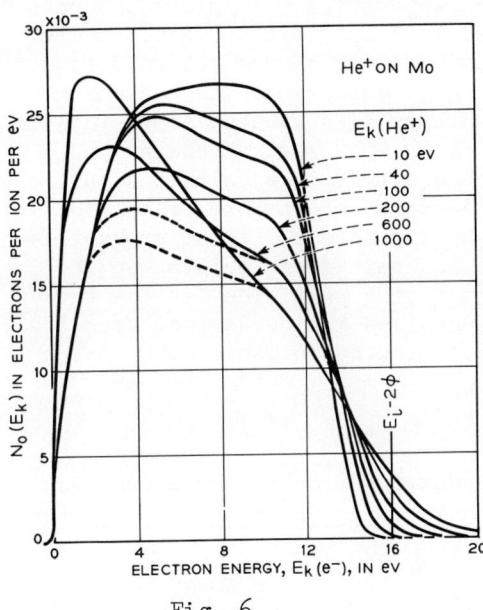

Fig. 6

detection of the electrons ejected from the solid. We can measure the total yield of electrons per incident ion, γ_i, as a function of incident ion energy (Fig. 5) or, what is more informative, the kinetic energy distribution, $N_o(E_k)$, of ejected electrons at different ion kinetic energies (Fig. 6). The data of Figs. 5 and 6 are for an atomically clean, polycrystalline molybdenum surface with incident singly-charged noble gas ions.(6) Evidence in these data for the onset of a process of electron ejection other than Auger neutralization is discussed in Sec. VI, and that enabling us to detect the occurrence of two-stage electron ejection is discussed in Sec. X.

It is clear that the resonance tunneling processes at the top of the triangle of Fig. 4 cannot be detected directly since no free electron is produced by them. When they occur they occur as precursors to Auger ejection processes thereby reducing the number of incident particles participating in a direct process. Resonance neutralization is the possible precursor of Auger de-excitation for incident ions and resonance ionization is the possible precursor of Auger neutralization for incident excited atoms. Whether one or the other of these two-stage processes is in fact possible relative to its companion direct process is determined by the position of the excited level relative to the Fermi level in the metal (Sec. IX). Each two-stage process has both a different γ_i and a different $N_o(E_k)$ from its companion direct process which facts enable us to detect its occurrence (Sec. X).

V. TRANSITION RATE AND PROBABILITY FUNCTIONS

The basic function specifying the rate at which a process occurs is the transition probability per unit time or the so-called transition rate $R_t(s)$. It is a function of the process and of the distance of the particle from the surface. For particles held at a specified separation s it gives the rate at which they become involved in a specific process. There are two probability functions that are convenient to define.(1) These have been discussed by many authors as far back as 1930.(3) They are the probability $P_o(s,v)$ that a particle of normal velocity v will reach s in its original charge and excitation state and the probability $P_t(s,v)ds$ that a particle of normal velocity v will undergo an electronic transition in the distance element ds at the distance s. These probabilities can be defined for incoming particles incident on the surface, P_o^i and P_t^i, and for outgoing particles leaving the surface, P_o^o and P_t^o. These functions may be derived from the basic rate function $R_t(s)$ as indicated in Table 1.

TABLE 1

Derivations of Probability Functions from the Rate Function

INCOMING	OUTGOING
$dP_o^i/ds = (R_t/v)P_o^i$	$dP_o^o/ds = -(R_t/v)P_o^o$
$P_o^i = \exp[-\int_s^\infty (R_t/v)ds]$	$P_o^o = \exp[-\int_0^s (R_t/v)ds]$
$P_t^i = (R_t/v)P_o^i$	$P_t^o = (R_t/v)P_o^o$

For the specific choice of an exponential rate function $R_t(s) = A\exp(-as)$ the resulting P_o and P_t functions are given for incoming particles in Table 2 and for outgoing particles in Table 3. The functions of Tables 2 and 3 are plotted, respectively, in Figs. 7 and 8. Note the quite different forms of these functions. For the incoming particle the P_t^i function is peaked at the separation $s = s_m$ with s_m given, for $R_t(s) = A\exp(-as)$, in Table 2. Thus as v is changed both P_o^i and P_t^i in Fig. 7 move relative to $R_t(s)$. The same is true for the P_o^o and P_t^o functions of Fig. 8. The specific velocity chosen in each case is that which for Fig. 7 makes $as_m = 6$, and for Fig. 8 makes the survival probability at $s = \infty$, $P_o^o(\infty,v) = \exp(-A/av)$, equal 0.1. In the case of an exponential rate function for incoming particles the functions for P_o^i and P_t^i written in terms of $(s-s_m)$ indicate that these two functions do not change form as v is varied but move "bodily" closer to

TABLE 2

Specific Functions for an Incoming Particle

	$s=0$	$s=s_m$	$s=\infty$
$R_t(s) = A\exp(-as)$	A	$A\exp(-as_m)$	0
$P_o^i(s,v) = \exp[-(A/av)\exp(-as)]$ $ = \exp\{-\exp[-a(s-s_m)]\}$	$\exp(-A/av)$	$1/e$	1
$P_t^i(s,v) = (A/v)P_o^i(s,v)\exp(-as)$ $ = aP_o^i(s,v)\exp[-a(s-s_m)]$	$(A/v)P_o^i(0,v)$	a/e	0

$$s_m = (1/a)\ln(A/av)$$

TABLE 3

Specific Functions for an Outgoing Particle

	$A=0$	$s=\infty$
$R_t(s) = A\exp(-as)$	A	0
$P_o^o(s,v) = \exp\{(A/av)[\exp(-as)-1]\}$	1	$\exp(-A/av)$
$P_t^o(s,v) = (A/v)P_o^o(s,v)\exp(-as)$	A/v	0

---

or farther away from the surface as v increases and decreases, respectively.

Figures 7 and 8 illustrate graphically a basic difference between the cases for incoming and outgoing particles. We note that the incoming processes are limited by the $P_t^i(s,v)$ function to distances where $R_t(s)$ is a small fraction (< 0.01) of its value at the surface. On the other hand, for outgoing particles transition processes can occur at all distances from the surface. This circumstance makes the exponential rate function a better approximation for incoming particles than for outgoing particles since $R_t(s)$ is closest to an exponential far from the surface and deviates most from an exponential close to the surface. We shall have occasion to calculate the parameters A/a for the Auger neutralization process in Sec. VII.

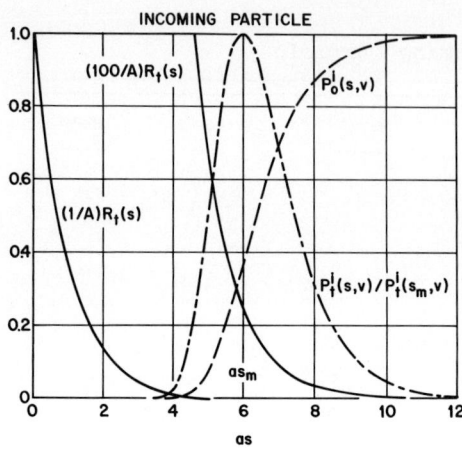

Fig. 7

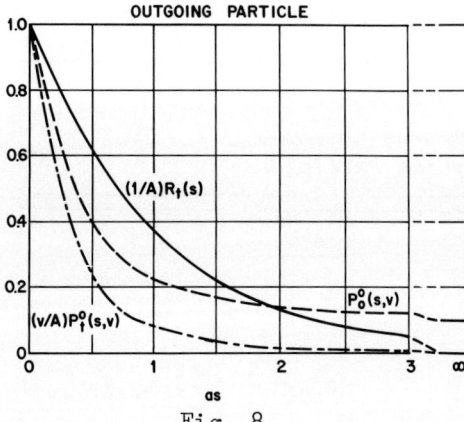

Fig. 8

VI. THE AUGER NEUTRALIZATION PROCESS

Perhaps the most important process of those diagrammed in Fig. 4 is the Auger neutralization process. It is the only process occurring for particle-solid combinations in which the work function ϕ is greater than the effective ionization energy, $E_i' - E_x'$, of the atom's lowest lying excited state near the surface. As we shall see in Sec. X an interesting case develops when these quantities are equal at some distance from the surface where these processes can occur.

An electron energy diagram for the Auger neutralization process is given in Fig. 2. Here we see that the kinetic energy of the electron ejected in a specific process is $E_k(e^-) = E_i' - \alpha - \beta$ when the initial states of the two electrons involved lie α and β below the vacuum level. The maximum of E_k, $(E_k)_{max} = E_i' - 2\phi$, occurs when $\alpha = \beta = \phi$ and both electrons lie initially at the Fermi level of the metal. The minimum kinetic energy is $(E_k)_{min} = E_i' - 2\varepsilon_0$ when both electrons lie initially at the bottom of the conduction band. The unknown quantity among these energies is the effective ionization energy of the parent atom, or the neutralization energy of the ion, at the distance from the surface where the electronic transition occurs. As is expected this energy does not equal its free space counterpart because of atom and ion interactions with the solid.

The energy level variations occurring in an atom near a surface is best depicted by the potential energy diagram. Figure 9 is such a diagram for Auger neutralization.(1) Here the energies of initial and final states are plotted as functions of separation of the atomic particle from the surface.

De-excitation and Neutralization Processes   9

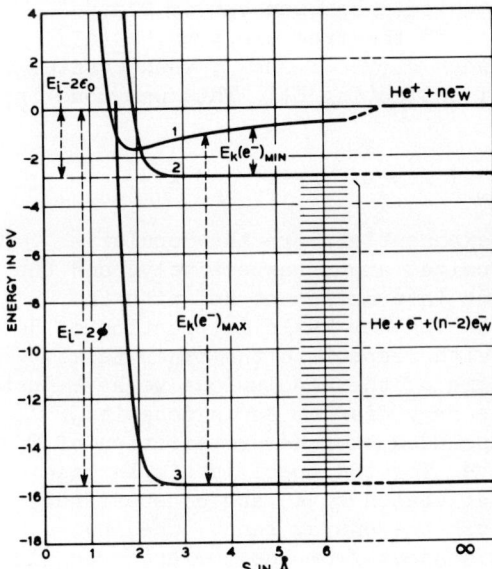

Fig. 9

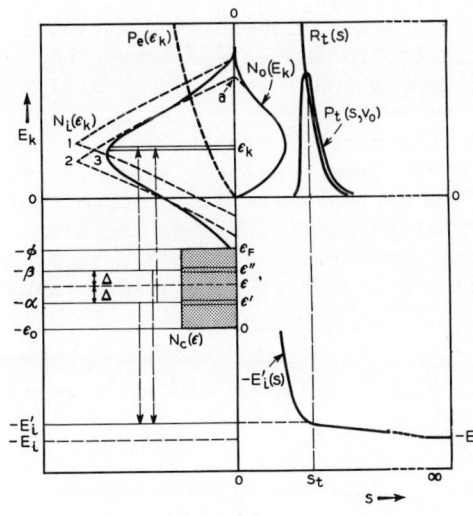

Fig. 10

The initial state, $He^+ + ne_{\overline{W}}$, specified as a He^+ ion plus n electrons in the metal here taken as tungsten W, varies at larger distances by virtue of the image force interaction, $-3.6(eV)/s(\overset{o}{A})$. Closer to the surface a repulsive interaction sets in. We do not know exactly where such interactions occur but those shown in Fig. 9 are estimates based on atom-atom interactions and particle diameters.(1) There is a band of final states in Fig. 9, $He + e^- + (n-2)e_{\overline{W}}$ corresponding to the removal of the 2 electrons from positions in the conduction band varying from ϕ to ε_0 below the vacuum level. This results in a band of final states $2(\varepsilon_0 - \phi)$ in width. The final state atom He interacts with the solid via a van der Waals interaction whose magnitude is so small as to be invisible on Fig. 9.

We expect electronic transitions between potential curves for initial and final states to obey the Franck-Condon principle. This states that during the electronic transition neither position nor momentum of the nucleus of the atomic particle can change appreciably. This means that transitions, as for diatomic molecules, must occur on the potential diagram vertically and that the kinetic energy of He after the electronic transition equals that of He^+ before the transition. From this fact we are forced to conclude that the energy distance between curve 1 and a curve in the manifold between curves 2 and 3 must equal the kinetic energy of the free electron, $E_k(e^-)$. $E_k(e^-)_{max}$, the distance between curves 1 and 3, when equated to $E_i' - 2\phi$ yields a means of estimating E_i'. This was done in Ref.[1] with the result:

$$E_i - E_i' = B_n exp(-b_n s) - B_i exp(-b_i s) + 3.6/s, \quad (1)$$

where the first and second exponentials are the repulsive terms for the neutral and ionized atom, respectively, and the third term is the image force interaction of He^+ with the solid in eV with s in Å. The van der Waals interaction of He with the solid is neglible with respect to the other terms.

We may now translate some of the conclusions we have just drawn back to the electron energy diagram as is done in Fig. 10.(1) Here we have included the energy variation of E_i' and energy distribution curves for the specific simple case of a constant density of initial states $N_c(\epsilon)$ and constant Auger transition probability through the energy band. The Auger transitions take place at distances from the surface specified by the $P_t(s,v_o)$ function, shown here with its maximum at s_t, the distance at which the transition most probably occurs.

If we neglect energy level change and energy broadening we obtain curve 1 for $N_i(\epsilon_k)$, the kinetic energy distribution of electrons inside the solid before any have crossed the surface barrier. $N_i(\epsilon_k)$ is proportional to the probability for producing electrons in the energy range $d\epsilon_k$ at ϵ_k. For the elemental process shown this must be proportional to the product of the initial state electron densities at $\epsilon = \epsilon'$ and ϵ'', that is, to $N_c(\epsilon')N_c(\epsilon'') = N_c(\epsilon+\Delta)N_c(\epsilon-\Delta)$. We note that there is an infinite set of paired initial states yielding electrons at ϵ_k, namely those symmetrically disposed $\pm\Delta$ from ϵ, the halfway point between the levels at ϵ_k and $-E_i$. Thus:

$$N_i(\epsilon) \propto \int_{-(\epsilon - \epsilon_F)}^{(\epsilon - \epsilon_F)} N_c(\epsilon + \Delta)N_c(\epsilon - \Delta) d\Delta, \quad (2)$$

with $N_i(\epsilon_k)$ obtained from $N_i(\epsilon)$ using the transformation:

$$\epsilon_k = E_i' - \epsilon_o + 2\epsilon. \quad (3)$$

Equation (3) results from equating the magnitudes of the transitions of the down and up electrons. Since $E_k = \epsilon_k - \epsilon_o$ we may also write:

$$E_k = E_i' - 2(\epsilon_o - \epsilon). \quad (4)$$

The function $N_c(\varepsilon)$, which is constant between $\varepsilon = 0$ and ε_F and is zero outside these limits, folds to a triangle of base width $2\varepsilon_F$. Curve 1 for $N_i(\varepsilon_k)$ in Fig. 10 is slightly distorted from a simple equilateral triangle by the variation of the free-electron-like density of final states which enters into the transition probability.(1)

If we now use E'_i at $s = s_t$ instead of E_i we obtain curve 2 for $N_i(\varepsilon_k)$. It lies lower in energy than curve 1 because of the reduced ion neutralization energy. Finally, if we broaden curve 2 by convolution with a Lorentzian or a Gaussian to account approximately for the energy broadening inherent in the Auger process(7) we obtain curve 3 as the internal energy distribution of Auger electrons. $P_e(\varepsilon_k)$ is the probability of electron escape over the surface barrier and $N_o(E_k)$ is the external distribution in kinetic energy measurable outside the solid.(Fig. 6) The total yield γ_i is, of course,

$$\gamma_i = \int_0^\infty N_o(E_k) dE_k. \qquad (5)$$

VII. EVIDENCE FOR ENERGY LEVEL VARIATION

The simple theory underlying Fig. 10 predicts that, as E'_i is reduced, characteristic features of $N_i(\varepsilon_k)$ that are reflected in the measured $N_o(E_k)$ will move to lower E_k, that is, will appear closer to the vacuum level, and that the total yield γ_i will decrease. Data for He^+ ions of varying kinetic energy incident on a germanium surface in Fig. 11 clearly show that the higher energy peak, a feature resulting from the form of $N_c(\varepsilon)$, moves to lower energy as $E_k(He^+)$ is increased.(8) We note also in Fig. 11 the greater energy broadening as $E_k(He^+)$ increases.

That total yield decreases with increasing incident ion energy is seen in the He^+ data of Fig. 5 in the range $0 < E_k(He^+) < 400$ eV.(6) The rise in γ_i for ion energies above this range is clearly seen from Fig. 6 to be attributable to the onset of another ejection mechanism.(6)

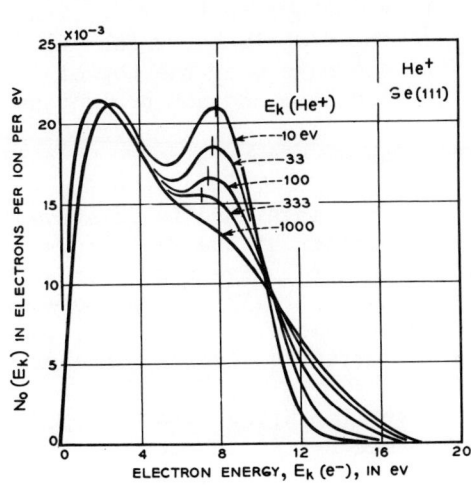

Fig. 11

For $E_k(He^+) > 400$ eV a new component of $N_o(E_k)$ is appearing made up of slower electrons than those in the Auger distributions. This is, in fact, the onset of so-called kinetic ejection in which electrons are ejected by the kinetic energy of the incident particle. If the Auger component of the distributions is extrapolated to the distributions for $E_k(He^+)$ = 600 and 1000 eV as suggested by the dashed lines and the Auger yields determined, we obtain the dashed-line extrapolation of γ_i shown on Fig. 5. The expected downward trend of γ_i for the Auger component is thus confirmed. The anomalous behavior of Ne^+ (Fig. 5) will be discussed in Sec. XI.

A second and more quantitative determination of energy level shift is obtained by fitting the measured kinetic energy distribution of electrons ejected by He^+ incident on Ge(111) starting from a reasonable valence band density $N_V(\epsilon)$ and including the known transition probability factors.(8) As seen in Fig. 12 we start from an $N_V(\epsilon)$ made up of four parabolas to simulate the s-p band between $\epsilon = 0$ and ϵ_V as well as the degenerate p band between the energies $\epsilon = p\epsilon_V$ and ϵ_V. As parameters we have p in the expression $\epsilon_V - p\epsilon_V$, the width of the degenerate p band, and r in the expression $(1-r)(\epsilon/\epsilon_V)N_i(\epsilon)$ which decreases the transition probability as one moves up in the band. This effect results from the fact that the p orbitals, which predominate near the top of the band, do not project as strongly from the surface as do s orbitals of the same energy. Thus p wave function magnitude is relatively smaller at the ion than s wave function magnitude resulting in a relatively smaller transition probability for electrons initially in a p state as compared to an s state. Two other parameters are $P_e(E_k)$, the probability of escape, and E_i^*, the effective ionization energy near the surface.

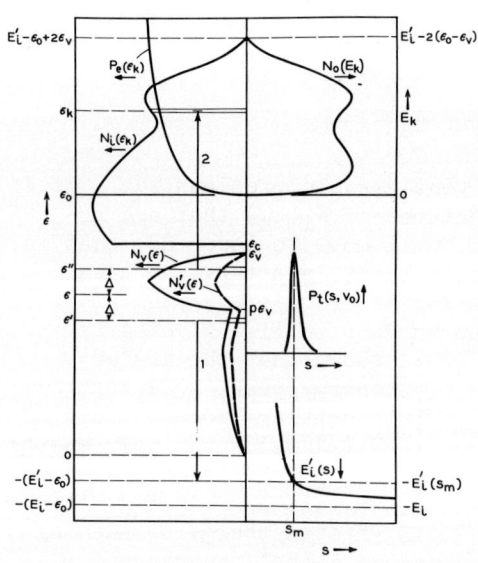

Fig. 12

The procedure was then to determine $N_i(\epsilon_k)$ from Eqs. (2) and (3) and to multiply by an assumed $P_e(\epsilon_k)$ to get $N_o(E_k)$. Several such theoretical $N_o(E_k)$ are shown in Fig. 13 where they are compared with experimental data points. Curve 1

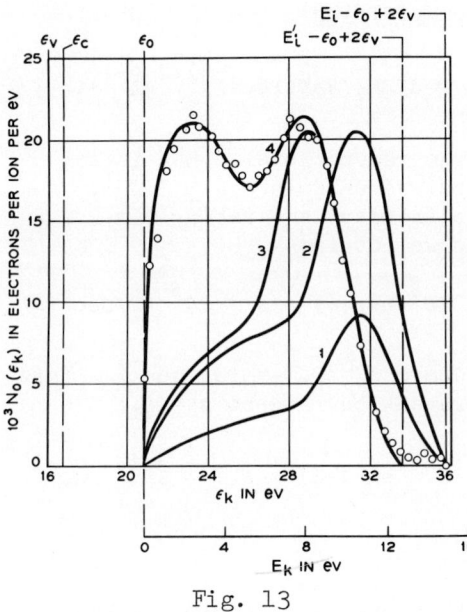

Fig. 13

corresponds to using $E_i'' = E_i$, the free-space ionization potential, and a $P_e(\varepsilon_k)$ calculated from an isotropic distribution of excited electrons. We see that it is too small and predicts faster electrons than are actually observed. Curve 2 is obtained by increasing $P_e(\varepsilon_k)$ appropriately and curve 3 by, in addition, using $E_i'' = E_i - 2$ eV. Finally, curve 4, in good agreement with the data points, is achieved by adjusting the parameter r discussed above. The important thing to note about this procedure is that the parameters are basically orthogonal to one another and that there is no way to bring coincidence between the high energy sides of the theoretical and experimental distributions without making $E_i'' = E_i - 2$ eV. Although $E_i - E_i''$ could depend on the specific particle-solid combination, 2 eV has turned out to be a remarkably universal figure for, say, 10 eV He^+ ions on a variety of solid surfaces.

VIII. TRANSITION RATE PARAMETERS FOR AUGER NEUTRALIZATION

The determination of the change in neutralization energy, ΔE_i, between infinite separation and that separation at which Auger neutralization takes place enables us to calculate approximate values for the parameters A and a in the expression $R_t(s) = A\exp(-as)$ for the transition rate of the process. If we equate $\Delta E_i = -2$ eV to the image force interaction at the separation s_m corresponding to the maximum of the $P_t^i(s,v)$ function we obtain

$$\Delta E_i = -2 \text{ eV} = -3.6/s_m(\overset{\circ}{A}); \quad s_m \overset{\sim}{=} 2 \overset{\circ}{A}. \quad (6)$$

The parameter a is determined by the rate of fall off of the wave functions outside the surface. If, at larger distances from the surface the wave function has the form $\psi \propto \exp(-\lambda s)$, then $a = 2\lambda$ because $R_t(s)$ involves the square of the wave function. λ is expressible in terms of the ionization energy,

E_i, of the wave function state, yielding:

$$a = 2\lambda = 2[2mE_i/\hbar^2]^{1/2} = 4[2E_i(Hartrees)]^{1/2} (\overset{\circ}{A}^{-1}). \quad (7)$$

For a state lying 7 eV below the vacuum level, that is for which E_i = 7 eV $\cong$ 0.25 Hartrees, we calculate a $\sim$ 3 $\overset{\circ}{A}^{-1}$. Using the expression $s_m = (1/a)\ln(A/av)$ and taking $v = 2.2 \times 10^6$ cm/sec for a 10 eV He$^+$ ion we obtain:

$$A/a = v \exp(as_m) = 2.2 \times 10^6 \exp(6) \cong 9 \times 10^8 \text{ cm/sec.} \quad (8)$$

With $a \cong 3 \overset{\circ}{A}^{-1}$ as calculated above $A \cong 2.7 \times 10^7$ sec^{-1}. This leads to an estimate of the transition rate at $s = s_m$:

$$R_t(s_m) = A \exp(-as_m) = 2.7 \times 10^7 \exp(-6) \cong 6.7 \times 10^{14} \text{ sec}^{-1}.$$

$$(9)$$

This is a rate close to that expected for a radiationless Auger process and to the value quoted in Ref. 1. In Table X of Ref. 1 theoretical values for A/a by Shekhter(2) and Cobas and Lamb(4) are given for the Auger and resonance processes. In general the experimental estimates are considerably larger than the theoretical. The theoretical estimates for s_m appear to be unrealistic, however.

Estimates of A/a have been obtained by others using the expression $P_t^0(\infty,v) = \exp(-A/av)$ for the survival probability at infinity of an excited or ionized particle leaving the solid.(TABLE 3) Van der Weg and Bierman(9) have obtained A/a $\cong 1.2 \times 10^6$ cm/sec from the survival probability of 30-90 keV reflected Ar$^+$ ions with respect to Auger neutralization on the outward trip. A/a $\cong 2 \times 10^6$ cm/sec is estimated by the same authors(10) from the line profile of CuI 3247 Å observed in the collision of 80 keV Ar$^+$ on Cu. The same number is estimated by White and Tolk(11) from radiationless de-excitation processes of receding particles of energies 10-3000 eV. However, higher values of A/a in the range 1.2 to 1.5 $\times 10^8$ cm/sec have been obtained by two groups(12)(13) studying Doppler broadened line shapes. No work involving the de-excitation of outgoing particles has yielded as high a value for A/a as that estimated above for the Auger neutralization of incoming slow ions.

IX. THE AUGER DE-EXCITATION PROCESS

The electron energy diagram for the Auger de-excitation process has already been presented in Fig. 3. The corresponding potential energy diagram for the de-excitation of a neon metastable is given in Fig. 14.(1) The variation of the final state energy with atom-solid separation varies as it did for the Auger neutralization process.(Fig. 9) The initial state behavior is considerably different. The potential curve for the state $Ne^m + ne_W^-$ is compounded of two terms: a van der Waals attraction, which, although larger for Ne^m than for Ne, is still small (~ 0.1 eV), and a repulsive interaction which, because of the larger size of the metastable, must cause the potential energy curve to rise above the asymptote at considerably larger distances than does that for Ne. The magnitude of this effect was estimated in Ref. 1 and is shown in Fig. 14.

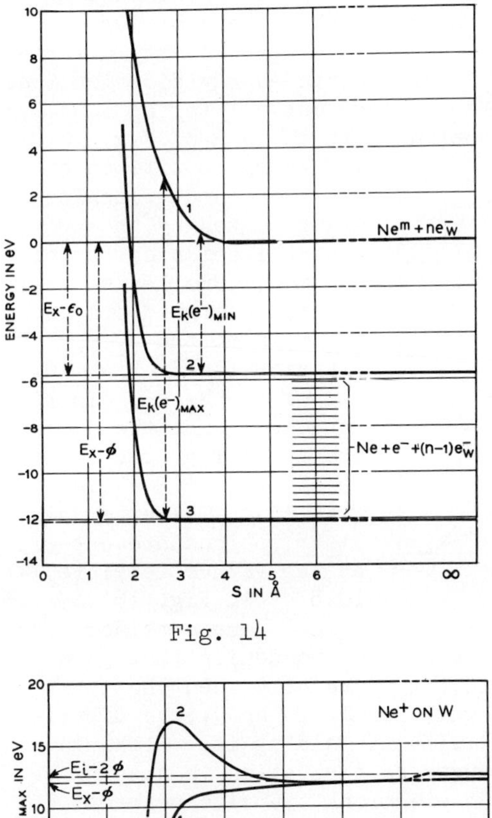

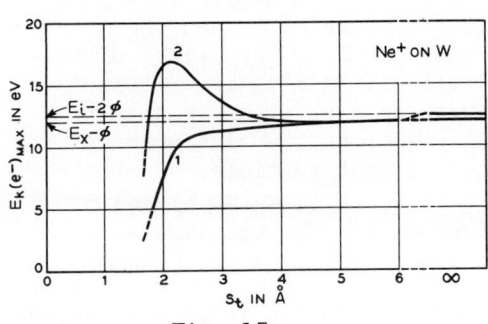

Fig. 14

Fig. 15

The form of the potential curves in Fig. 14 indicate that, in contradistinction to Auger neutralization, the separation of the curves for the initial state (curve 1) and a final state (a curve between curves 2 and 3) initially increases as s decreases. Since the kinetic energy of the Auger electron outside the solid is the separation of such curves we expect the maximum kinetic energy $E_k(e^-)_{max}$ to increase above the free space value, as curve 2 of Fig. 15 shows, before decreasing close to the surface. Curve 1 in Fig. 15 shows that $E_k(e^-)_{max}$ for Auger neutralization decreases monotonically as the atom-surface separation at which the process occurs decreases.(1)

The differences in electron kinetic energy maxima just discussed will form the basis for an experimental demonstration of the occurrence of Auger de-excitation.(Sec. XI) In anticipation of this we can investigate for the simplified solid of Fig. 10 what an admixture of a small percentage of Auger de-excitation with the predominant process of Auger neutralization will do to the observed kinetic energy distribution of ejected electrons. The kinetic energy distributions in Fig. 16 for He, Ar, Kr, Xe ions and curve 1 for Ne ions correspond to $N_i(\varepsilon_k)$ distributions like curve 3 of Fig. 10.

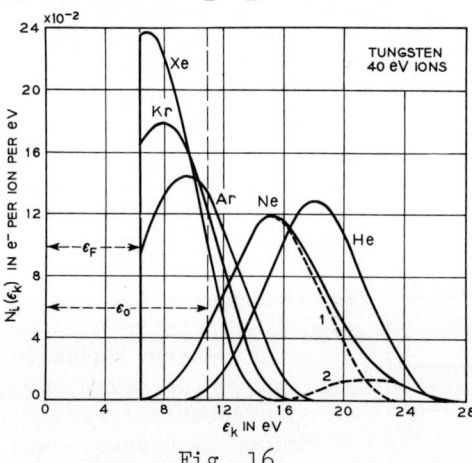

Fig. 16

The energy positions of the broadened triangles are dictated by the neutralization energy of the ion. Smaller neutralization energy drops the position of the $N_i(\varepsilon_k)$ distribution in Fig. 10. All of the $N_i(\varepsilon_k)$ distributions inside the solid are cut off at the Fermi level ($\varepsilon_k = 6.3$ eV in Fig. 16). The external distributions are cut off at the vacuum level ($\varepsilon_k = 10.8$ eV in Fig. 16).

If one considers the case of no energy level shifts and no broadening that gave curve 1 for $N_i(\varepsilon_k)$ in Fig. 10, it is readily seen that the Auger de-excitation process will result in an $N_i(\varepsilon_k)$ distribution that reproduces $N_c(\varepsilon)$ without folding it. Although a two-electron Auger process Auger de-excitation is quasi one-electron since one of the two electrons transits between specific atomic levels in the incident particle. Furthermore for no level shift or broadening the triangular $N_i(\varepsilon_k)$ of base width $2\varepsilon_F$ from neutralization and the rectangular $N_i(\varepsilon_k)$ of base width ε_F from de-excitation have the same maximum point.

When one admits energy level variation in each process we see by Fig. 15 that the de-excitation distribution moves up in energy whereas the neutralization distribution moves down. When broadening is also included we see how the component $N_i(\varepsilon_k)$ functions 1 and 2 in Fig. 16 are obtained for an 8% admixture of the de-excitation process. We would thus predict that such an admixture would distort the high energy end of the distribution producing a measurable shift of its maximum to higher energy. It also increases the yield because the faster electrons in curve 2 leave the solid with greater probability than they would if distributed over a distribution such as curve 1.

X. RESONANCE TUNNELING PROCESSES

Finally, among the individual processes shown in Fig. 4 we discuss the resonance tunneling processes. These are shown in the electron energy diagram of Fig. 1 where the circumstance that determines which of the two processes will occur is also suggested. The position of the excited (or ground) level in the atom relative to the Fermi level and its occupancy determine whether resonance ionization or neutralization occurs.

A potential energy diagram appropriate to the resonance tunneling processes is given in Fig. 17. For the initial state of ion X^+ and n electrons in the solid, $X^+ + ne_S^-$, the potential curve appropriate to resonance neutralization is shown partially imbedded in the manifold between curves 2 and 3 that describe the final states $X^m + (n-1)e_S^-$. The states between curves 2 and 3 differ in the position in the filled band of the hole left by the neutralizing electron. It is clear that for resonance neutralization to occur at all distances from the solid curve 1 must lie between curves 2 and 3 at all distances from the surface. This requirement arises because of the resonance character of the electronic transition. There is no ejected electron, as in the Auger processes, to carry off excess potential energy. Thus transitions must occur at the crossing of the initial-state potential curve and one of the manifold of curves between 2 and 3.

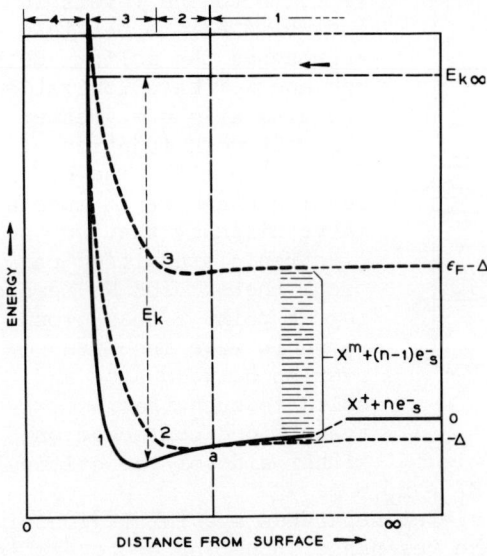

Fig. 17

In Fig. 17 curve 2, one of the extremals of the manifold of final states in which a hole is resident in the filled band, and curve 1 cross at the point a. This means that resonance neutralization can occur only for distances greater than the crossing point a. The energy Δ in the diagram is

$\Delta = E_i - E_x - \phi$, the energy of the asymptote of curve 2. The energy level of curve 2 at any distance relative to the zero energy defined in Fig. 17 is $E_i' - E_x' - \phi$. At the crossing point at a this energy is zero. From this we derive the condition $E_i' - E_x' - \phi = 0$. Since $E_i' - E_x'$ is the ionization energy of the metastable state, point a thus defines the condition at which the metastable level lies at the Fermi level of the solid. At distances less than that of point a, where curve 1 is outside the manifold between curves 2 and 3, the metastable level lies above the Fermi level.

Translating energies from potential diagrams such as that of Fig. 17 back to an electron energy diagram yields Fig. 18.

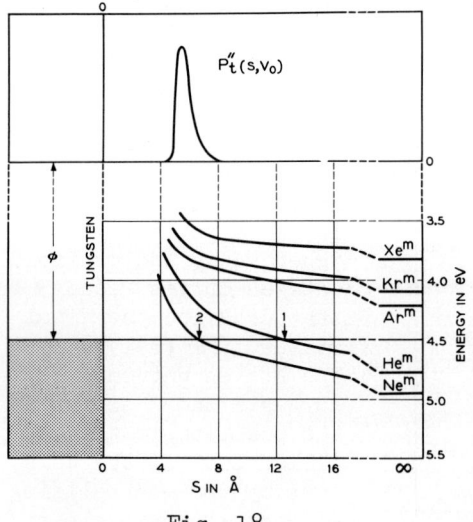

Fig. 18

Here we see that the metastable energy levels of the noble gas atoms rise relative to the levels in the solid as the particle approaches the solid. Both He^m and Ne^m have ionization energies at $s = \infty$ greater than the work function of $W(\phi_W = 4.5 \text{ eV})$. The He^m level crosses ε_F at such a large distance that no electronic transition can occur there. The Ne^m crossing at point 2 does cross ε_F where wave function overlap between atom and solid is such that transitions can occur on either side of the critical distance $s = s_c$ defined by $E_i' - E_x' = \phi$.

The critical distance $s = s_c$ separates spatial regions in which one or the other of the resonance tunneling processes is possible. For $s > s_c$, that is in region 1 shown at the top of Fig. 17, resonance neutralization is possible but not resonance ionization. In regions 2 and 3, for $s < s_c$, resonance ionization is possible but not resonance neutralization. Region 3 is separated from region 2 as the region in which the Auger processes become probable. The particle is reflected at the potential curve barrier and so in principle does not enter region 4. E_k and $E_{k\infty}$ in Fig. 17 are ion kinetic energies.

XI. TWO-STAGE ELECTRON EJECTION PROCESSES

We are now prepared to discuss the two-stage electron

ejection process in which an incident ion is neutralized to an excited state followed by Auger de-excitation of the excited atom thus formed with the ejection of an electron. This process takes us across the top of the triangle in Fig. 4 and down the right-hand side. In our discussion of Sec. IX we saw that Auger de-excitation has a significantly different energy distribution from Auger neutralization and in Fig. 16 we saw the effect of an 8% admixture of the de-excitation process. In Sec. X we concluded that for Ne^+ the critical point at s_c most likely lies close enough to the surface to separate regions in which the two resonances process can occur.

Our ability to detect the partial occurrence of the Auger de-excitation process depends on the above facts and, importantly, on the further fact that the partition of processes for Ne^+ incident on W, for which s_c is properly placed, is a function of incident ion velocity. How this comes about is as follows. If we plot the probability $P_0(s,v)$ that an incident ion remains an ion (the survival probability of Table 1 and 2) for various incident velocities we get a graph like Fig. 19.(1) Here s_c is the critical distance discussed above, and $a^{\sim}$ is the parameter in the rate $R_t^{\sim}(s) = A^{\sim} \exp(-a^{\sim} s)$ of the resonance neutralization and ionization processes. For a sufficiently slow ion curve 1 of Fig. 19 would be appropriate. In this case all Ne^+ ions would become Ne^m metastable atoms before s_c is reached from larger distances. After the critical separation s_c is crossed these metastables can revert to ions via

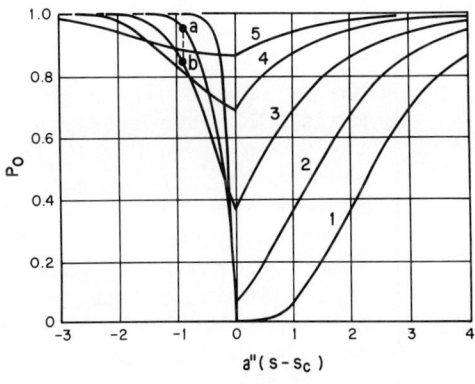

Fig. 19

the resonance ionization process. As ion velocity toward the surface is increased we progress from curve 1 to curve 5 on Fig. 19. We note that at $s = s_c$ a larger fraction remains as ions as v is increased. However, the space rate of reversion to metastables is similarly slowed for the faster ions so the curves cross each other in the region $s < s_c$. If we assume that the Auger process, whichever is possible, occurs at $a^{\sim}(s-s_c) \cong -0.9$ the partition between the Auger processes for curve 2 is given by point a, and for curve 3 by point b. Thus for faster particles a smaller number of ions have converted to metastables at $s = s_c$ but a greater number remain as metastables at the separation where the Auger processes occur. Thus we expect the partition between the Auger processes to shift from essentially 100% Auger neutralization at very low velocities to an increasing percentage of Auger de-excitation, near 10% say, as velocity increases. According to our discussion in Sec. IX we expect one result of this to be an increase in the yield γ_i since the yield per ion of de-excitation is greater than that for neutralization. The former process produces, on the average, faster electrons more of which can leave the solid. This is seen to occur for Ne^+ in Fig. 5 and explains the Ne^+ anomaly. A second consequence of this shifting partition is to increase the maximum kinetic energy of ejected electrons as predicted by Fig. 16. Figures 20 and 21 demonstrate that this occurs.(6) The vertical lines along the $E_k(e^-)$ ordinate axis indicate the values $E_i(X^+) - 2\phi$ for each ion. The Ne^+ distribution in Fig. 20 is normal for 10 eV ions for which the yield γ_i in Fig. 5 is also normal. At 40 eV ion energy the Ne^+ distribution in Fig. 21 clearly has an anomalously high maximum kinetic energy and the yield γ_i in Fig. 5 is also anomalously high. Thus we are able to understand in essentially complete detail the nature of the Auger neutralization and de-excitation processes and how they are partitioned by the kinetics of the resonance processes that are the precursors of de-excitation in the two-stage ejection mechanism.

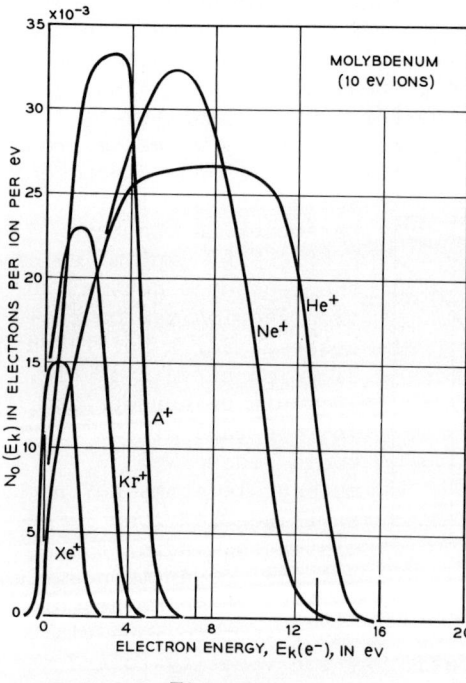

Fig. 20

XII. PROCESSES INVOLVING INCIDENT $He^+(2s)$

We have, to this point, been discussing a family of electronic transition processes which involve singly charged unexcited ions and excited neutral atoms. Another interesting family of transition processes involves metastably excited ions and doubly excited neutral atoms.(14) We shall discuss these as our final example of what can be learned from such studies.

De-excitation and Neutralization Processes

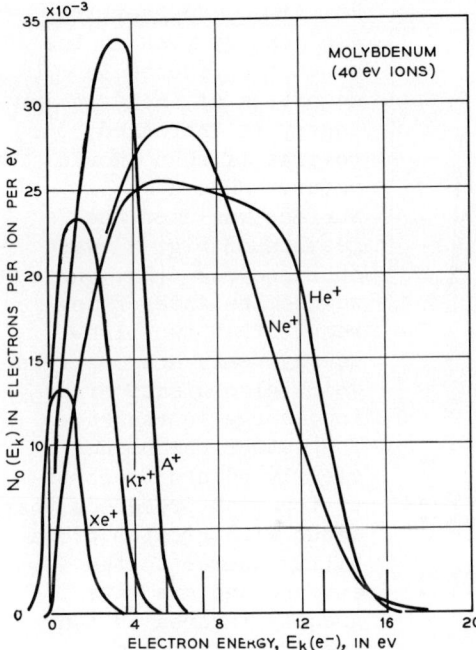

Fig. 21

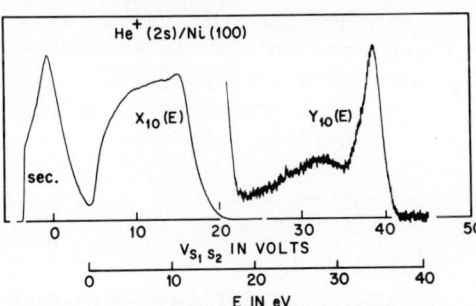

Fig. 22

In these studies the transition we had hoped to see was the ionic de-excitation process:

$$He^+(2s) + ne_s^- \to He^+(1s)$$
$$+ (n-1)e_s^- + e^-, \quad (10)$$

in which a metastably excited helium ion is de-excited to its ground state with the release of a metal electron. The process is energetic – the excitation energy is 40.8 eV – and like all de-excitation processes it is quasi two-electron so that the kinetic energy distribution is not a fold but reflects directly the local density of states. (Sec. IX) The data were taken with a beam of He^+ ions, 99.9% being $He^+(1s)$ and 0.1% $He^+(2s)$. The possible processes for these ions produce electrons in quite different energy ranges so they could be studied separately. When this mixed beam was incident on the clean Ni(100) surface the observed electron energy spectrum was that of Fig. 22. Near $V_{S_1 S_2} = 0$ a peak of secondary electrons from the grids of the analyzer is observed. Electrons ejected from the Ni surface are accelerated by 4 eV to separate them from these secondaries. This results in the electron energy scale labelled E in Fig. 22. The $X_{10}(E)$ distribution is that ejected by 10 eV $He^+(1s)$, the $Y_{10}(E)$ distribution that ejected by 10 eV $He^+(2s)$. The sensitivity of the apparatus was increased 300 times for the $Y_{10}(E)$ measurement.

The $Y_{10}(E)$ distribution of Fig. 22 looks superficially like what one would expect if the process were that given in

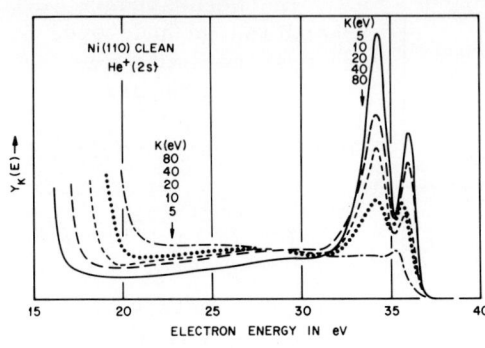

Fig. 23

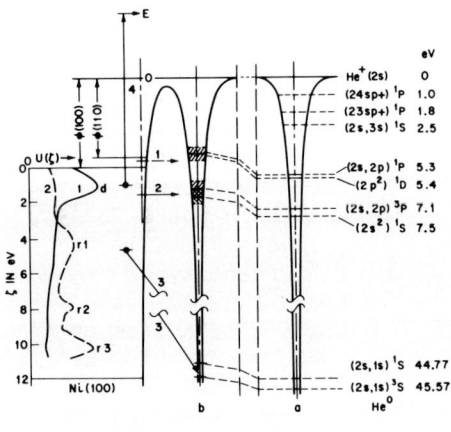

Fig. 24

Eq. (10). However, the high energy peak is too broad and reduces in magnitude as incident ion energy is increased, the reverse of what should occur. When a Ni(110) surface was used the results of Fig. 23 were obtained for Y(E). Here we see the interesting result that two high-energy peaks are observed which also disappear as ion energy increases. The latter phenomenon clearly points to competition with processes that occur with greater probability nearer to the surface and are thus greatly favored as ion energy is increased.

The processes that are in fact being observed are the following. First, the He$^+$(2s) is neutralized to a doubly-excited state by a resonance tunneling process such as:

$$He^+(2s) + ne_S^- \rightarrow He^{o**}(2s^2) + (n-1)e_S^-. \quad (11)$$

As can be seen from Fig. 24 there are four possible levels, $(2s^2)^1S$, $(2s,2p)^3P$, $(2p^2)^1D$, and $(2s,2p)^1P$, into which electrons near the Fermi level can tunnel. These amalgamate into two broader groups of states near the surface. Tunneling into these is indicated by arrows 1 and 2 of Fig. 24. After the formation of the He^{o**} particle in any one of these states autoionization occurs, an example of which is:

$$He^{o**}(2s,2p) \rightarrow He^+(1s) + e^-. \quad (12)$$

This is an Auger de-excitation process in which all the electrons are in the atom. It occurs near the surface where the atomic levels are shifted as is evidenced by the fact that the electrons ejected in each of the high energy peaks of

Fig. 23 are 0.6 to 0.7 eV faster than those observed when these states autoionize in free space.(14)

Figure 24 indicates why we observe one autoionization peak for Ni(100) and two for Ni(110). In the former case only one of the two groups of excited He^{o**} states lies below the Fermi level. In the latter case at least part of the upper of the two groups lies below the Fermi level where tunneling from filled states in the solid is possible. Curves 1 and 2 at the left of Fig. 24 are experimental local densities of states used in the detailed discussions of Ref. 14. Transitions 3 and 4 are those of the process of Eq. (10).

The reason why the Auger electrons are faster when the process occurs near the surface can be seen from the potential energy diagram of Fig. 25. Here the resonance neutralization process occurs at the crossing of an initial potential curve, 2, 3, 4, or 5, with one of the potential curves representing the interaction of He^{o**} with the surface, 6, 7, 8, or 9. The subsequent autoionization process involves a transition from one of these curves to the final state curve, 10, representing the de-excited ion and an ionized electron. An electron produced in this process near the surface is faster than one produced in a process in free space because of the nature and configuration of the potential curves of Fig. 25. Here, unlike Auger neutralization (Fig. 9), the ionic state is the final state. Its potential curve (curve 10) lies below the flatter potential curve (curve 9, say) for the neutral initial state. Nearer the surface these curves lie farther apart, not

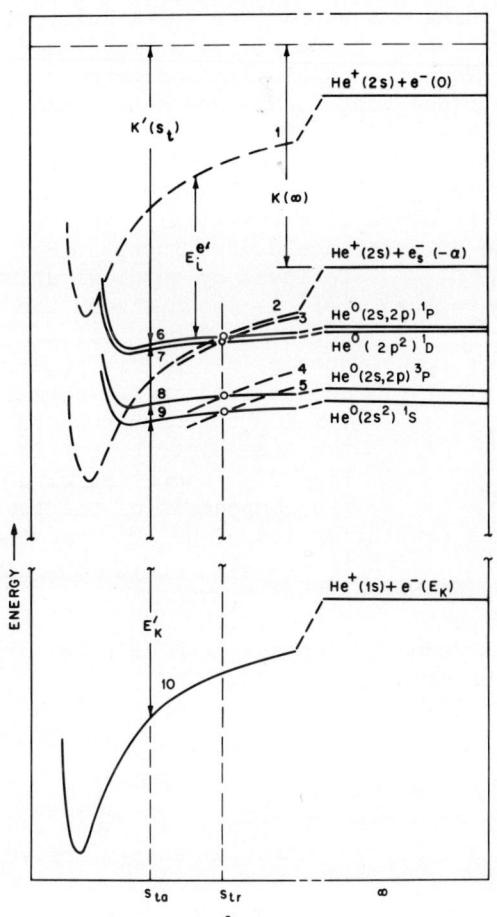

Fig. 25

closer together as in Fig. 9, resulting in a faster electron, that is, greater E_K^* in Fig. 25.

The amount by which the ejected electron is faster than that from a free-space process is the result, principally, of the image force interaction of the ion with the surface. The same statement can be made about the magnitude of the upward shift of the doubly excited levels (Fig. 24) into which the tunneling electron transits in the first stage of the process. This shift must be just enough at the particle solid separation where the tunneling process occurs to put the ^1P and ^1D states above the Fermi level for the Ni(100) surface and at or just below the Fermi level for the Ni(110) surface. This consistency requirement between the energetics of the two components of the two-stage process is a particularly good test of the general picture of the role of particle-solid interaction we have been using. Equating the level shift to the image force and van der Waals terms yields a transition separation of about 5 Å. Thus the autoionization process occurs farther from the surface than does Auger neutralization.

The family of transition processes involved in the de-excitation and neutralization of metastably excited helium ions near a solid surface is shown diagrammatically in Fig. 26. The two-stage process of resonance neutralization (RN) followed by autoionization (AI) is shown across the top. The final ionic species $He^+(1s)$ is Auger neutralized in the normal fashion to contribute to the $X(E)$ distribution in Fig. 22. Competing with the components of the two-stage process are the Auger neutralization of $He^+(2s)$ and the Auger de-excitation of He^{o**} shown vertically in Fig. 26. As incident ion energy increases the partition among these processes shifts in the direction of the curved, dashed-line arrows, accounting for the reduction in the peak or peaks of electrons from the autoionization process.

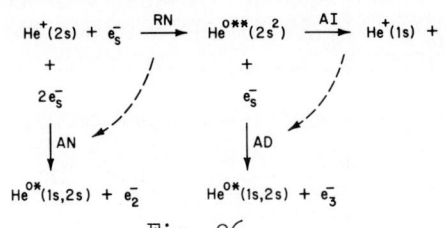

Fig. 26

XIII. SUMMARY

In this paper we have discussed the processes which can occur when atomic particles carrying potential energy arrive in the vicinity of a solid surface. We have shown how these processes are related to each other in families and how their energetics can be described in terms of potential energy diagrams involving particle-solid interaction energies. Some aspects of the kinetics of these processes have been discussed in terms of probability functions dependent upon particle-solid separation and particle velocity.

XIV. REFERENCES

1. Hagstrum, H.D., Phys. Rev. 96, 336 (1954).
2. Shekhter, S.S., J. Exptl. Theor. Phys. (U.S.S.R.) 7, 750 (1937).
3. Massey, H.S.W., Proc. Cambridge Phil. Soc. 26, 386 (1930).
4. Cobas, A. and Lamb, W.E. Jr., Phys. Rev. 65, 327 (1944).
5. Oliphant, M.L.E., and Moon, P.B., Proc. Roy. Soc. (London) A127, 388 (1930).
6. Hagstrum, H.D, Phys. Rev. 104, 672 (1956).
7. Hagstrum, H.D., Takeishi, Y., and Pretzer, D.D., Phys. Rev. 139, A526 (1965).
8. Hagstrum, H.D., Phys. Rev. 122, 83 (1961).
9. Van der Weg, W.F., and Bierman, D.J., Physica 44, 177 (1969).
10. Van der Weg, W.F., and Bierman, D.J., Physica 44, 206 (1969).
11. White, C.W., and Tolk, N.H., Phys. Rev. Letters 26, 486 (1971).
12. Baird, W.E., Zivitz, M., Larsen, J., and Thomas, E.W., Phys. Rev. A 10, 2063 (1974).
13. Kerkdijk, C.B., Smits, C.M., Olander, D.R., and Saris, F.W., Surf. Sci. 49, 45 (1975).
14. Hagstrum, H.D., and Becker, G.E., Phys. Rev. B 8, 107 (1973).

neutralization and inelastic energy losses in low-energy ion scattering

W. Heiland[1]
Bell Laboratories

E. Taglauer
Max-Planck-Institut fur Plasmaphysik

I. INTRODUCTION

The elastic part of the interaction of rare gas ions with solid surfaces at energies below a few keV is well understood in terms of binary collisions[1]. Even though the cross sections involved, the influence of thermal vibrations, and the effects of surface structure are not yet known sufficiently, good qualitative agreement has been achieved between experiments and ion scattering theories or models based mainly on successive single binary collisions. By comparison in the field of inelastic surface particle interaction, at best qualitative understanding has been achieved, owing to the strong influence of the electronic structure of the solid on, for example, the neutralization of impinging and scattered ions. In this paper we will outline the experimental results related to the inelastic ion-surface interaction and discuss the mechanisms involved. The phenomena of interest will be electronic stopping, electron emission, excitation and neutralization of the projectiles. Part of these problems will also be covered by other authors in these proceedings, so we will be rather short on, for example, potential emission of electrons[2].

II. INELASTIC ENERGY LOSSES - GENERAL REMARKS

A particle colliding with another particle, a surface or a solid loses energy by the "elastic" collisions of the nuclear masses and by the inelastic interaction of the electrons. If the scattered particles are the subjects of interest the

---

[1]On leave from Max-Planck-Institut fur Plasmaphysik 8046 Garching, Germany.

processes are defined as nuclear and electronic energy loss. For a given ion-solid combination, the phenomenon is described by the nuclear and the electronic stopping cross section S which is related to the energy loss Q via

$$S = 2\pi \int_0^{r_c} pQ(p,E)\,dp \qquad (1)$$

where p is the impact parameter, E the particle energy and r_c the maximum value of the impact parameter, which is defined by the structure and density of the solid. The general assumption is that nuclear loss and electronic loss can be treated separately and simply add up to the total loss[3]. The evidence from the large amount of high energy (>20 keV) experiments proves this basic assumption to be correct.

The electronic energy losses are on the other hand the source of a variety of other phenomena like secondary electron emission and emission of electromagnetic radiation (light, X-rays). Secondary electrons from the solid are observed if sufficient kinetic energy is transferred to an electron in order to overcome the surface barrier. Photon and X-ray emission are observed from excited atoms (projectile and target) and from direct electron-hole recombination in insulators (for reviews see (4,5)). Owing to the complexity of the phenomena, there is no complete theoretical description of these processes possible, there are rather mosaic-like pieces of theoretical approaches for different single processes, for example, Auger-emission of electrons[2]. Some of them we will discuss in this paper. A verbatim description of the fate of an ion scattered by a solid in order to list the processes involved is given in the following paragraph.

A particle beam hitting a solid may be classically thought of as being divided into three parts, backscattered, trapped and transmitted[6]. In our context we consider the backscattered particles only, since they are the most important probe at low energies, even though measurements of the trapping coefficient provide the same information as measurements of the reflection coefficient[7]. But both quantities are based on the average fate of many particles and give rather little information on the single processes involved. The trajectory of the probing ion is governed by repulsive screened coulomb potentials in our energy range. The attractive parts of the interaction due to dispersive forces between projectile and surface can be neglected at beam energies above about 50 eV; at lower energies they contribute appreciably to the scattering process[8]. Hence the probing ion will undergo a series of single binary collisions accumulating an energy loss of

$$\Delta E_N = E_o \sum_{n=1}^{N} \left[(1-f_n) \prod_{m=1}^{n} f_{m-1} \right]$$

with

$$f_n = f\left(\frac{M_1}{M_2}, \theta_n\right) \text{ and } f_o = 1$$

where

$$f\left(\frac{M_1}{M_2}, \theta\right) = \frac{M_1^2}{(M_1+M_2)^2} \left\{ \cos\theta \pm \left[\left(\frac{M_2}{M_1}\right)^2 - \sin^2\theta \right]^{1/2} \right\}^2 \quad (2)$$

describes the conservation of energy and momentum in a collision between two mass points (M_1 = mass of the projectile, M_2 = mass of the target), where the projectile is scattered into an angle θ in the laboratory system. The scattering angle, the impact parameter p and the interaction potential V(r) are related by

$$\theta_{com} = \pi - 2 \int_{R_{min}}^{\infty} \frac{p\,dr}{r^2 (1-(p/r)^2 - V(r)/E)^{1/2}} \quad (3)$$

in the center-of-mass system, where R_{min} is the distance of closest approach.

During the whole trajectory the electronic interactions occur. The first interaction may be an Auger-neutralization process[2], whereby an electron tunnels into the ground level of the approaching ion. The change of the potential energy of the electron is balanced by the excitation of a second electron, which may leave the solid, thus causing the so-called potential emission of electrons. In principle we have now two classes of impinging particles, neutrals and ions, which may see slightly different interaction potentials, but in the case of especially light ions and metals the coulomb forces between the free electrons and the outer-shell electrons or rather energy levels of the atom or ion will lead to an appreciable shift of these levels[9]. At close distances to the solid or inside the solid outer-shell electrons may essentially end up in non-binding states, leaving the projectile in an ionic state screened by the electrons of the target. Hence the question of the charge state of the particle inside the solid has no final answer for all cases at all energies. Nevertheless the moving particle will cause electronic excitation and eventually leave the surface and relax to a final state, which may be the ground state or an excited state of the neutral, the positive or negative ion, doubly ionized state, etc. Most experiments seem to indicate this state is mainly determined by the last inelastic interaction, when the particle leaves the surface, which may be an electron loss or electron pickup process, or merely an excitation-de-excitation process. In many cases the final velocity (i.e. the velocity component perpendicular to the surface) is the important parameter and so the trajectory of the particle and the energy loss bear on the final state. Deviations from the determination of the charge state or the state of excitation at the surface have been observed in charge state measurements of scattered rare gas projectiles[10]. Molecular effects on the charge state of backscattered hydrogen[11] and on the excited states of backscattered hydrogen[12] indicate an interaction between the excited electrons in the solid, created by the traveling particle, and the energy levels of the particle leaving the surface region. With these complications in mind, we will outline some of the models for single processes involved in the whole, complex phenomenon (Fig. 1).

III. ELECTRONIC ENERGY LOSS

The theories for the electronic energy loss are in fact not very specific as far as the actual electron states are concerned. They are based upon a Thomas-Fermi description of the atom[13,14]. This reflects the assumption that the excitation of the atomic shells will be distributed among all electrons, hence leading to an averaging process after several atom-atom collisions. On the other hand, a complete quantum mechanical approach is very complicated too. In the Firsov theory[13] the calculation starts from a model, where during the collision of two atoms an electron flux occurs through a virtual plane separating the two atoms. The velocity distribution of the electrons in each atom is assumed to be spherical symmetric so the electron flux is proportional to $nv_e/4$. This flux exerts a force on the atom and by proper integration, the energy to slow down the atom is obtained. The relation between electron density and velocity is given by the Thomas-Fermi model ($v_e \propto n^{1/3}$). The final result relates the energy for a single collision Q to the atomic numbers Z_1 and Z_2, the projectile velocity v and the impact parameter p by

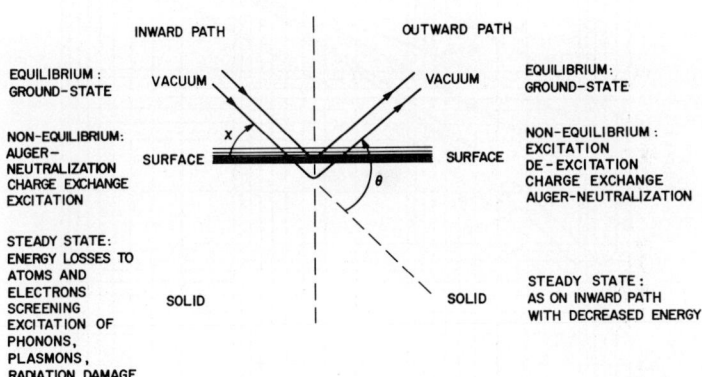

Fig. 1. Scheme of the ion-surface interaction. Impact angle ψ and laboratory scattering angle θ are defined by this figure.

$$Q = \frac{(Z_1+Z_2)^{5/3} \; 4.3 \cdot 10^{-8} \; v}{(1+3.1(Z_1+Z_2)^{1/3} \; 10^7 p)^5} \qquad (4)$$

where Q is obtained in eV, for v in cm/s and p in cm. The important result is the proportionality between energy loss and $\sqrt{E}$, which is slightly modified by the energy dependence of the impact parameter for a given scattering angle.

Lindhard and Scharff[14] without giving too much detail give the same energy dependence for the stopping power $S \sim \sqrt{E}$, without any dependence on the impact parameter. Their model is also based on a Thomas-Fermi atom. Recently Oen and Robinson[15] gave a semi-empirical approach to the problem, i.e. by comparing results from an elaborate computer simulation (MARLOWE[16]) with experiment they found that indeed an impact parameter dependence has to be taken into account. This is especially important at low energies (Fig. 2). The

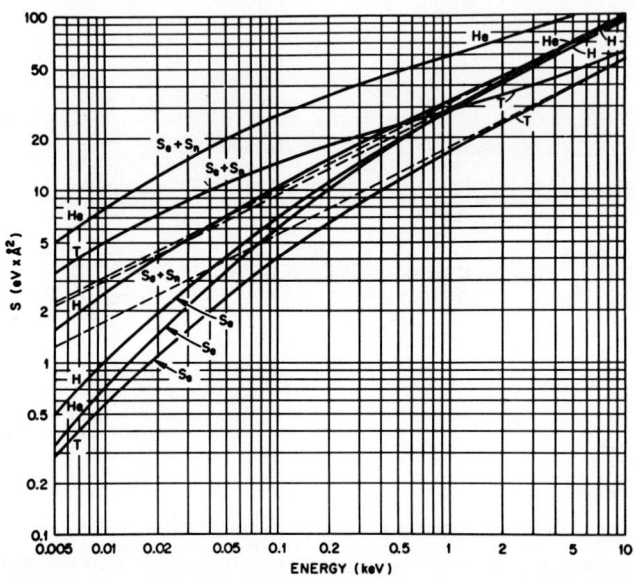

Fig. 2. Electronic S_e and total ($S_e + S_n$ = electronic plus nuclear) stopping cross section for light ions (H, T and He) in Cu. (Oen and Robinson[15]).

impact parameter dependence is taken into account by an exponential function to model approximately the electron density around the atom, thus

$$Q(p,E) = (0.045\ KE^{1/2}(\pi a_{12}^2)\ \exp[-0.3R(p,E)/a_{1,2}] \quad (5)$$

where K is a constant, a_{12} the screening length in the Molière approximation of the Thomas-Fermi potential and R the distance of closest approach.

An experimental test of these models at low ion energies is difficult to obtain. The method of measuring the energy loss by passing an ion beam through thin foils is not practicable, since good foils which are thin enough are hard to make. Measurements of particle reflection[17-19] or trapping[20,21] allow a comparison with computer simulation including the electronic energy loss. Figure 3 shows calculated and experimental results for a number of light

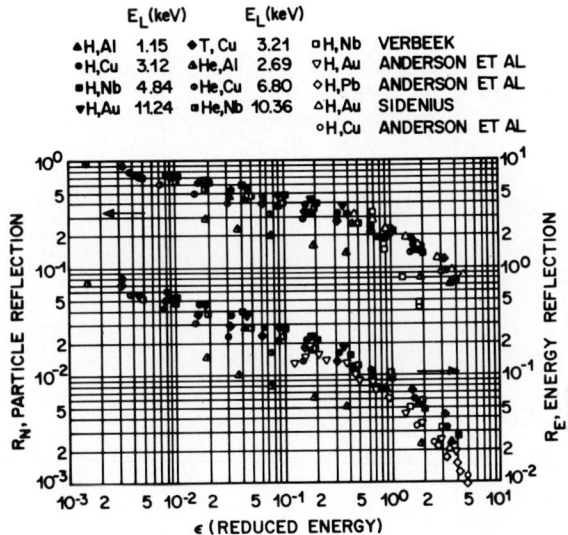

Fig. 3. Particle and energy reflection coefficients as a function of reduced energy ($\varepsilon = aE_r/Z_1Z_2e^2$ with $a = 0.8853\ a_o\ (Z_1^{1/2} + Z_2^{1/2})^{-2/3}$, $E_r = M_2E/(M_1 + M_2)$ and a_o is Bohr's radius) from computer calculations (filled symbols) and experiments (open symbols). (Oen and Robinson[15]).

projectiles in different targets. The calculated values are from the program Marlowe[15,16], which include the electronic loss according to Eq. (5). Even though energy and particle reflection coefficients are rather complex quantities and contain averages over a large number but different ion trajectories more direct approaches have not been very successful owing to the experimental problems.

Figure 4 is an example from the scattering below 2 keV[22]. The energy loss has been calculated from the experimental backscattering spectra using

$$Q = E_o \left[\frac{2M_1}{M_2} \left(\frac{E_1}{E_o} \right)^{1/2} \cos\theta - \left(1 - \frac{M_1}{M_2}\right) \frac{E_1}{E_o} + \left(1 - \frac{M_1}{M_2}\right) \right] \quad (6)$$

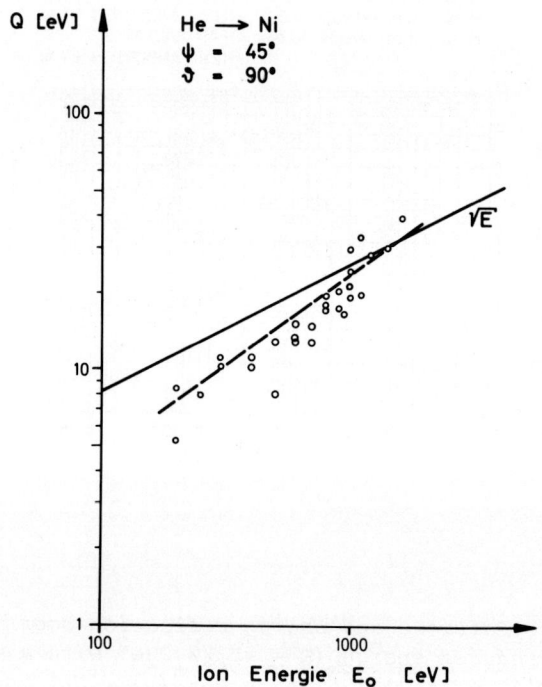

Fig. 4. Inelastic loss of He^+ scattered from Ni under single scattering condition. Solid line corresponds to $\sqrt{E}$ dependence, dashed line is drawn to aid the eye.

where E_0 is the primary ion energy and E_1 the energy of the scattered ion. This contains the assumption that under the given geometrical conditions single binary scattering prevails. The result shows the expected behavior but in point of view of the experimental peak width, which has about the same magnitude as the estimated loss, it is not convincing. Similar results have been recently reported for He and He^+ scattering at somewhat higher energies (2-8 keV)[23].

An analogous attempt under multiple scattering conditions (Ne→Ni, Ne→Ag, Ne→W)[24] showed good general agreement between a Marlowe calculation and the experiment, e.g. Fig. 5. But for the problem in question the influence

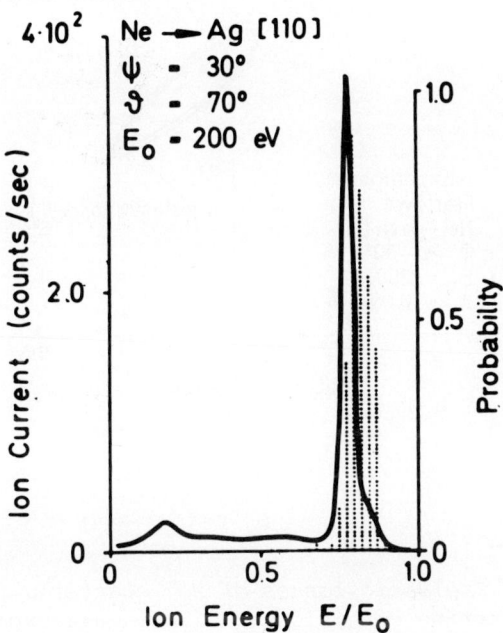

Fig. 5. Comparison of measured energy spectrum (solid line) and a spectrum (histogram) generated by the computer program MARLOWE (Heiland, Taglauer and Robinson[24]).

of the electronic losses on to the backscattering turns out to be too small to be essentially tested by this type of experiment (Fig. 6). From these results we might conclude that these energy losses are not important, but on the other hand the losses in the order of 10-50 eV are sufficient for outer-shell excitations. These in turn can be expected to influence neutralization effects and they are of course the source for optical radiation[5,25]. So quantitative measurements of the energy loss of scattered particles would contribute to the understanding of these phenomena.

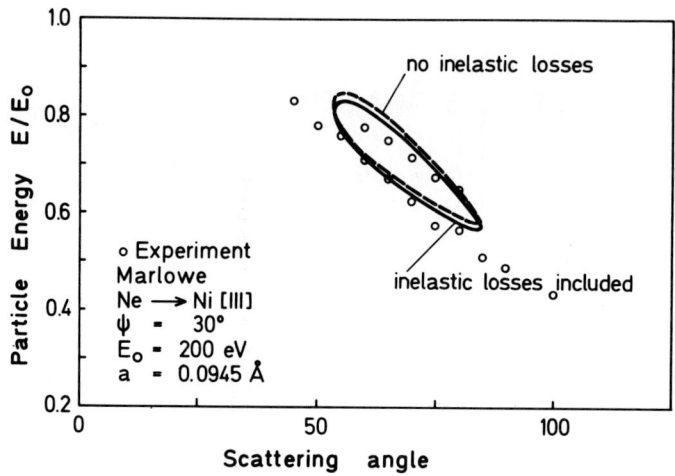

Fig. 6. Inelastic losses of Ne^+ scattered from Ni under multiple scattering conditions. Circles are experimental, the solid and dashed lines generated by MARLOWE (Heiland, Taglauer and Robinson[24]).

IV. EXCITATION

From the discussion of the electronic energy losses, the observation of ions or neutrals backscattered with sizable energy losses can be expected, as are found in ion-gas scattering experiments[26], especially in the single scattering regime which is a good approximation to the ion-gas case. But in general the experimental evidence from the two types of experiments is quite different. In the ion-gas scattering (usually small angle scattering $< 10°$), the inelastic events dominate, i.e. for a given impact parameter at a given energy certain energy levels of the target or the projectile are excited, resulting in numerous but distinct peaks in the spectra. In ion-surface collisions (usually large angle, $> 10°$) the general result is a single peak spectrum (e.g. Fig. 5). The expected inelastic events are only found at rather small scattering angles (as expected)

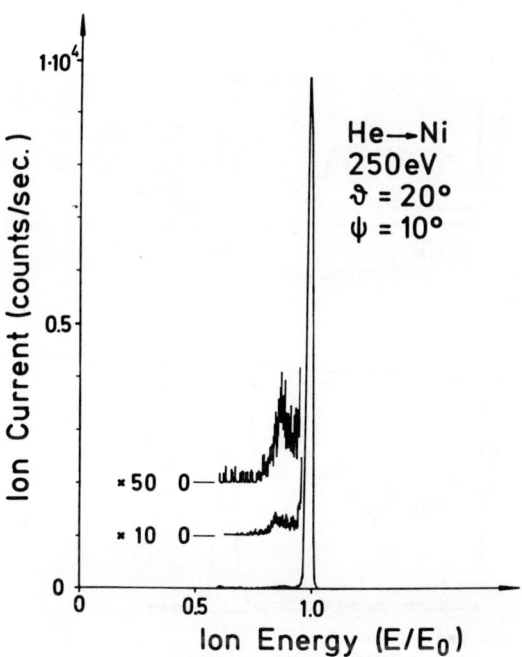

Fig. 7a. *Energy spectrum of He$^+$ scattered from Ni.*

but with very low intensity. For a distinction from peaks due to adsorbates or due to multiple scattering careful energy and angular measurements are necessary, i.e. it has been tested whether the peaks follow Eq. (2) or (6). Figures 7a and b show examples for He→Ni and Ne→Ag scattering[22]. Figure 8 shows the evaluation of the energy dependence of the Ne peak with about 45 eV energy loss according to Eq. (6). The equivalent result is obtained for Q as a function of the scattering angle at a constant energy. In Fig. 7b comparison is made with a Ne→Ne spectrum[27] at about the same $E \cdot \theta$ value. A similar evaluation and comparison has been made for He. It shows that the interpretation of these peaks as due to inelastic excitation may be correct.

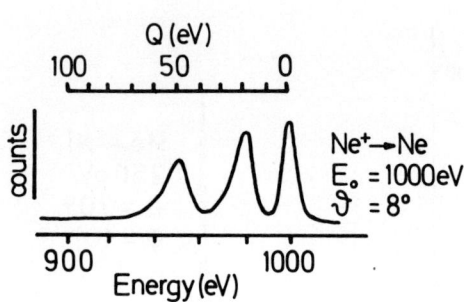

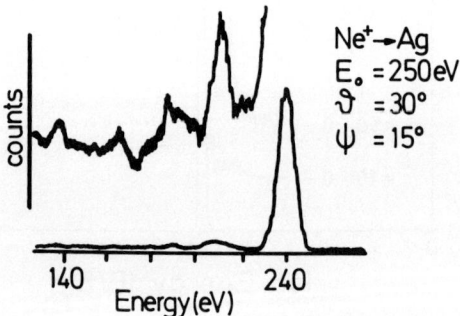

Fig. 7b. Energy spectrum of Ne^+ scattered from Ag. Comparison is made to a spectrum from $Ne^+ \to Ne$ scattering (Gerber, Niehaus and Stephan[27]). The low energy peaks are due to excitation mainly into autoionizing states.

In order to get the observed peak, e.g. for Ne, the following scheme of processes should occur

$$Ne^+ + \text{surface} \to Ne^\circ + e^- \quad \text{via Auger-Neutralization}$$
$$Ne + Ag \to Ne^{**} - Q \quad \text{Excitation} \tag{7}$$
$$Ne^{**} \to Ne^+ + e^- \quad \text{Auto-Ionization}$$

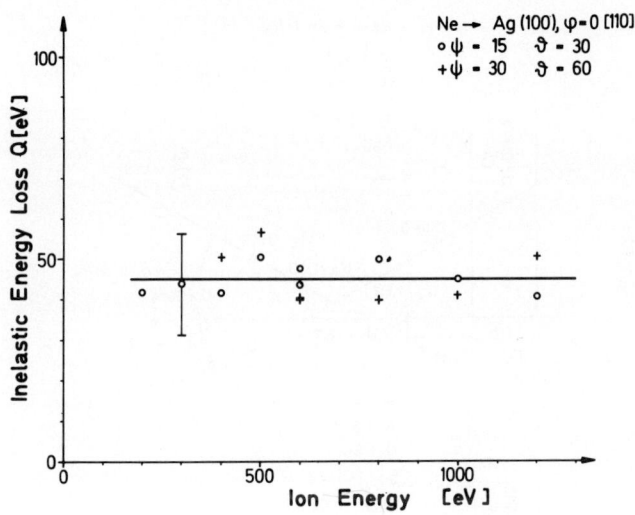

Fig. 8. Inelastic loss for one of the low energy peaks in Fig. 7 as a function of the primary energy for two sets of impact and scattering angles.

where $Q = 49.5$ eV would correspond to $Ne^{**}(\ldots 2s\ 2p^4({}^3P, {}^1D, {}^1S))$ and $Q = 20$ eV (Fig. 7) to the singly excited ion $Ne^{+*}(\ldots 2s\ 2p)$ or to the autoionizing state of the neutral $Ne^*(2s\ 2p^6\ 3s$ or $4s)$. A model for the actual trajectory of a Ne^+ ion scattered from a Ag single crystal surface can be obtained from a string-model[27] (Fig. 9), a model for the electronic interaction from Hagstrum's work[2] (Fig. 10). An estimate for the survival probability of the ion as an ion based on the Auger-neutralization process is inserted in Fig. 9. It shows that indeed the first step in our "reaction" scheme (7) is very probable. In the "single" scattering trajectory a sufficient small impact parameter may be reached to make the second step. In order to survive as an excited neutral, the particle faces a de-excitation process (Fig. 10), which may be very efficient depending on the velocity and the actual distance of the particle from the surface, since the surface barrier for the excited states is shallow or even not existent. On the other hand the lifetime of the excited Ne^{**}

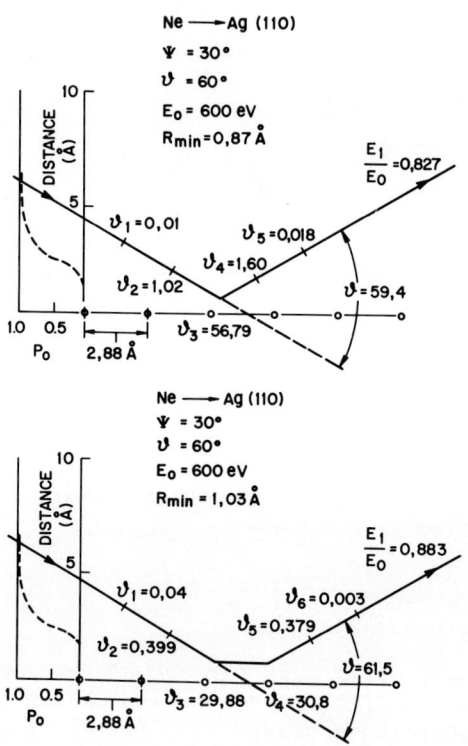

Fig. 9. Trajectories of Ne scattered from an Ag[110]-chain in a Ag(110) surface into a total scattering angle of about 60. R_{min} is the distance of closest approach, in both cases it is reached for θ_3. Insert on the left shows the probability P_0 of the incoming ion to survive as an ion as a function of the distance. At about 2 A all incoming ions will be neutralized with high probability.

($\sim 10^{-6}$ s) is large enough[27] to get away from the surface, so a few ions are possibly observed.

Further evidence comes from experiments on electron emission. For He→Cu the proper electrons have been observed[29] and recently results for Ne$^+$→Mg and Al have been reported[30] which show the electrons resulting from the decay of the excited Ne (Fig. 11). In conclusion we find that

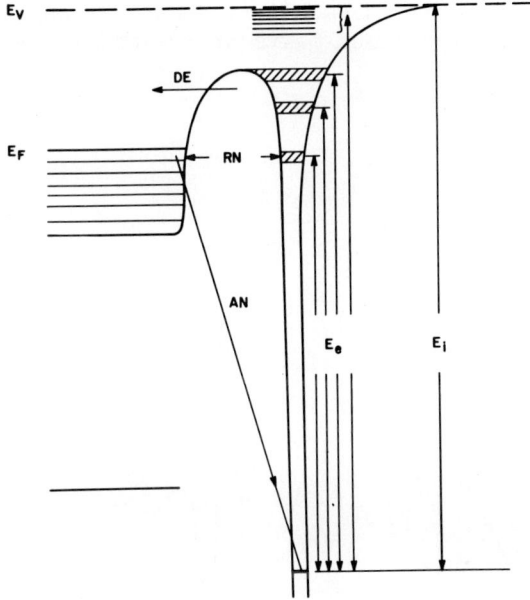

Fig. 10. Model for the charge exchange processes between an ion (atom) near a metal surface. E_v is the vacuum level, E_F the Fermi-energy, E_i the ionization energy and E_e are excitation energies. E_i and E_e are functions of the distance between the atom and the surface. AN is the Auger-neutralization, RN is a resonance-neutralization, DE is a de-excitation (or ionization).

the inelastic processes known from gas phase experiments are possible at surfaces, but they are heavily influenced by a de-excitation process, whereby the electrons from the excited states tunnel into the empty, but allowed states (conduction band and above) of the solid. Similar losses from target atoms are not to be expected (at least from metals) since the outer-shell electrons are energetically broadened into the valence band. Lower level excitation is possible (Fig. 11) but decreasingly less so with decreasing energy.

V. NEUTRALIZATION

The neutralization process is also discussed in the contributions of Hagstrum[2], Buck[10], Rusch[31] and Tully[32] in this book. The important parts of the model involved are

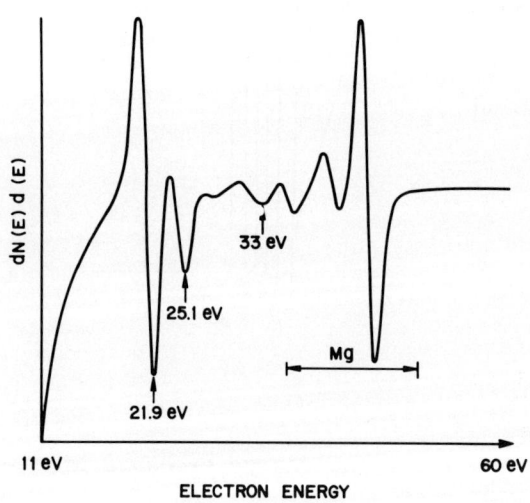

Fig. 11. Secondary electron spectrum (second derivative) due to Ne^+ impact at 3 keV onto a Mg surface, the peaks at 21.9, 25.1 and 33 eV are due to autoionizing states of the Ne created during the collision, the Mg-structure is due to an Auger-process of the Mg excited by the impinging Ne^+. (Ferrante and Pepper[30]).

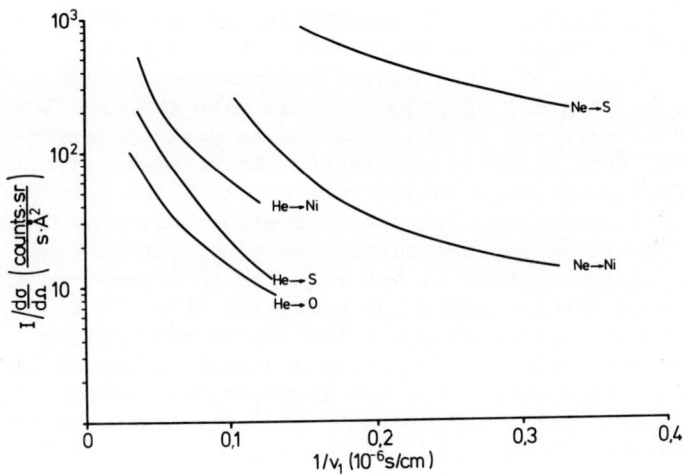

Fig. 12. Ion yield normalized to the scattering cross section for different projectile-target combination as a function of the inverse projectile velocity. (Taglauer and Heiland[34]).

shown in Fig. 10. Not shown is the dependence of the atomic energy levels on the distance between surface and atoms[9]. If resonance neutralization between sharp atomic levels is excluded (see[31,32]) a monotonic dependence of the ion yield on the velocity is to be expected[2]. Both for the Auger-neutralization and the resonance tunneling from the valence band into atomic states a dependence of the ion survival probability $P \propto \exp(-A/av)$ is predicted. This is observed for most ion-solid combinations[1,33], i.e. if the ion yield for a given scattering angle as a function of energy is normalized by the scattering cross section (Thomas-Fermi Moliere) and plotted versus 1/v monotonic decreasing functions are found (Fig. 12)[34]. Only in a few cases a strictly linear behavior is found indicating that more than one neutralization path is possible. It also cannot be excluded that part of the neutralization occurs on the outgoing path of the ion (some may survive the inward path, some are excited, reionized, etc.), where the particle has a lower velocity due to the elastic loss. Taking this into account the experimental findings can be fitted assuming only one neutralization mechanism, i.e. Auger-neutralization[23,35,36]. For heavier projectiles (higher Z) the possibility of neutralization of the particles inside the solid cannot be ruled out totally as shown by T. Buck[10] thus adding a further complication to the discussion. A still open question is also the the "low energy tailing" of rare gas ion backscattering peaks observed especially from oxidized metals[1]. Under these circumstances the surface properties change in a way to allow particles scattered from deeper layers and/or with additional inelastic excitations to escape the surface as ions. In general invoking neutralization is not required to understand the existence of single scattering peaks (as shown by Fig. 5). The model calculation does not include neutralization. For each ion-solid combination exists a certain range of impact angles and energy where the top layer shadows completely all deeper layers (total reflection). Towards higher energies and/or larger impact angles (= towards normal incidence) there exists a regime where ions can penetrate into the solid, but reflection from deeper layers is still highly improbable due to the energy losses encountered. When the regime of reflection from deeper layers is reached, neutralization becomes an important factor to shape the ion spectra. These regimes are obviously structure dependent and the thermal vibrations have to be considered. Again it can be concluded that an accurate understanding of ion yields has to include the ion trajectory in any specific case.

VI. SECONDARY ELECTRON EMISSION

Secondary electron emission is mainly discussed within two models, i.e. potential-emission (essentially covered by Hagstrum[2]) and kinetic emission (for reviews see (4,37,38): A further source of electrons are excited autoionizing projectiles (see above). Potential emission is schematically contained in Fig. 10; if the Auger-neutralization occurs, the electron tunneling from the valence-band into the ground state of the atom gains potential energy. This energy is carried away by a second electron as kinetic energy. The energy has to be high enough that the electron can reach empty states above the Fermi-level or the continuum above the vacuum level. If it starts close enough to the surface, it may in the latter case be observed as a secondary electron. The kinetic electron emission is also understood in terms of an Auger-process[39]. But at higher energies holes in the valence band and in deeper levels can be created by direct energy transfer (electron-electron collision). These processes are directly related to the energy loss discussed above and consequently the Firsov-model[13] has been used[39] to estimate the ionization cross sections involved.

Unfortunately virtually no experiments have been done which connect ion scattering and secondary electron emission under comparable conditions.

VII. ACKNOWLEDGMENT

We thank W. Eckstein (Garching) and N. H. Tolk (Murray Hill) for helpful discussions.

VIII. REFERENCES

1. Heiland, W. and Taglauer, E., Nucl. Instr. Meth. 132, 535 (1976).
2. Hagstrum, H.D., in "Inelastic Ion-Surface Collisions" (these proceedings).
3. Lindhard, J. and Scharff, M., Kgl. Danske Vidensk. Selsk. Math. Fys. Medd. 33, No. 14 (1963).
4. McCracken, G.M., Rep. Prog. Phys. 38, 241-327 (1975).
5. Tolk, N.H., Simms, D.L., Foley, E.B. and White, C.W., in "Ion Surface Interaction, Sputtering and Related Phenomena" (eds. R. Behrisch, W. Heiland, W. Poschenrieder, P. Staib, H. Verbeek) Gordon and Breach, London, 1973, p. 265.
6. Hou, M. and Robinson, M.T., Nucl. Instr. Meth. 132, 641 (1976).
7. Robinson, M.T., in "Proc. 3rd Int. Conf. on Atomic Collisions in Solids", Kiew, USSR (1974).
8. Hulpke, E., Surf. Sci. 52, 615 (1975).
9. Brandt, W., in "Atomic Collision in Solids" (eds. S. Datz, B.R. Appleton, C.D. Moak) Plenum Press, New York (1975) p. 261.
10. Buck, T.M., in "Inelastic Ion-Surface Collisions" (these proceedings).
11. Verbeek, H., Eckstein, W. and Datz, S., J. Appl. Phys. 47, 1785 (1976).
12. Tolk, N.H., Tully, J.C., Heiland, W., Kraus, J., Leung, S., and Hill, P., to be published.
13. Firsov, O.B., Sov. Phys. JETP 36, 1076 (1959).
14. Lindhard, J. and Scharff, M., Phys. Rev. 124, 128 (1961).
15. Oen, O.S. and Robinson, M.T., Nucl. Instr. Methods 132, 647 (1976).
16. Robinson, M.T. and Torrens, I.M., Phys. Rev. B9, 5008 (1974).

17. Sidenius, G., Phys. Letter 49A, 409 (1974).
18. Verbeek, H., J. Appl. Phys. 46, 2981 (1975).
19. Andersen, H.H., Lenskjaer, T., Sidenius, G. and Sorensen, H., J. Appl. Phys. 47, 13 (1976).
20. Kornelsen, E.V., Can. J. Phys. 43, 364 (1964).
21. Bohdansky, J., Roth, J., Sinha, M.K. and Ottenberger, W., in "Proc. 9th Symposium Fusion Tech.", Garmisch 1976.
22. Heiland, W. and Taglauer, E., Bull. Am. Phys. Soc. 20, 854 (1975).
23. Verhey, L., Thesis, Groningen 1976.
24. Heiland, W., Taglauer, E. and Robinson, M.T., Nucl. Instr. Meth. 132, 655 (1976).
25. Thomas, E., Van der Weg, W.F. and White, C.W., in "Inelastic Ion Surface-Collisions" (these proceedings).
26. Massey, H.S., Burhop, E.H.S. and Gilbody, H.B., in "Electronic and Ionic Impact Phenomena", Oxford University Press, London 1974.
27. Gerber, G., Niehaus, A. and Stephan, B., J. Phys. B6, 1836 (1973).
28. Taglauer, E. and Heiland, W., Surf. Sci. 33, 27 (1972).
29. Soszka, W. and Lipiec, J., Surf. Sci. 45, 27 (1972).
30. Ferrante, J. and Pepper, S.V., Surf. Sci. 57, 420 (1976).
31. Rusch, T.W., in "Inelastic Ion-Surface Collisions" (these proceedings).
32. Tully, J.C. and Tolk, N.H., in "Inelastic Ion-Surface Collisions" (these proceedings).
33. Rusch, T.W. and Erickson, R.L., J.V.S.T. 13 (1976).
34. Taglauer, E. and Heiland, W., Surf. Sci. 47, 234 (1975).
35. Verhey, L.K., Poelsema, B. and Boers, A.L., Rad. Eff. 27, 47 (1975).
36. Brongersma, H.H., et al., J.V.S.T. 13, 670 (1976).
37. Parilis, E.S., in "Atomic Collision Phenomena in Solids", (eds. D.W. Palmer, M.W. Thompson and P.D. Townsend), North Holland Publ. Co., Amsterdam 1970, p. 513.
38. Krebs, K.H., Fortschritte Physik 16, 419 (1968).
39. Parilis, E.S. and Kishinevski, L.M., Sov. Phys. Solid St. 3, 885 (1960).

neutralization behavior in medium energy ion scattering

T. M. Buck
Bell Laboratories

ABSTRACT

This paper is a review of some of the experimental literature on the neutralization of ions having primary energies of 5-200 keV, as they are reflected from solid surfaces or pass through thin foils. Most of the data are presented in the form of ion-fractions (Y^+/Y) as functions of energy of the scattered particles. A transition is observed: from the behavior of H and He at 30-200 keV in which there is no direct evidence of neutralization inside the solid, either because it does not occur or because equilibration lengths are too short to be resolved, to that found in the 5-30 keV range where H, He, Ne, and Ar have all shown evidence of enhanced neutralization of ions which penetrate the surface. Present evidence does not permit a confident choice between the theoretical models for protons above 10 keV which assert that neutralization inside the solid is prevented by electron screening, and another (proposed primarily for protons at higher velocities) which accepts the older concept of alternate electron capture and loss inside. Neutralization and scattering results demonstrate that the sharply peaked ion spectra of low energy ion scattering result from a combination of two factors: a) large scattering cross-sections which cause depletion of the ion beam as it enters and leaves the solid and b) more efficient neutralization of ions which penetrate beyond the surface.

I. INTRODUCTION

The aim of this chapter will be to summarize the neutralization behavior of ions with *medium energies* as they scatter from solid surfaces or are transmitted through thin foils. In addition it will be appropriate to describe a change in scattering behavior which occurs in this energy region, in order to distinguish scattering from neutralization effects. Results for H and He at 30-200 keV and for H, He, Ne and Ar at 5-32 keV will be emphasized. In recent years the interest in this energy range has arisen mainly from the use of electrostatic analyzers in surface analysis by ion scattering (1-3), and for investigation of plasma-wall interactions in high-temperature plasma technology (4-6). In these contexts it is important to know the charge states of scattered particles, and how they depend on such factors as the incident or scattered energy or velocity, on target material, depth of penetration, surface cleanliness, exit angle, and possibly other factors. Regarding depth dependence, one may ask whether particles beckscattered from a solid surface have the same ion-fraction for a given energy as those transmitted through foils, i.e. whether "equilibration" of charge state occurs in surface scattering. A further question which is implicit in most of the foregoing and which has received considerable theoretical and experimental attention is: where does the neutralization step occur, inside or outside the target surface? The sharply peaked spectra of low-energy ion scattering ($\leq$ 5 keV) with noble gas ions and electrostatic energy analysis, have frequently been attributed to a neutralization effect in which ions penetrating beyond the first layer or two of atoms are neutralized much more completely than those which are reflected from the surface. On the other hand, at higher energies, $\gtrsim$ 100 keV, theoretical arguments and experimental evidence have suggested that the neutralization step occurs outside the surface as the ion is leaving. In the results to be reviewed here there is indeed evidence of a transition from one type of behavior to the other; i.e., within the experimental resolution, no depth dependence of charge state has been detected for H and He in the 30-200 keV range while such a dependence is apparent for H, He, Ne, and Ar below 30 keV. Complicating the matter somewhat, there is also a transition in scattering behavior-toward sharply peaked energy spectra. At present there is no theory which explains the medium-energy neutralization results in quantitative detail, but mention will be made of several models which have been proposed and which predict qualitative trends.

II. 30-200 keV H AND He

A. <u>Transmission Through Thin Foils</u>

Although this chapter is concerned primarily with charge states of particles back-scattered from solid surfaces, some reference should be made to earlier experiments on ion beams transmitted through thin foils, since it now appears that the ion-fraction values for surface scattering agree closely with those obtained in transmission experiments, although it was not obvious, a priori, that this should be true. Much of the early work on H and He has been reviewed by Allison (7). Hall (8) studied proton beams with energies between 20 and 400 keV passing through foils of Be, Al, Ag and Au, finding that the ratios of charged to neutral components were the same for the light elements but slightly lower for gold above 100 keV. Phillips (9) found significant variations in ion fractions of 3-200 keV hydrogen beams transmitted through thin metal foils when a clean exit surface was maintained by evaporation. A strikingly similar effect in surface scattering will be described in section IIC2. Dissanaike's (10) results for ^{4}He (0.13-1.1 MeV) traversing thin metal foils showed average charge state dependent only on the emergent ion energy or velocity, not on target material. Armstrong et al. (11) also found no dependence on foil material (C, Al, Ni, Ag, or Au) in the range 0.2-6.5 MeV for ^{4}He but they suggested that carbon contamination might have been present on the metal foils.

B. Neutralization Mechanisms

The early experimental results for H and He traversing thin foils were usually discussed in terms of electron capture and loss in the foil, and described by cross-sections dependent on particle velocity (12,13). An average charge state equilibrium was supposed to be established at a particular velocity, after passage through some minimum thickness of material. For the simple case of neutrals and singly charged positive ions, the formalism is indicated by

$$\frac{df^+}{dx} = n[(1-f^+)\sigma_{01} - f^+ \sigma_{10}] \tag{1}$$

in which $f^+ \equiv Y^+/Y$ is the ion fraction, x is path length, n is target atom density, and $\sigma_{10} = \sigma_c$ and $\sigma_{01} = \sigma_\ell$ are the electron capture and loss cross-sections for the particle. Setting $df^+/dx = 0$ gives the equilibrium ion fraction:

$$f^+_{eq.} = \left(\frac{Y^+}{Y}\right)_{eq.} = \frac{\sigma_{01}}{\sigma_{01} + \sigma_{10}} = \frac{1}{1 + \left(\frac{\sigma_{10}}{\sigma_{01}}\right)} \tag{2}$$

Since equilibrium distances could not be measured but were evidently very small, the ratios of capture and loss cross-sections were measured, rather than absolute values of either (8,10,11). Comparisons of the velocity dependence of these ratios for medium energies with theoretical predictions (13) showed some rather unconvincing qualitative agreement. The lack of dependence on target Z was troublesome.

A few alternatives to this model of electron capture and loss inside the solid have been published in the last ten years, all arguing that Debye-type screening by the electron gas prevents bound states on the ion while it, a proton or deuteron in particular, is inside the solid, and therefore that any electron-pick-up must occur outside the surface as the ion leaves. Yavlinskii et al. (14) considered as one possible neutralization step outside, the recombination of ions with electrons in the surface distribution. Trubnikov and Yavlinskii (15) calculated also the probability of resonant tunneling of electrons from the metal out to the ion, and it was suggested that both these mechanisms, and also possibly Auger neutralization, might contribute to the total neutralization probability. Berkner et al. (16) have pointed out that these models (14,15) predict a strong influence by conduction electron density, e.g., as between Au and C, which is not observed experimentally. Brandt and Sizmann (17) have also argued against neutralization inside the solid, on the basis of bound states being prevented by electron screening and also, at high energies, by "collision broadening" (18). They proposed neutralization of emerging protons in the tail of the electron distribution at the surface and gave a preliminary expression for the neutral fraction,

$$\Phi_o \approx (1 + 40 E_p)^{-1} ,$$

applicable in the energy range $10^{-2} < E_p < 10^{-1}$. E_p is the kinetic energy of the emerging particle in MeV per atomic mass unit. The energy range is 10-100 keV for protons and 40-400 keV for He^+. All three of the models mentioned above were compared with data of Phillips (9) and found to predict the general trend of the data, but without very close quantitative agreement. Comparisons with back-scattering data will be shown in the following section.

Recently, Cross (19,20) has re-investigated the question of neutralization inside the solid through electron capture into and loss from bound states on the proton, i.e., discrete hydrogen atom eigenfunctions. For "fast" protons, with $V \gg V_o$ (21), he argues that bound states are not prevented by

electron screening since the collective screening charge is
"left behind" a distance of many Bohr radii. Collision
broadening, i.e. the uncertainty in the energy of the bound
state due to its short lifetime at high velocities, was also
shown to be unimportant. Furthermore, he demonstrates that
electron capture and loss cross-sections from ion-gas measurements (taken on hydrocarbon vapors) can be used quite successfully to predict the neutral fractions of hydrogen emerging
from carbon foils at high velocities and even those of hydrogen scattered from "dirty" gold surfaces at velocities in the
"medium" range, conjecturing as Armstrong et al. (11) had
that carbon was a common contaminant on practical metal
surfaces.

Carbon layers do indeed grow on surfaces by means of ion-
beam-induced decomposition of hydrocarbon layers, as can
easily be observed at $p \approx 10^{-7}$ Torr in 100 keV He^+ or H^+ back
scattering from silicon (2). For various reasons (19) the
intermediate range is more difficult to deal with theoretically than is the high energy ($V \gg V_o$) region, e.g., the
collective electron screening cannot be ignored. Nevertheless,
the high-energy approach may "retain some usefulness at the
intermediate energies". It should be noted that according
to this model the equilibration depth should be very short
at medium energies, e.g. $\approx 3\text{Å}$ for 100 keV, so that exceptionally good depth resolution would be required to distinguish between neutralization inside and outside the solid.

C. Back-Scattering from Solid Surfaces

In addition to their direct relevance in ion scattering for
surface analysis and plasma wall interactions, charge state
measurements on *back-scattered* particles offer interesting
possibilities for finer depth selectivity which may help to
resolve the question of where the neutralization occurs.
Transmission experiments rely on defect-free films of various
discrete thicknesses, which are not easily prepared for many
materials, and studies of energy dependence require a series
of fixed beam energies. On the other hand a pair of backscattering energy spectra taken at one incident energy can
provide charge state information corresponding to a continuous range of energy and depth, as will be shown. Furthermore, charge states of particles scattered from impurities
which are known to be very close to the surface may be
observed.

1. Practical Surfaces

A distinction will be made here between "practical" surfaces, sometimes called "dirty", and "clean" surfaces, since pronounced differences in neutralization behavior at such surfaces have been observed. By "practical" we mean surfaces, e.g. of Cu, Au, and Si which have been carefully cleaned by etching and washing before mounting in the chamber but which have not been cleaned in ultrahigh vacuum by ion bombardment or high-temperature annealing, as would be required for a "clean" surface. The clean surface is presumably free of adsorbed gases while the practical surface is not.

A typical back-scattering charge state experiment (22,26) in the 30-200 keV range consists essentially of two measurements: (A) An energy spectrum of the total yield, ions and neutrals, back-scattered from a thick target into a silicon surface-barrier detector behind a small aperture. (B) A second measurement, similar in all respects to the first except that an electric field is applied to deflect all ions out of the scattered beam, so that only the neutrals enter the detector. Letting A and B represent the respective scattering yields, the ion fraction is then given by $(A-B)/A \equiv Y^+/Y$ which is effectively the ratio of singly charged positive ions to the total yield in the case of H or He, since H^- and He^{++} fractions are known to be very small, < 0.01, in the energy range under discussion (9,10,2,23). A pair of these A and B spectra are shown in Fig. 1. The vertical separation between

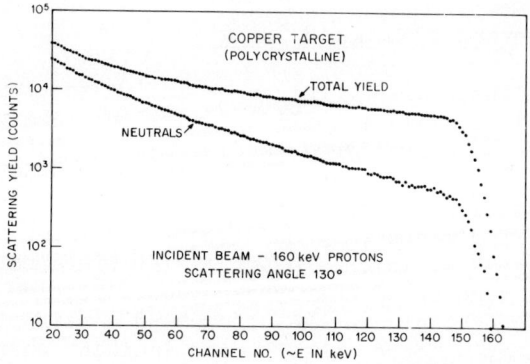

Fig. 1. Energy spectra of hydrogen backscattered from Cu - ions and neutrals (upper curve); neutrals only (lower curve). Equal ion doses used for the two spectra. (Ref. 22).

the two curves at any energy represents the ion fraction for
that energy. The decrease in separation at lower energies
must indicate a dependence of ion fraction on back-scattered
energy or penetration depth, or both. This uncertainty is
resolved by taking spectra at different incident energies.
Ion-fraction data for three incident energies are plotted in
Fig. 2. The close intermingling of points for the same

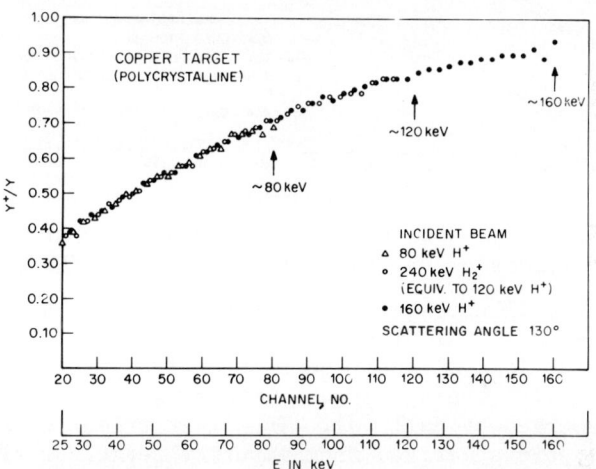

*Fig. 2. Composite of ion-fraction, (ions)/(ions plus
neutrals), versus scattered energy for H^+ incident on polycrystalline Cu at 3 different energies. Practical surface,
not cleaned in UHV (Ref. 22).*

scattered energy but different incident energy is clearly
evident, i.e., there is no evidence of any dependence of Y^+/Y
on the depth from which H^+ and H^o scattered to emerge at a
given energy. If particles scattered from the near-surface
region had higher ion fractions this should be manifested by
an upward inflection of points at the high energy end of a set
for a particular incident energy. It may also be observed
that a lower average charge state of the incoming H, in the
form of H_2^+, did not change the ion-fraction of the backscattered particles. At lower energy such an effect has been
observed (24).

Depth resolution was not especially good in the measurements
of Fig. 2, no better than ~175 Å as estimated from the detector
energy resolution of 7.5 keV. A finer discrimination of the
depth from which ions emerge is afforded by the data for ^{4}He
in Fig. 3 (25). The dashed and dotted curves give ion-
fractions of ^{4}He scattered, respectively, from silicon and

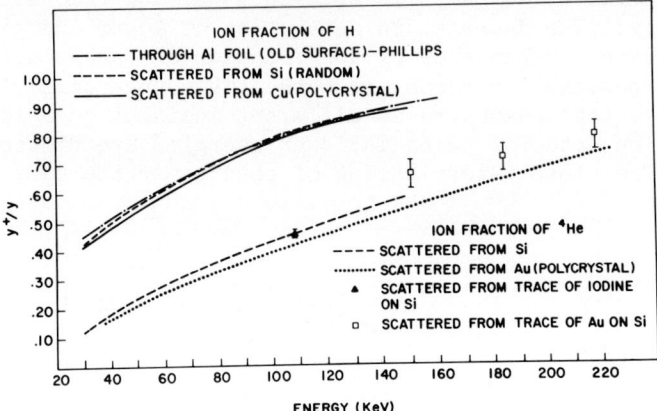

Fig. 3. Charge state data for H and ^{4}He scattered from practical surfaces of Cu, Au, and Si. $\theta_L = 130°$ (Ref. 22, 25).

gold, and they illustrate a small but reproducible influence by the target material, i.e., a "practical" gold surface neutralizes He ions slightly more effectively than a silicon surface does with the difference extending to lower energy than was observed by Hall (8). The black triangle and open squares are points derived from impurity peaks for traces of iodine ($\sim 2 \times 10^{13}/cm^2$) and gold ($\sim 1.4 \times 10^{13}/cm^2$) on the silicon surface. The gold points represent three incident energies. The iodine point lies on the silicon curve, and the iodine atoms were on the surface or very close to it since it had been adsorbed from an iodine-methanol solution at room temperature. Similarly, the gold was deposited from a very dilute cyanide solution. It seems unlikely that these trace impurities were covered by more than a few Angstroms of oxide or carbon and yet the He scattered from them exhibits Y^+/Y values which lie on the silicon curve in the iodine case and close to an extrapolation of it in the gold examples.* These trace impurity results suggest that He$^+$

---

* At higher energies, 0.3-2 MeV Lurio and Ziegler (in "Beam-Foil Spectroscopy" Vol. II p. 665, I. A. Sellin and D. J. Pegg Eds. Plenum Press, New York, 1976) have determined charge state fractions for He scattered from a monolayer of tungsten on sapphire by applying a 30 kV bias to the target so that separate peaks for He$^\circ$, He$^+$, and He^{++} could be resolved by a surface barrier detector. Charge state distributions agreed with those found for He transmitted through carbon foils (11), and it was concluded that equilibration occurred in 10-15 Å of carbon which had been deposited on the gold layer at a chamber pressure of 2×10^{-6} Torr.

neutralization at medium energies is dominated by the substrate surface electron distribution rather than the particular identity of an isolated target atom on the surface.

Ion fractions of H lie higher than those of ^{4}He because exit velocity has the dominant influence, and the H velocity is twice that of ^{4}He at the same energy.

Additional data for H are shown in Fig. 4 (26) which includes experimental results from different groups and also predictions by several theoretical models. There is close agreement

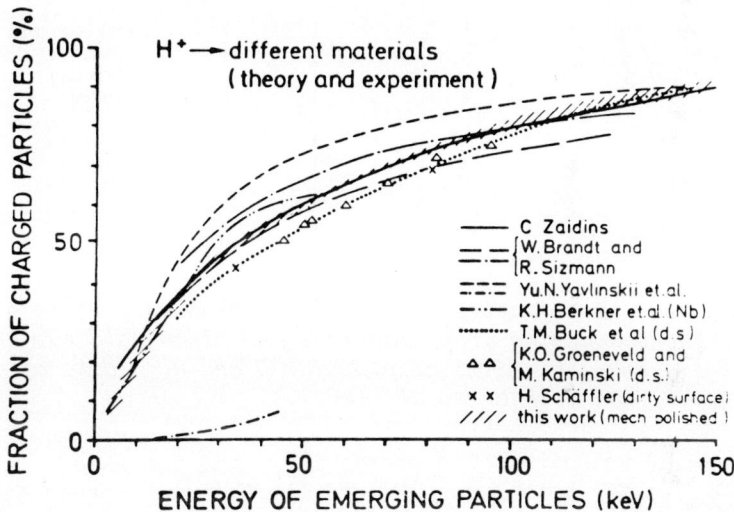

Fig. 4. *Charge state data for H scattered from different materials. Experimental results of Berisch et al. compared with results of other groups and with theoretical predictions. (Berisch et al., Ref. 26).*

between results on different "dirty" surfaces and slightly higher values for "mechanically polished" surfaces. The theoretical predictions generally do not fall very close to the experimental curves although the semi-empirical predictions of Zaidins (27) which combine experimental results with probability considerations of Dmitriev (28), agree closely with the data for a mechanically polished surface.

2. *Clean Surfaces*
Effects of surface cleaning on neutralization are illustrated in Fig. 5a in which Y^+/Y values for ^{4}He and H are plotted against exit velocity. The curves demonstrate the primary

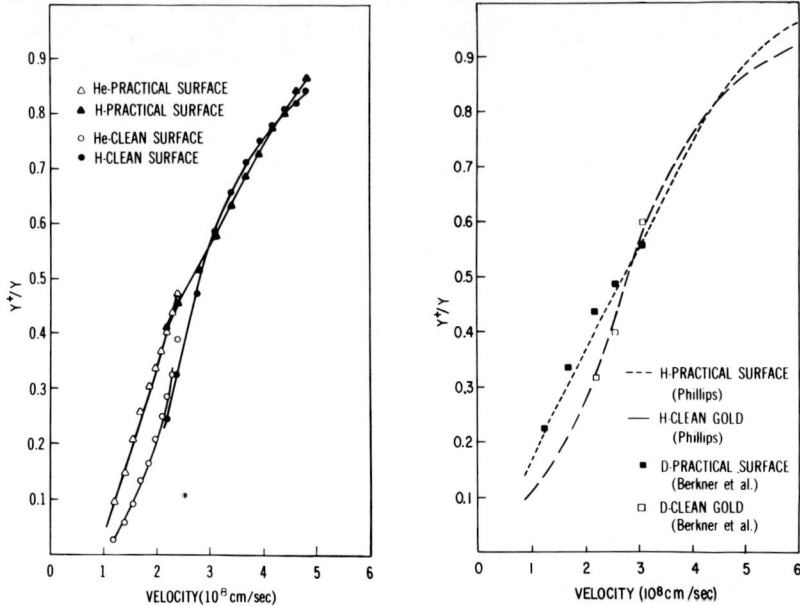

Fig. 5a. Ion-fraction vs particle velocity for ^{4}He and H after backscattering from practical and clean gold surfaces. (Ref. 25). Fig. 5b. Similar plot for D and H. (data of Phillips (9) and Berkner et al. (16)).

influence of velocity. The energy range was the same (25-120 keV) for both H and ^{4}He and the H curves appear as extensions of the He curves, on a velocity plot. In both cases measurements taken before and after cleaning in UHV are shown. Cleaning consisted of heating to 800°C followed by argon ion bombardment. Pressure in the chamber after bombardment was about 2×10^{-8} Torr. Considering first the ^{4}He it can be seen that cleaning of the gold surface lowered the Y^+/Y curve or increased the neutralization probability throughout the velocity range covered. The H results are more complicated: at the low-velocity end the clean surface curve is lower than the practical curve but it rises more steeply, crosses over the proctical curve at 3×10^8 cm/sec, and then crosses again at 4.2×10^8 cm/sec. This intricate behavior matches very closely the data of Phillips (9) on protons and Berkner et al. (16) on deuterons which are plotted in a similar way in Fig. 5b. In Phillips' experiments proton beams were transmitted through Al foils and the practical or dirty condition was that of the Al foil after long standing in a rather poor vacuum. The clean surface condition was achieved by evaporating a fresh layer of gold on the exit surface of

the Al foil shortly before the scattering experiment. In the
measurements of Berkner et al. with deuterons a similar
evaporation procedure was used and the points shown were
obtained for freshly evaporated gold. It is interesting that
on a velocity plot the two isotopes H and D produce curves
nearly identical with those formed by the elements H and ^{4}He
except for a small difference in slope between the H and He
data which does not appear between H and D. In other words,
on either kind of surface, velocity rather than identity of
the particle is the dominant factor governing neutralization.
Similar effects of cleaning by high temperature annealing
have been observed by Behrisch et al. (26) with H scattered
from V, Be, and Mo, the two surface conditions being the
mechanically polished and high-temp annealed states. Thus,
Figs. 5a and 5b show a difference between clean and con-
taminated metal surfaces which is quite reproducible for
several target materials and cleaning conditions.

The neutralization models discussed in section IIB do not
explain these cleaning effects. However, it seems reasonable
to speculate that the increased neutralization efficiency on
a clean gold surface, for velocity from 1 to 3 10^8 cm/sec,
might be associated with a tunneling process occurring as
the ion leaves the surface, which would be inhibited by a
surface film. The cross-overs at higher velocity, however,
would have to be explained by one or more additional mechanisms. On the other hand, equilibration in a carbon contamination layer can also account rather well for the
practical curves (19), while the clean curves may represent
equilibration in gold. There may be contributions from both
internal and external neutralization mechanisms in the medium
energy range. However, one should emphasize that in the
30-200 keV energy region for light ions there is no direct
evidence of neutralization inside the solid, either because
it does not occur or because equilibration depths are too
short to be resolved. We can conclude that although understanding is incomplete the neutralization behavior of light
ions scattered from solid surfaces with energies of 25-200 keV
is rather simple and reproducible, depending primarily on the
exit velocity and only slightly, if at all, on other factors
listed in section I, except for surface cleanliness.

III. 6-32 keV H, He, Ne, Ar

A. Transition in Energy Spectra of Ions

It was mentioned in the introduction that neutralization of

ions inside the solid has been proposed to account for the sharp peaks which are typical of low-energy ion scattering spectra of noble-gas ions obtained by electrostatic energy analysis (29). The transition from medium to low energy behavior is illustrated in Fig. 6 where ESA spectra for ^{4}He scattered from gold are shown (30). Four incident energies

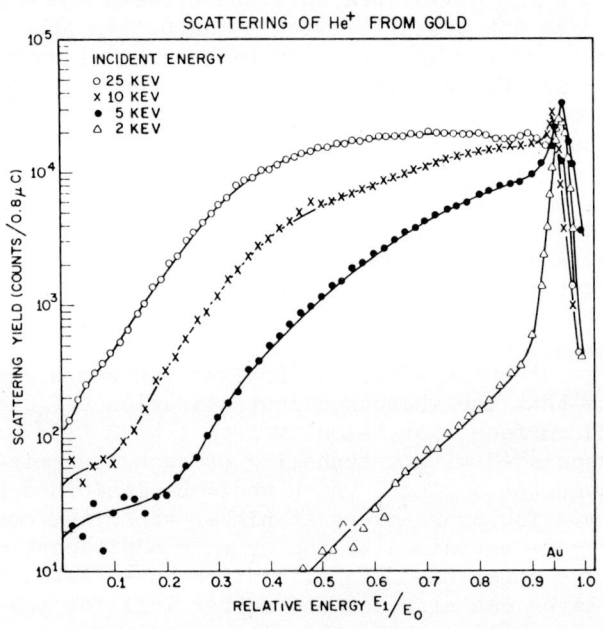

Fig. 6. *Energy spectra for ^{4}He ions backscattered from polycrystalline gold. Four different primary energies. Measured by ESA at 120° scattering angle. (Ball et al., Ref. 30)*

are represented, from 25 keV down to 2 keV, and scattered intensities are plotted on a log scale against relative energy E/E_0. For E_0 = 25 keV, there is a broad background, or low-energy tail, due to ions which penetrated and lost energy on their inward and outward paths, in addition to that lost in the back-scattering collision ($\theta_L = 120°$). This background decreases rapidly as the primary energy is reduced, and a sharp surface peak develops. To explain such behavior, one is forced to accept either that there is more efficient neutralization of ions which penetrate, as has been supposed, or that the total scattering yield of ions and neutrals from beneath the surface has decreased in some way. The experimental results in the 6-32 keV energy region which are presented in this section show that both mechanisms actually contribute to the transition in spectral shape illustrated in Fig. 6.

B. Experimental Techniques for Neutralization Studies

For particle energies below about 30 keV the experimental technique (section IIC), utilizing a silicon detector to measure energy spectra of ions and neutrals from which to derive ion-fraction values, cannot be used owing to the dead-layer thickness. Two other methods have been used successfully, however. In one, a known fraction of the scattered neutrals are re-ionized or "stripped" in a calibrated gas cell after which the ions pass through an electrostatic analyzer (26,24,31). Success with this technique depends on accurate knowledge of the stripping efficiency as a function of energy. The other method is a **time-of-flight** (TOF) technique (32) in which a pulsed beam of ions strikes the target and the scattered particles travel down a flight path of 1 meter length, for example, to an electron multiplier which detects both ions and neutrals. Pulses from the beam pulsing system and the multiplier are fed to a time-to-amplitude convertor which in turn supplies pulses to a multichannel pulse height analyzer (PHA). Spectra for ions plus neutrals and for neutrals alone are accumulated concurrently by applying an intermittent field normal to the scattered beam to deflect all ions out of the beam during half of each 2-second cycle. Fig. 7 (33) shows both spectra for 8 keV Ar^+ scattered from gold; they resemble mirror images of energy spectra. The sharp peak at 5.8 μsec is assumed to represent single scattering of Ar from gold surface atoms. The shoulder on the short-time or high-energy

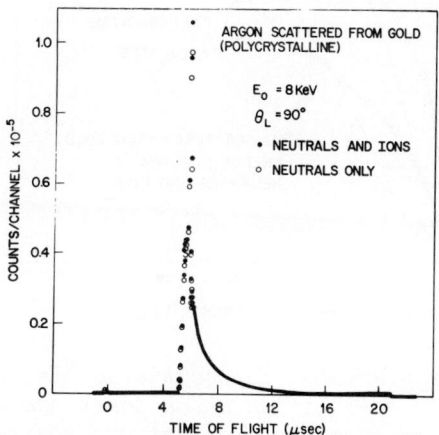

Fig. 7. Time-of-flight spectra for Ar scattered from polycrystalline gold (Ref. 33).

side of the peak represents double scattering, i.e. from two
atoms in sequence, near the surface while the long-time or
low-energy tail beyond the peak is due to scattering from
deeper inside the target. The small separation of the two
spectra is evidence of a high degree of neutralization.
These TOF spectra are readily transformed to the more usual
energy spectra, and to curves of ion-fraction as function
of scattered energy (33).

C. Transition in Energy Spectra of Neutrals and Ions

The reason or reasons for the transition in shape of the ion
spectra, going from medium energy to low energy (Fig. 6), can
be investigated using either of the two techniques described
above. Evidence from TOF experiments for one contributing
factor is shown in Fig. 8. Although it is a scattering
rather than a neutralization phenomenon, discussion of it
here seems justified in order to put neutralization effects
at low energy in proper perspective in regard to their
influence on spectral shape. Figure 8 shows energy spectra

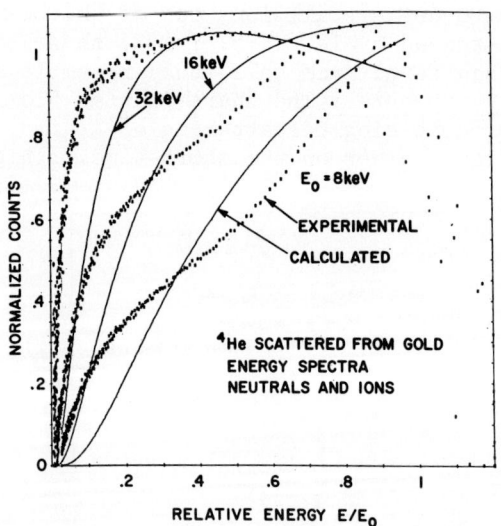

*Fig. 8. Energy spectra for scattered ^4He, neutrals plus
ions, compared with calculation which takes account of the
removal of particles from the incident and scattered beams
as they enter and leave the solid. Calculated curves normalized to experimental points at the maxima. (Ref. 33).*

for He, neutrals plus ions, scattered from gold which were
obtained by transformation of TOF spectra. Three incident
energies are represented, and the experimental data show a
distinct trend as the incident energy decreases: the low
energy yield drops and the spectrum begins to develop a peak
at the high energy end, the same trend exhibited by the ion
spectra of Fig. 6. In this case, however, preferential
neutralization of the ions which penetrate beyond the surface
cannot be invoked as an explanation since neutrals are
included in the spectrum. The Ne and Ar spectra of Fig. 9,
taken at 8 keV exhibit peaks which are progressively sharper
than the 8 keV He. Chicherov (31) had observed similar

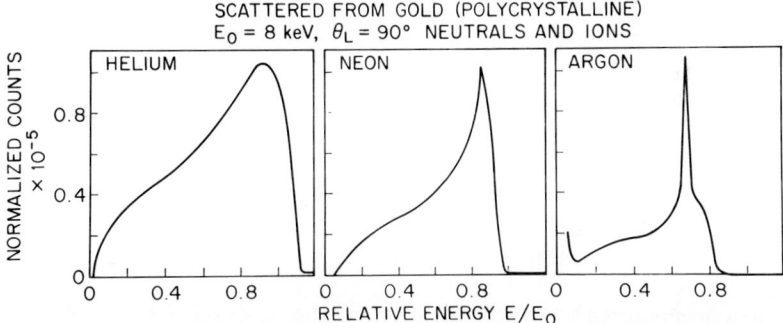

Fig. 9. Energy spectra of He, Ne, and Ar, neutrals plus ions, scattered from gold. $E_0 = 8$ keV. (van der Weg et al., Ref. 39).

spectra for neutral Ar scattered from gold, using the stripping
method. The energy trend of Fig. 8 and the Z trend of Fig. 9
both suggest an influence of the scattering cross-section,
which increases with decreasing energy and increasing Z. The
scattering cross-sections evidently become so large that the
ion beam is depleted significantly as it enters and leaves
the solid. This reduces the yield of ions and neutrals
scattered from inside and emphasizes the peaks due to scattering from surface atoms. In the calculation of the solid
curves of Fig. 8, which obviously follow the trend of the
experimental data, the beam attenuation effect was explicitly
taken into account. Simplifying assumptions were made which
were reasonable for He, but not for Ne or Ar. Energy loss was
derived from electronic stopping, ignoring nuclear stopping,
and plural scattering sequences were not included, only single
events, in calculating the spectra. The failure to reproduce
the inflections of the experimental He curves may be due to
neglect of plural scattering. Fig. 10 shows spectra for
H (H^0, H^+, and H^-) scattering from stainless steel. These

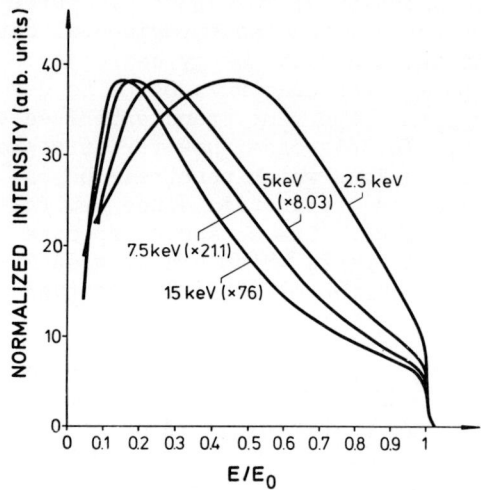

Fig. 10. Energy spectra of backscattered hydrogen (neutral, positive, and negative) for various primary energies. $\theta_L = 135°$. (Eckstein et al., Ref. 23).

spectra were obtained by the stripping method (23). Here again the maximum in the spectrum shifts toward higher relative energy as incident energy decreases, although the maximum has not reached the high energy edge of the spectrum even for $E_o = 2.5$ keV. Presumably, a surface peak would develop at still lower primary energy. This downward shift in the energy required for peaking behavior for H scattering from stainless steel, as compared with He, Ne, or Ar on Au, is to be expected on the basis of smaller Z's and σ's. These authors also derived absolute reflection coefficients, R_N, the ratios of reflected to incident particles, from the back scattering spectra and found R_N increasing at lower primary energy, from ∼0.05 at 15 keV to ∼0.2 at 2 keV. Computer simulations of back-scattered hydrogen spectra (34,35) show a similar trend toward pronounced peaks at low energy, as illustrated in Fig. 11. The simulations facilitate inclusion of plural scattering sequences which become important when energies are low and scattering cross-sections large.

Thus, it is now established that the energy distributions of back-scattered H, He, Ne and Ar, including both neutral particles and ions, develop peaked shapes when primary energies are low enough, owing to scattering effects. However, this alone does not account for the sharpness of ion scattering spectra obtained with an ESA; neutralization also plays a part.

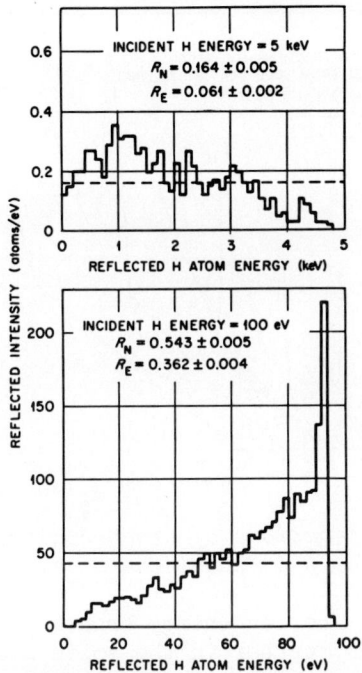

Fig. 11. Energy distributions of H reflected from copper (E_O = 5 and 0.1 keV) calculated by the MARLOWE binary collision cascade simulation program. (Oen and Robinson, Ref. 34).

D. Ion-Fractions of Ar, Ne, He, and H as Functions of E and E_o.

Evidence of neutralization inside the solid, which further enhances the peaks of ion spectra is shown in Fig. 12 for the case of Ar scattering from gold. The ion fractions of scattered Ar derived from TOF data are plotted against scattered energy, for several incident energies. These composite curves are quite different from the smooth, overlapping data of H and He at higher energies, in Fig. 2, for example. In the Ar case there is clearly an influence of distance traveled or time spent in the solid, on the charge state. For each incident energy the curve is peaked at an energy corresponding to single scattering from gold surface atoms (E/E_o = 0.67 for 90° scattering). By comparison with the Ar energy spectrum in Fig. 9, one sees that the high energy branch of each Y^+/Y curve represents ions which were scattered twice near the surface, evidently causing more neutralization than occurs in

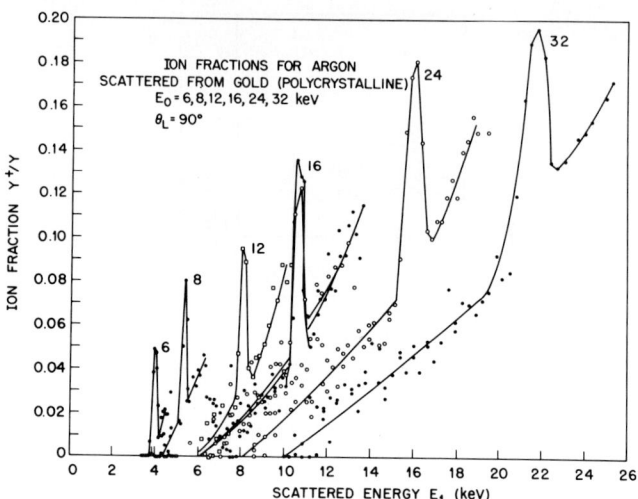

Fig. 12. Ion fractions of argon scattered from polycrystalline gold, for 6 incident energies. (Ref. 33).

single scattering. The low energy branch corresponds to those which traveled farther in the solid losing more energy and becoming still further neutralized. The peak values of Y^+/Y rise with increasing energy, a trend which is expected for neutralization of ions by the Auger process or resonance transfer (36) as they leave the surface. The distinct separation between the low energy tails of the Y/Y^+ curves for different E_o is rather surprising and difficult to explain. It implies that charge equilibrium is not reached after an energy loss of at least 5 keV which, assuming a stopping power of 10^3 keV/micron corresponds to a distance traveled of 50 Å. However, electron pick-up cross-sections for Ar in gases, of $\sigma_c \sim 10^{-15}$ cm^2 (37), suggest the distance should be $\lambda = 1/n\sigma_c = 1$-$2$ Å, n being the target atom density or 5.9×10^{22}/cm^3 in gold. Inner shell ionization might explain the apparent long life of the argon ion, 2.5×10^{-14} sec. for 50 Å at $V = 2 \times 10^7$ cm/sec, but known lifetimes do not seem encouraging, e.g. 2×10^{-15} for the Ar 2P vacancy at 400 keV (38).

Ion-fraction curves for Ne scattering from gold are shown in Fig. 13 (39). For each incident energy the curve has a steep branch corresponding to surface scattering and a lower branch for bulk penetration and neutralization. The lower branches which are defined by least-squares curves show no clear separation, i.e., the equilibration depths are evidently short, in accord with the capture cross-section argument

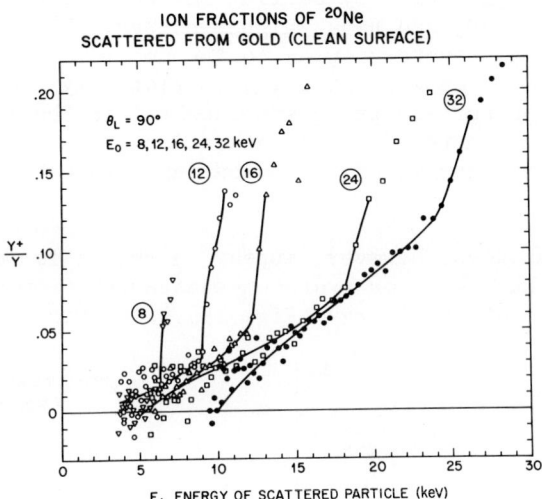

Fig. 13. Ion fractions of neon scattered from polycrystalline gold, for 5 incident energies. (Ref. 39).

which was unsuccessful in the Ar case above. The end-points of the drawn curves correspond to surface scattering peaks in the energy spectra and the points beyond show no evidence of increased neutralization for doubly scattered particles, as in the Ar curves. The reason for this difference is not known. The surface peak values of Y^+/Y for Ne and Ar fall in the same range although the Ne velocity is 40% greater for a given energy, and the He values in Fig. 14 are considerably smaller even though the He velocity is 2-3 times higher.

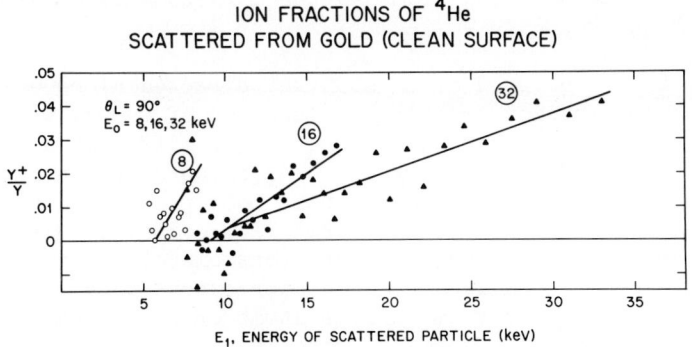

Fig. 14. Ion fractions of helium scattered from polycrystalline gold, for 3 incident energies (Ref. 39).

There is no simple velocity scaling for He, Ne, and Ar below 30 keV as for H, D, and He at higher energy. The data for He are rather unsatisfactory owing to the small ion-fractions being measured, poor statistics, and relatively poor energy resolution. No surface peaks are resolved in the ion fraction curves although there is clear separation of curves for different incident energy, again indicating more neutralization of deeper penetrating particles.

A surface peak does, however, appear in recent ion-fraction data of Matschke (40) for hydrogen scattering from a gold single crystal at low energy (Fig. 15). In this work the

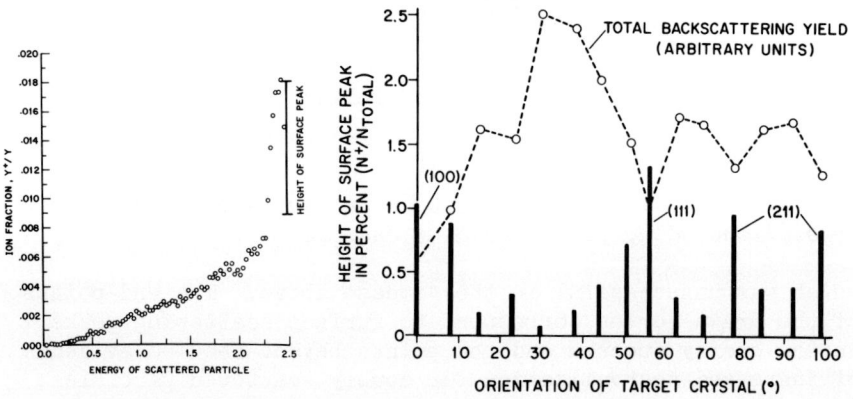

Fig. 15a. Ion fraction of hydrogen scattered from single crystal gold surface (110) at a specific azimuthal orientation. Primary beam was H_2^+ incident at 5 keV. Fig. 15b. Height of surface peak, as defined in 15a, as function of azimuthal scattering direction. Directions indicated are the planes defined by the incoming beam and the direction of observation. (after Matschke, Ref. 40).

primary beam consisted of H_2^+ at 5 keV directed along a <110> axis. Values of ion-fraction as function of energy were determined by the stripping technique for different azimuthal orientations of the gold crystal. As illustrated in Fig. 15b the ion-fraction curves showed their most pronounced peaks when the direction of the scattered beam coincided with a planar channel direction, i.e., protons emerging from planar channels were neutralized to a greater extent than were those which were scattered from surface atoms or in a random direction from inside. It seems reasonable to suggest that this interesting and rather surprising result is due to the lower electron density in the channel which reduces the electron screening which is supposed to prevent bound states on the proton (14,15,17). It is also interesting to note that the neutralization inside the solid reviewed in this section for H, He, Ne, and Ar has been observed only at energies corresponding to $V \gtrsim V_F$, V_F being the Fermi velocity for electrons in the solid, 1.4×10^8 cm/sec for gold (Fig. 16). Thus, available bound states and favorable velocities both appear to be required for electron pick-up by ions in solids.

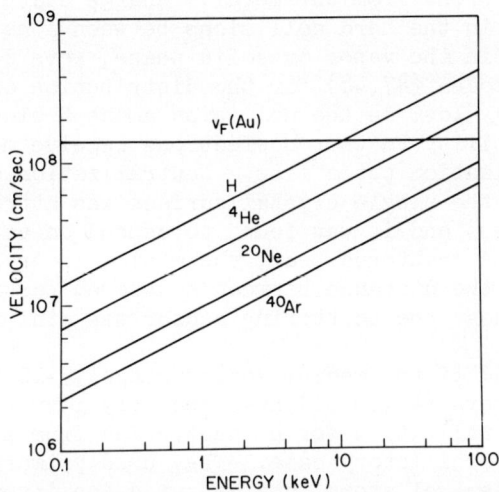

Fig. 16. *Ion velocity as function of energy for H, He, Ne and Ar, compared with Fermi velocity of electrons in gold.*

IV. CHARGE EXCHANGE BEFORE, DURING, AND AFTER THE PARTICLE-SURFACE ATOM COLLISION.

Considerable attention has been given in this chapter to the question of where the neutralization step occurs, inside or outside the solid. Little has been said about the specific details of charge exchange as a particle approaches the surface, collides with a surface atom and is reflected. Other chapters in these proceedings deal with the recently discovered oscillations in neutralization probability with incident energy, which occur for certain ion-atom combinations at low energy (41,42) and with other non-oscillatory processes at low energy (43).

In the medium energy range van der Weg et al. (44,45) studied the interactions between ion and solid in surface scattering by measuring the multiple charge-states of Ar and Cu ions resulting from collisions of 30-90 keV Ar^+ ions with Cu atoms in the vapor state (44) and also in the mono-crystalline state (45) with a channeling direction aligned with the ion beam to emphasize surface scattering by suppressing scattering from deeper layers. It was known (46) that the mean charge of Ar ions was lower after scattering from solid Cu than from the vapor and this difference was attributed to neutralization of the highly charged ions formed in the ion-atom collisions by capture of electrons from the metal. Charge state distributions produced in the hard collisions between ions and Cu atoms, whether in the vapor or solid phase, were described by a statistical model (47,48) for the distribution of the inelastic energy lost in the collision among N electrons to find the probability Pn that n electrons receive more energy than their ionization potentials. Neutralization of the ions scattered from the single crystal surface was attributed to the Auger process and it was found to depend on azimuthal direction of the scattered beam, the minima in Ar^+ ion yield occurring when the distance between scattered particles and surface atoms near the scattering center was minimal.

At the lower end of the medium energy range (3-10 keV) Verhey, Poelsema and Boers (49,50,51) have recently investigated the details of neutralization for He scattering from a Cu (100) surface, a non-oscillatory case. They used primary beams of ions and also neutral atoms, and measured ion intensity at scattering angles of 30° or 60° as a function of incidence angle (between ion beam and target surface) and primary energy. They also proposed a model which assumes that the trajectory of the reflected particles can be divided into three parts: the incoming path, the region of violent collision with a surface atom, and the outgoing path. Along the incoming and outgoing trajectories Auger neutralization may occur while ionization or neutralization may take place during the collision. It is not assumed that all particles start off as ions after the collision. In this respect the model differs significantly from that of van der Weg and Bierman (45) and the possible sequences of electron capture and loss which contribute to an average charge state of the scattered particles are more numerous. Only neutrals and singly ionized particles were considered in the model. Since neutral yields were not measured directly, only ions, an assumption about scattering behavior was required. The differential scattering cross-section for a fixed scattering angle was assumed to be a symmetric function of the incidence angle around specular reflection. This was confirmed by computer simulations. When needed, absolute cross-sections were derived from a suitable interaction potential, the Moliere approximation of the Thomas-Fermi. The dependence of ion yield on incidence angle at scattering angles of 30° and 60° was predicted successfully for primary beams of both ions and neutral He atoms. The characteristic velocity for Auger neutralization ($V_c \simeq 1.6$–2.1×10^7 cm/sec) and also ionization probabilities for He atoms colliding with Cu atoms (rising from 0 below 3 keV to ~20–30% at 10 keV) were derived (50,51). Neutralization probabilities during collision also increased with energy, maintaining a constant ratio of ~3.3 to the ionization probability (51). Further clarification of the situation was believed possible by direct measurements of ion fractions and of scattering angle influence.

V. ACKNOWLEDGMENT

The author wishes to thank J. E. Robinson, W. F. van der Weg, and H. Verbeek for helpful discussions and cooperation during the workshop, and F. E. P. Matschke for permission to include some of his experimental results in this review before publication.

VI. REFERENCES

1. van Wijngaarden, A., Miremadi, B., and Baylis, W.E., Can. J. Phys. 49, 2440 (1971).
2. Buck, T.M., and Wheatley, G.H., Surface Sci. 33, 35 (1973).
3. Feuerstein, A., Grahmann, H., Kalbitzer, S., and Detzmann, H., in "Ion Beam Surface Layer Analysis" (O. Meyer, G. Linker, F. Kappeler, Eds.) p. 471. Plenum Press, 1976.
4. McCracken, G.M., and Freeman, N.J., J. Phys. B Atom. Molec. Phys. 2, 661 (1969).
5. Eckstein, W., and Verbeek, H., J. Vac. Sci. Tech. 9, 612 (1972).
6. Verbeek, H., J. Appl. Phys. 46, 2981 (1975).
7. Allison, S.K., Rev. Mod. Phys. 30, 1137 (1958).
8. Hall, T., Phys. Rev. 79, 504 (1950).
9. Phillips, J.A., Phys. Rev. 97, 404 (1955).
10. Dissanaike, G.A., Phil. Mag. 7, 1051 (1953).
11. Armstrong, J.F., Mullendore, J.V., Harris, W.R., and Marion, J.B., Proc. Phys. Soc. (London) 86, 1283 (1965).
12. Brinkman, H.C., and Kramers, H.A., Proc. Akad. Amsterdam 33, 973 (1930).
13. Bohr., N., Kgl. Danske Videnskab. Selskab, Mat-fys. Medd. 18, No. 8 (1948).
14. Yavlinskii, Yu.N., Trubnikov, B.A., and Elesin, V.F., Izv. Akad. Nauk SSSR Ser. Fiz 30, 1917 (1966).

15. Trubnikov, B.A., and Yavlinskii, Yu.N., Sov. Phys. JETP 25, 1089 (1967).
16. Berkner, K.H., Bornstein, I., Pyle, R.V., and Stearns, J.W., Phys. Rev. A 6, 278 (1972).
17. Brandt, W., and Sizmann, R., Phys. Lett. 37A, 115 (1971).
18. Brandt, W., in "Atomic Collisions in Solids" (S. Datz, B.R. Appleton, and C.D. Moak, Eds.), Vol. 1, p. 261. Plenum Press, New York, 1975.
19. Cross, M.C., in "Inelastic Ion-Surface Collisions" (N.H. Tolk et al., Eds.).
20. Cross, M.C., Phys. Rev. B15, 602 (1977).
21. Chateau-Thierry, A., Gladieux, A., and Delaunay, B., in "Atomic Collisions in Solids" (F.W. Saris and W.F. van der Weg, Eds.), p. 553 North-Holland Pub. Co., Amsterdam, 1976.
22. Buck, T.M., Wheatley, G.H., and Feldman, L.C., Surface Sci. 35, 345 (1973).
23. Eckstein, W., Matschke, F.E.P., and Verbeek, H., J. Nuclear Materials, in press (1977).
24. Verbeek, H., Eckstein, W., and Datz, S., J. Appl. Phys. 47, 1785 (1976).
25. Buck, T.M., Feldman, L.C., and Wheatley, G.H., in "Atomic Collisions in Solids" (S. Datz, B.R. Appleton, and C.D. Moak, Eds.), Vol. 1, p. 331. Plenum Press, New York, 1975.
26. Berisch, Eckstein, W., Meischner, P., Scherzer, B.M.U., and Verbeek, H., in "Atomic Collisions in Solids" (S. Datz, B.R. Appleton, and C.D. Moak, Eds.), Vol. 1, p. 315. Plenum Press, New York, 1975.
27. Zaidins, C.S., quoted by J.B. Marion and F.C. Young in "Nuclear Reaction Analysis Graphs and Tables", p. 34. North-Holland Pub. Co., Amsterdam, 1968.
28. Dmitriev, I.S., Soviet Phys. JETP 5, 473 (1957).
29. Smith, D.P., J. Appl. Phys. 38, 340 (1967).
30. Ball, D.J., Buck, T.M., MacNair, D., and Wheatley, G.H., Surface Sci. 30, 69 (1972).
31. Chicherov, V.M., Sov. Phys. JETP Letters 16, 231 (1972).
32. Chen, Y.-S., Miller, G.L., Robinson, D.A.H., Wheatley, G.H., and Buck, T.M., Surface Sci. 62, p. 133 (1977).
33. Buck, T.M., Chen, Y.-S., Wheatley, G.H., and van der Weg, W.F., Surface Sci. 47, 244 (1975).
34. Oen, O.S., and Robinson, M.T., in "Atomic Collisions in Solids" (F.W. Saris and W.F. van der Weg, Eds.), p. 647, North-Holland Pub. Co., Amsterdam, 1976. See also Heiland, W., Taglauer, E., and Robinson, M.T., p. 655.
35. Robinson, J.E., Kwok, K.K., and Thompson, D.A., in "Atomic Collisions in Solids" (F.W. Saris and W.F. van der Weg, Eds.), p. 667, North-Holland Pub. Co., Amsterdam, 1976.

36. Hagstrum, H.D., Phys. Rev. 96, 336 (1954), and in "Inelastic Ion-Surface Collisions"(N.H. Tolk, et al., Eds.).
37. Lo, H.H., and Fite, W.L., Atomic Data 1, 305 (1970).
38. Feldman, L.C., Silverman, P.J., and Fortner, R.J., in "Atomic Collisions in Solids" (F.W. Saris, and W.F. van der Weg, Eds.), p. 29. North-Holland Pub. Co., Amsterdam, 1976.
39. van der Weg, W.F., Buck, T.M., and Wheatley, G.H., abstract in "Atomic Collisions in Solids" (F.W. Saris, and W.F. van der Weg, Eds.), p. 571. North-Holland Pub. Co., Amsterdam, 1976. Complete paper to be published, Surface Sci.
40. Matschke, F.E.P., Ph.D. Thesis, Technical University Munich, 1977, to be published.
41. Rusch, T.W., and Erickson, R.L., in "Inelastic Ion-Surface Collisions" (N.H. Tolk et al., Eds.).
42. Tully, J.C., and Tolk, N.H., in "Inelastic Ion-Surface Collisions" (N.H. Tolk et al., Eds.).
43. Heiland, W., and Taglauer, E., in "Inelastic Ion-Surface Collisions" (N.H. Tolk et al., Eds.).
44. van der Weg, W.F., Bierman, D.J., and Onderdelinden, D., Physica 44, 161 (1969).
45. van der Weg, W.F., and Bierman, D.J., Physica 44, 177 (1969).
46. Snoek, C., van der Weg, W.F., Geballe, R., and Rol, P.K., in Proc. 7th Int. Conf. on Phenomena in Ionized Gases, Belgrade, p. 145, 1966.
47. Russek, A., Phys. Rev. 132, 246 (1963).
48. Everhart, E., and Kessel, Q.C., Phys. Rev. 146, 27 (1966).
49. Verhey, L.K., Poelsema, B., and Boers, A.L., Rad. Effects 27, 47 (1975).
50. Verhey, L.K., Poelsema, B., and Boers, A.L., Nuc. Inst. and Methods 132, 565 (1976).
51. Verhey, L.K., Thesis, Groningen State Univ., The Netherlands (1976).

# oscillatory scattered ion yields in low energy ion-surface scattering

T. W. Rusch and R. L. Erickson
3M Company

Recent observations of structure in the energy dependence of scattered ion yields indicate that electronic interactions during the scattering process behave very much like ion-atom collisions. Oscillatory structure observed for the specific cases of He^+ scattering from Ga, Ge, As, In, Sn, Sb, Tl, Pb, and Bi is attributed to quasiresonant charge exchange between the vacant He^+ 1s state and the outer d states of these atoms. This structure is observed to be sensitive to the chemical environment of the target atom. In addition, Fourier analysis of the oscillatory structure suggests that more than one charge transfer mechanism may be present.

I.   INTRODUCTION

Low energy (0.1 - 5 keV) ion scattering from solid surfaces is receiving increasing attention as a tool for surface studies. (1-3) Surface properties such as elemental composition (4), defect formation (5), and surface and adsorbate structure (6) can be examined using the energy and angular distributions of scattered ions. These investigations have relied mainly upon the kinematics of the scattering process which can be described by classical models. The recent observation of oscillatory elastic scattered ion yields in ion-surface scattering (7) has stimulated interest in electronic interactions accompanying the scattering process. In this paper we will review some of the experimentally observed scattered ion yield characteristics, their implications for modeling low energy ion-solid interactions, and areas for further experimental and theoretical investigations.

II. SELECTED PRINCIPLES

In the discussion which follows, we will be interested in both the energy and number of ions scattered from a solid surface. Hence, we will briefly review the principles and experimental techniques relating to these quantities.

The scattering of an ion from a surface is considered to be the result of a single binary elastic collision between the ion and a surface atom (4) or of a sequence of binary elastic collisions (2). For a single collision, conservation of energy and momentum gives the relationship

$$\frac{E_1}{E_0} = \frac{1}{(1+\mu)^2} [\cos\theta + (\mu^2 - \sin^2\theta)^{1/2}]^2 \quad (1)$$

for $\mu = m_2/m_1 \geq 1$. In this equation, m_1 and E_0 are the incident ion mass and energy, respectively, m_2 is the target atom mass, E_1 is the energy of the elastically scattered ion, and θ is the laboratory scattering angle with respect to the incident ion direction. For a typical ion scattering measurement, E_0, m_1, and θ are known; E_1 is determined experimentally from the measured scattered ion energy distribution. These four quantities are used in Eq. (1) to infer m_2. An example of a scattered ion energy distribution is presented in Fig. 1 for 2.0 keV ^{4}He$^+$ scattering from Pb with $\theta = 90°$. Atoms of different mass on the surface will each produce an elastically scattered peak in the scattered ion energy distribution (within the mass resolution of the technique). In addition, the scattered ion energy distribution normalized with respect to E_0 is substantially independent of E_0 in the range from 0.1 to 5.0 keV (1,8).

The number, or yield, of singly scattered ions is of considerable interest for quantitative surface analysis. In this discussion, we are interested in particular in the dependence of the yield on incident ion energy or velocity. The measured

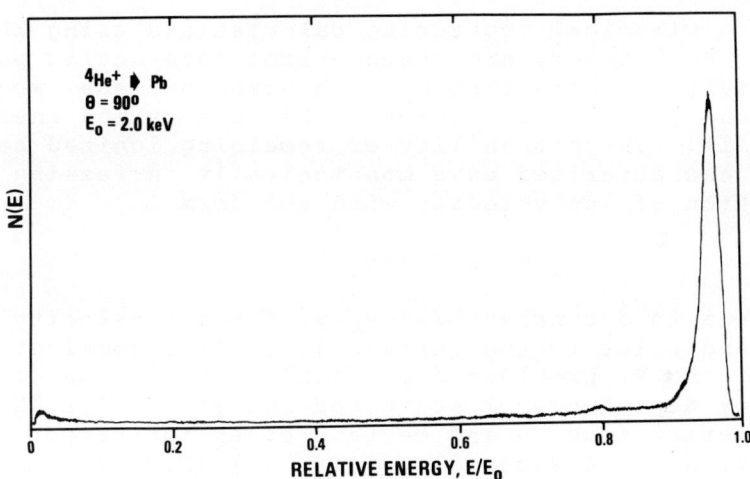

Fig. 1. Energy distribution of secondary and scattered ions which indicates the prominent peak for $^4He^+$ elastically scattered from Pb.

yield of elastically scattered ions, Y_i, from the ith constituent of a surface can be expressed as

$$Y_i(E_o,\theta) = k\, N_i\, I_o\, T\, D\, \Delta\Omega\, \sigma_i(E_o,\theta)\, P_i(E_o,\theta) \quad (2)$$

where N_i is the surface density of the ith constituent, I_o is the primary ion current, T is the analyzer transmittance, D is the detector sensitivity, $\Delta\Omega$ is the analyzer acceptance angle, $\sigma_i(E_o,\theta)$ is the differential scattering cross section, $P_i(E_o,\theta)$ is the probability that the ion remains ionized after scattering, and k contains the appropriate conversion factors (9). For a given experimental configuration, $\Delta\Omega$ is constant and the energy dependencies of T, D, and I_o are measurable. When using elemental samples, the value of N_i should be roughly independent of E_o. Hence, if the yield is normalized with respect to these factors,

$$Y_i(E_o,\theta) \propto \sigma_i(E_o,\theta)\, P_i(E_o,\theta) \quad (3)$$

and the energy dependence of the normalized scattered ion yield represents the energy dependence of the product $\sigma_i P_i$.

Classical scattering calculations using the Bohr, Born-Mayer, and Thomas-Fermi interaction potentials indicate that σ_i is a structureless, monotonically decreasing function of primary ion energy (10,11). The probability of remaining ionized has been characterized as a monotonically increasing function of ion velocity with the form

$$P_i = \exp(-a/V_\perp) \qquad (4)$$

where a is a constant and $v_\perp$ is the ion velocity perpendicular to the surface (1). This combination of σ_i and P_i provides a plausible explanation of the energy dependence of scattered ion yields for $^4He^+$ scattering from Cu and certain other ion-target combinations as discussed by Smith (1) (Fig. 2).

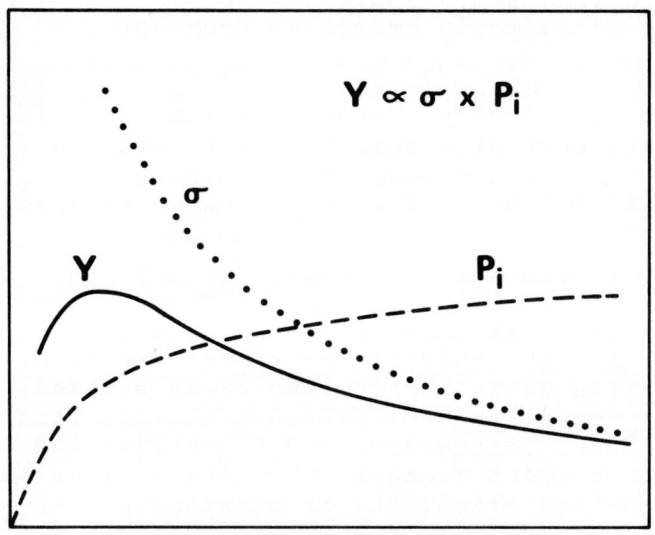

Fig. 2. *Qualitative energy dependencies of the elastic scattering cross section, σ, the probability of the scattered particle remaining ionized, P_i, and the resultant ion yield, Y.*

The experimental results of present interest depart significantly from this model. For example, the dependence of the scattered ion signals on E_o for $^4He^+ \rightarrow Pb$ is presented in Fig. 3. The broad dashed line highlights the pronounced oscillation in the

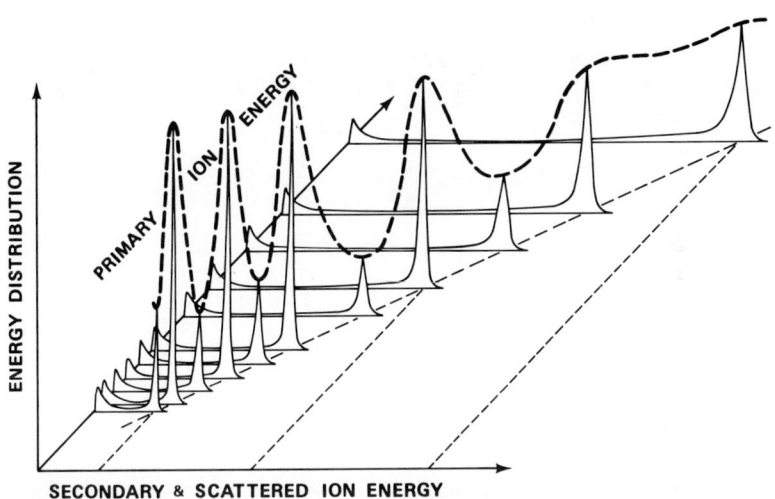

Fig. 3. Energy distributions of the secondary and scattered ions at selected values of primary ion energy. The dashed curve highlights the oscillatory variation of the elastically scattered ion signal with primary ion energy.

elastically scattered ion signal which occurs as a function of incident ion energy. Note that these energy distribution curves are normalized with respect to primary ion current but not energy analyzer transmittance. These results and others to be presented in Section III indicate that the relatively simple scattered ion yield curve observed for $He^+ \rightarrow Cu$ is not generally obtained. Some structure, often non-oscillatory, is present in the yield curves for many ion-surface combinations. These observations, as well as those of Hagstrum and Becker, (12) suggest that multiple neutralization processes are operative. If we define the ion neutralization probability P_n to be

$$P_n = 1 - P_i \tag{5a}$$

then multiple neutralization processes can be expressed as

$$P_n(E_o, \theta) = \sum_j P_{nj}(E_o, \theta) \tag{5b}$$

where the summation is over the j individual neutralization processes.

To relate these probabilities more directly to experimentally tractable quantities, it seems to be appropriate to define P_{nj} as

$$P_{nj}(E_o, \theta) = Q_j(E_o, \theta)/\sigma_i(E_o, \theta). \tag{6}$$

In this equation, Q_j is the cross section for the jth neutralization process and σ_i is the differential scattering cross section for the ith element as noted previously. Substituting Eqs. (5a), (5b), and (6) into Eq. (3),

$$Y_i(E_o, \theta) \propto \sigma_i(E_o, \theta) - \sum_j Q_j(E_o, \theta). \tag{7}$$

With the yield expression reformulated in this manner, structure in the scattered ion yield curves can be directly assigned to structure in the charge transfer (i.e., neutralization) cross sections. This also provides a format for relating Q_j values between solid and gaseous targets.

III. CLASSIFICATION OF SCATTERED ION YIELDS

Scattered ion yields as a function of primary ion energy have been measured for He^+, Ne^+, and Ar^+ scattering from elements and compounds in selected regions of the periodic table. These data were collected either manually (7) or automatically by electronically scanning E_o and synchronously tracking the maximum value of the scattered ion signal (13,14).

For convenience of presentation and handling, the energy dependencies of these scattered ion yields have been classified according to four general shapes (Fig. 4). Some physical significance can be attributed to the designation of classes I and II, but classes III and IV have been defined purely

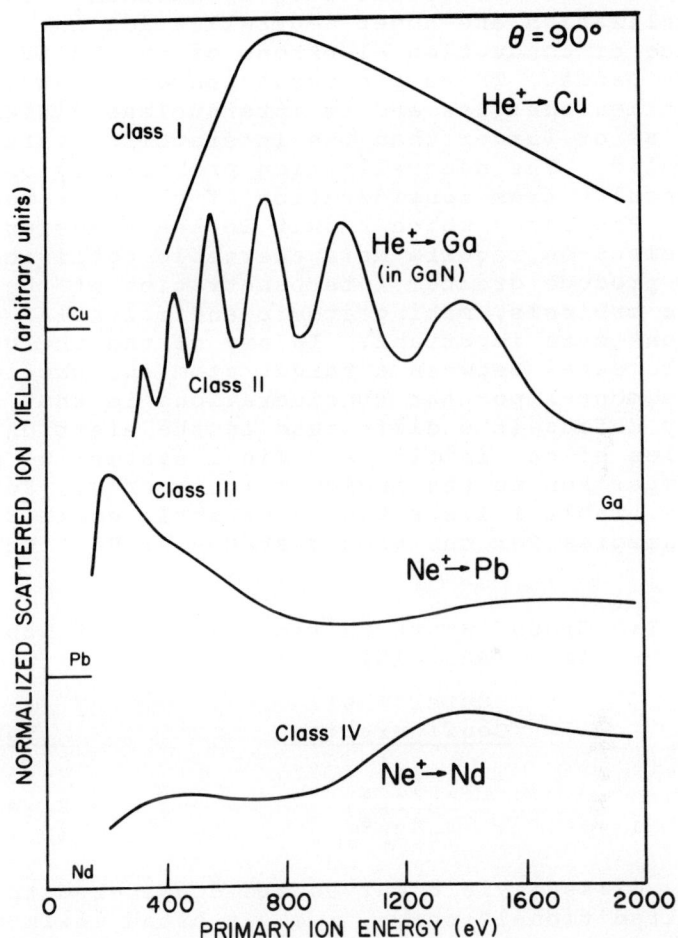

Fig. 4. Representative normalized ion yield curves illustrating the four general classes observed for He^+, Ne^+ and Ar^+ scattering from solid surfaces. (After Rusch and Erickson (13).)

on the basis of curve shape. Also, the classes are subject to the limited range of energy (100-2500 eV) so that a different classification scheme might be necessary if the data were obtained over a larger energy range.

The factors that govern the yield curve shapes are not all understood, but it is believed that several different neutralization processes may be important. Processes such as resonance tunneling neutralization and Auger neutralization involve the valence or conduction electrons of the solid state energy bands. These processes can occur at low incident ion energies and at internuclear distances as large as or larger than the interatomic distances of the solid. The neutralization probability of Eq. (4) results from consideration of these processes (40). Processes which result in the transfer of a core electron require more energetic collisions which produce greater interpenetration of the electronic orbitals, making atomic and molecular considerations more important. In any of the charge transfer processes between a target atom and the ion, one of the most important considerations is that the energy defect (the difference in the electronic energies of the initial and final states) be small in comparison to the incident ion energy. For reference, Table I lists the outer shell configurations and energies for the ground states of He^+, Ne^+, and Ar^+.

Table I: Ground State Characteristics of Noble Gas Ions (Ref. 15)

Ion	Outer Shell Configuration	Energy
He^+	$1s^1$	24.580 eV
Ne^+	$2s^2 2p^5$	21.559 eV
Ar^+	$3s^2 3p^5$	15.755 eV

Class I yield curves are considered to have the "traditional" shape, i.e., a broad maximum followed by a monotonically decreasing yield. As discussed in the previous section, this energy dependence can be explained using a monotonically decreasing differential scattering cross section and a monotonically increasing probability of the scattered ion remaining ionized. Class I yield curves have been observed for He^+ scattering from Al, Si, Cu, Zn, Zr, Ag, Cd, and Ta.

For the example of $He^+ \rightarrow Cu$ (Fig. 4), the outermost states of Cu are the 3p and 3d states having binding energies (16) of 78 and 6 eV, respectively, which differ significantly from the 24.58 eV energy of He^+. (A 4 eV work function has been used to refer the tabulated binding energies for solid targets to the vacuum level.) It seems unlikely that the 3p electron of Cu would be excited to the He$^+$ ground state at the energies used in these measurements, and so one might expect that neutralization takes place by processes involving only the higher energy states. In the case of Ta, there are 4f states of 29 eV and 31 eV. Although these states do not pose a large energy defect, no direct evidence has been found for core electron transfer processes.

Class II yield curves have a rapidly oscillating amplitude with an envelope similar to a class I curve. Such curves have been observed when He$^+$ ions are scattered from elements having d-electron states close in energy to the He$^+$ ground state. Erickson and Smith (7) have attributed this oscillatory structure to quasiresonant charge exchange between the vacant He$^+$ 1s state and a d-state in the solid. Class II yield curves have been observed for He$^+$ scattering from Ga, Ge, As, In, Sn, Sb, Tl, Pb, and Bi, as well as for Ne$^+$ scattering from Ga. The characteristics of these oscillatory yields will be discussed in greater detail in the following section.

Class III yield curves have a maximum at energies below 1 keV with a broad secondary maximum at higher energy. Often there is some fine structure in the curves but it is not generally of an oscillatory nature. Class IV yield curves have the general characteristic of slowly varying or increasing yield with increasing primary ion energy. Subtle oscillatory structure or other irregular structure as illustrated in Fig. 4 is observed. It is not possible at present to assign a mechanism to the structure observed in the Class III and Class IV yield curves. A number of neutralization processes could be possible and determination of a particular process for a given system will have to await further work (13).

Table 2 summarizes the classification of scattered ion yield curves exhibited by ion-solid combinations examined thus far. Sections IV and VI will review the characteristics of the oscillatory yields in more detail and point out additional considerations not incorporated into this simple classification scheme.

IV. OSCILLATORY SCATTERED ION YIELDS

The discussion in Section II indicated that

Table 2: Preliminary Classification of Experimental Scattered Ion Yields, $\theta = 90°$.

Class	Incident Ion		
	He^+	Ne^+	Ar^+
I	Al, Si, Ni, Cu, Zn, Zr, Nb, Pd, Ag, Cd, Ta, Pb, Au	Zn, Sb, Te, W	
II	Ga, Ge, As, In, Sn, Sb, Tl, Pb, Bi	Ga	
III		Si, S, Ni, Cu, Ge, Pd, Ag, In, Sn, Hf, Pt, Au, Pb	In, Tl, Pb
IV	S, Te, Sc, La, Ce, Nd, Sm, Gd, Dy, Er, Yb, Hf	Cd, La, Ce, Nd, Sm, Gd, Dy, Er, Yb, Tl	Cu, Ge, Pd, Ag, Cd, Sn, Sb, Te, La, Ce, Nd, Sm, Gd, Dy, Er, Yb, Hf, Pt, Au

the scattered ion yield is proportional to a kinematic parameter (the differential scattering cross section) and an electronic parameter (the probability of the incident ion remaining ionized after the collision; which can be related to charge transfer cross sections). In this section we will discuss these parameters and the experimental results which have allowed the general mechanism producing oscillatory yields in ion-solid scattering to be inferred from models of oscillatory charge transfer cross sections in gas phase ion-atom scattering.

Processes other than those associated with the ion-solid interaction (e.g., charge transfer to the ambient gas) are most unlikely because the scattered ion yields depend specifically upon the target material. Also, kinematic processes have been dismissed as the cause of the oscillations in the yield for reasons discussed below.

As noted in Section II, classical scattering

calculations indicate that the differential scattering cross section, σ, is a structureless, decreasing function of primary ion energy. Furthermore, σ is a slowly varying function of target atom mass as shown in Figure 5 for the screened Coulomb potential (11).

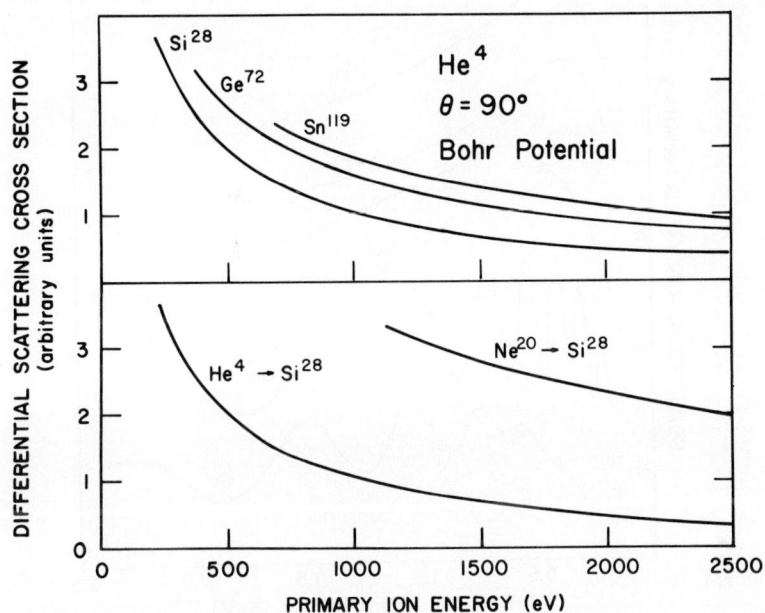

Fig. 5. Differential scattering cross section as a function of primary ion energy calculated using the Bohr potential.

Other commonly used potentials exhibit similar behavior (10). Experimentally, however, it is observed that both the occurrence and the details of oscillatory structure vary significantly in adjacent elements. For example, oscillatory yields are observed for Ga but not Zn and for In but not Cd. In addition, the oscillatory structure differs in the sequences Ga-Ge-As, In-Sn-Sb, and Tl-Pb-Bi (Figs. 6-8).

Other possible kinematic contributions to the oscillatory yield might be multiple scattering or

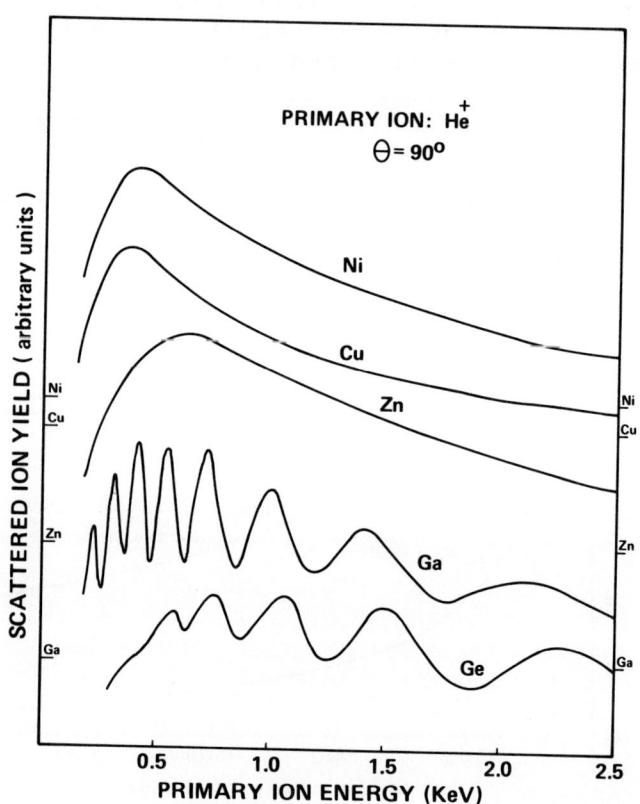

Fig. 6. Scattered ion yield curves for $^4He^+$ scattering from a sequence of elements which illustrates the effect of 3d-electron energy on the presence of oscillatory structure.

channeling. These possibilities are discounted by examining the scattered ion yield from an element in different crystalline environments. For $^4He^+$ scattering from indium in metallic In, and InP and $ZnIn_2S_4$ crystals, the yield curves exhibit important subtle differences (Fig. 9) which appear to relate to the chemical environment of the In (Section V). In addition, changing the target angle with respect to the incident ion beam does not alter the oscillatory structure (17). Hence, it is concluded that multiple scattering or channeling which are dependent

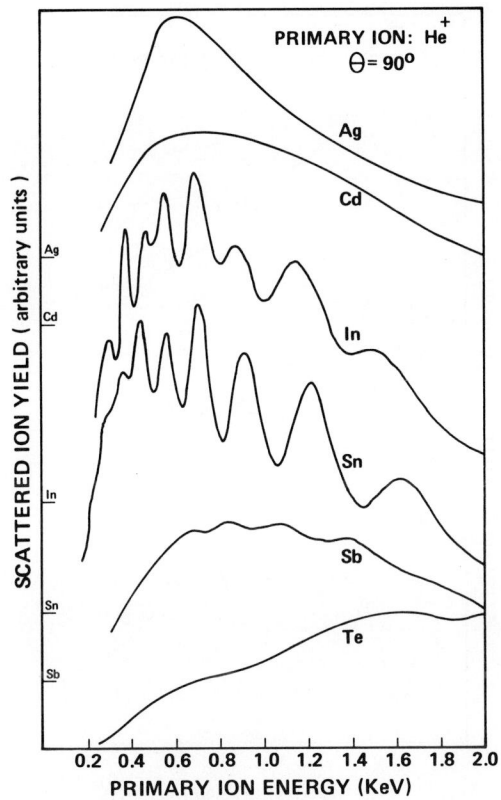

Fig. 7. Scattered ion yield curves for $^4He^+$ scattering from a sequence of elements which illustrates the effect of 4d-electron energy on the presence of oscillatory structure.

upon crystalline structure and angle of incidence do not contribute significantly to the oscillatory scattered ion yield.

Rather than being kinematic in origin the oscillatory scattered ion yields probably originate from electronic charge transfer processes associated with the ion-solid interaction. Within the mathematical framework of the ion yield, these processes are contained in the ion neutralization probability, P_n, or, as discussed in Section II, in the individual

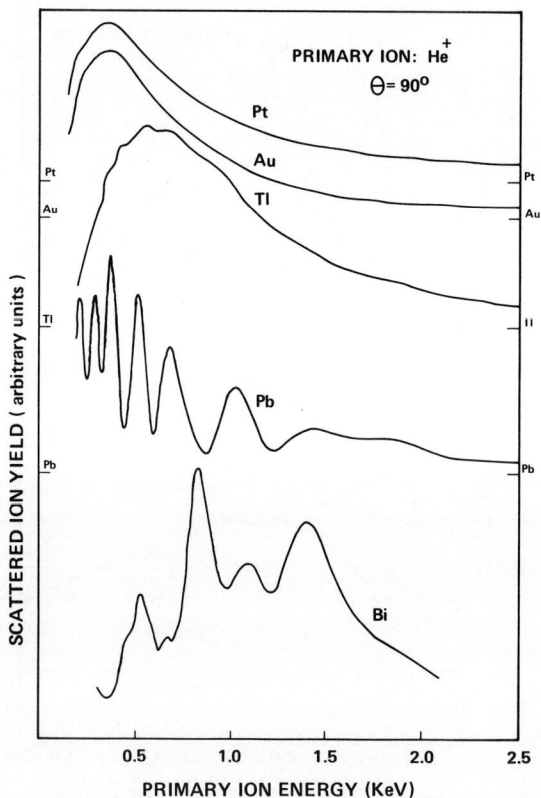

Fig. 8. Scattered ion yield curves for $^4He^+$ scattering from a sequence of elements which illustrates the effect of 5d-electron energy on the presence of oscillatory structure.

cross section terms, Q_j. In the gas phase, oscillatory charge transfer cross sections (both differential and total) have been observed for a number of ion-atom and ion-molecule collisions (18-20). The accepted mechanism for these observations is resonant (21) or quasiresonant (22) charge exchange.

A basic theoretical model for oscillatory charge transfer cross sections assumes that a single electron is transferred between two resonant or quasiresonant states of the collision complex. In

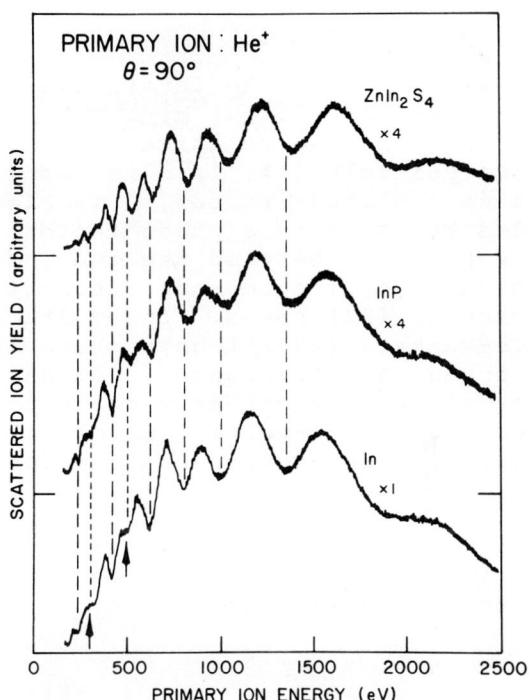

Fig. 9. Scattered ion yield curves for $^4He^+$ scattering from indium in samples of metallic In, and single crystals of InP and $ZnIn_2S_4$. The dashed lines reference the energies at which minima occur for the case of metallic In. These curves have not been normalized with respect to analyzer transmittance. (After Rusch and Erickson (13).)

the particular case of ion-surface scattering, the collision complex appears to be formed by the incident ion and the target atom on the surface. Quantum mechanical phase interference between the two states leads to an oscillatory differential charge transfer cross section of the form

$$Q(p,v) = A_1(p,v) + A_2(p,v)\sin^2[\beta/\hbar - \delta] \quad (8)$$

where

$$\beta = \int_{r_o}^{\infty} \frac{\Delta E(r)}{v(r)} \, dr. \tag{9}$$

In these equations, r is the internuclear separation, r_o is the distance of closest approach, p is the impact parameter, v is the relative velocity between the colliding particles, A_1 is a non-oscillatory term, A_2 is an amplitude factor, $\hbar$ is Planck's constant divided by 2π, δ is a phase factor, and ΔE is the energy separation between the two interacting states. Eq. (8) applies to both resonant (23,24) and quasiresonant (22) cases. In addition, F. J. Smith and co-workers (25,26) have shown that for two classes of potential difference functions, $\Delta E(r)$, (i.e., those exhibiting either a maximum as a function of r or a strong repulsive core), the <u>total</u> charge transfer cross-section has the form of Eq. (8).

When the relative velocity is approximately constant during the collision, as is the case for scattering through small angles, Eq. (9) becomes

$$\beta = \frac{1}{v} \int_{r_o}^{\infty} \Delta E(r) \, dr = \frac{<Er>}{v}. \tag{10}$$

Substituting this result into Eq. (8),

$$Q(p,v) = A_1(p,v) + A_2(p,v) \sin^2\left\{\frac{<Er>}{\hbar v} - \delta\right\}. \tag{11}$$

This equation indicates that the period of the oscillatory structure should be uniform when plotted as a function of 1/v. In addition, when using different isotopes for the primary ions (e.g., ^{3}He and ^{4}He), the oscillatory structure should be essentially identical when plotted as a function of 1/v.

Under the conditions of relatively large velocity v and large scattering angle θ, the parameter r_o is small and slowly varying with v and θ so that the integral of Eq. (10) is relatively constant. Hence, the velocity dependence of the oscillating structure should be nearly independent of scattering angle. This behavior has been observed for gas phase resonant charge exchange (27).

Although the idealized conditions discussed above are not, in general, satisfied in the experiments considered here, no large deviations from predicted behavior are expected. This is borne out by the experiments discussed below where results of the isotope and angular dependencies of oscillatory cross sections in ion-solid interactions are in good agreement with the resonant charge exchange model.

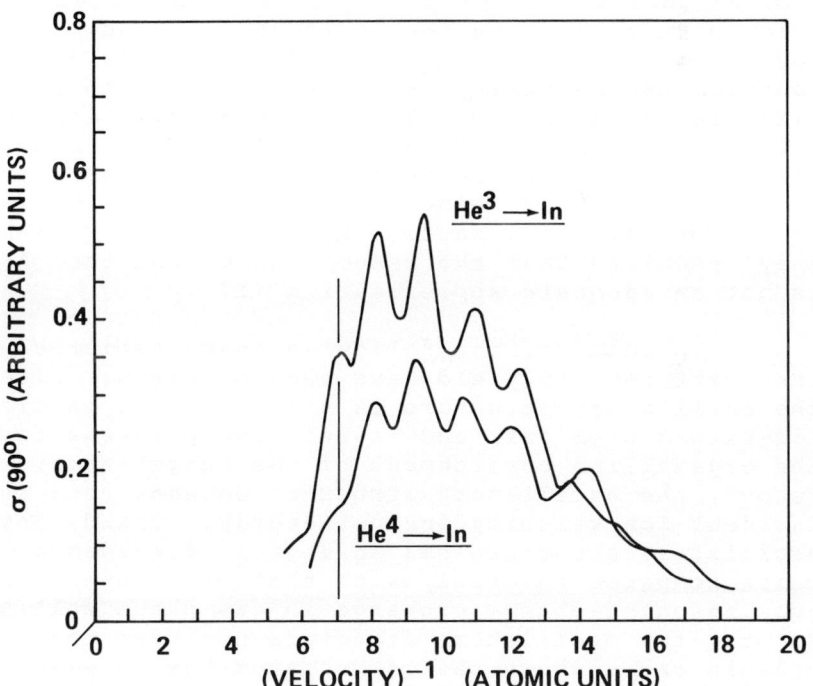

Fig. 10. Scattered ion yield versus the reciprocal incident ion velocity for $^3He^+{\rightarrow}In$ and $^4He^+{\rightarrow}In$. (After Helbig and Adelman (28).)

Erickson and Smith (7) and Helbig and Adelman (28) measured the scattered ion yields for ^3He and ^4He ions scattering from several solids (Fig. 10). In accordance with the above model, the positions of the maxima and minima are found to depend on ion velocity and not energy. The deviations from the expected behavior are being investigated by Helbig, who has postulated that the observation of a peak (or lack of it) near seven inverse velocity units may be related to differences in the distance of closest approach of the two isotopes. Note also in Fig. 10 that the spacing of the minima is not constant but increases with increasing inverse velocity. This is a result of the breakdown in the assumptions leading to Eq. (10).

Tolk et al (17) measured the scattered ion yields for $^4\text{He}^+ \rightarrow \text{Pb}$ and $^4\text{He}^+ \rightarrow \text{Ga}$ (in GaP) as a function of laboratory scattering angle and target angle with respect to the incident ion beam. The positions of maxima and minima in the yield as a function of scattering angle are observed to behave differently in the two cases. The angular dependence for $^4\text{He}^+ \rightarrow \text{Ga}$ can be fit quite nicely using a simplified model for quasiresonant charge exchange which assumes a constant value for $\Delta E(r)$ in Eq. (10), within an estimated interaction distance. For $\text{He}^+ \rightarrow \text{Pb}$, they speculate that the assumption of constant $\Delta E(r)$ is not an adequate approximation (17).

In summary, the following characteristics of the scattered ion yield have been observed. First, the oscillatory structure is specific to a particular ion-target atom pair and is relatively insensitive to the crystalline environment of the target atom. Second, the oscillatory structure depends upon the incident ion velocity and not energy. Third, the oscillatory structure has an angular dependence which compares favorably with that of resonant or quasiresonant charge exchange in gas phase collisions. Fourth, the oscillatory structure has been observed only in cases where the target atom has an electronic state close in energy to the He^+ ground state. It is concluded, therefore, that the predominant mechanism producing the oscillatory scattered ion yields for ion-solid scattering is resonant or quasiresonant charge exchange in agreement with the conclusion of Erickson and Smith (7).

V. CHEMICAL EFFECTS

As noted in the previous section, subtle

differences are observed in the scattered ion yields for ^{4}He$^+$ scattering from In in different crystalline environments. Referring to Fig. 9, two notable differences exist: the oscillatory structure in the compound yield curves is shifted to higher primary energy values with respect to the elemental In yield curve, and shoulders in the elemental In yield curve at 300 and 600 eV are distinct minima in the compound curves.

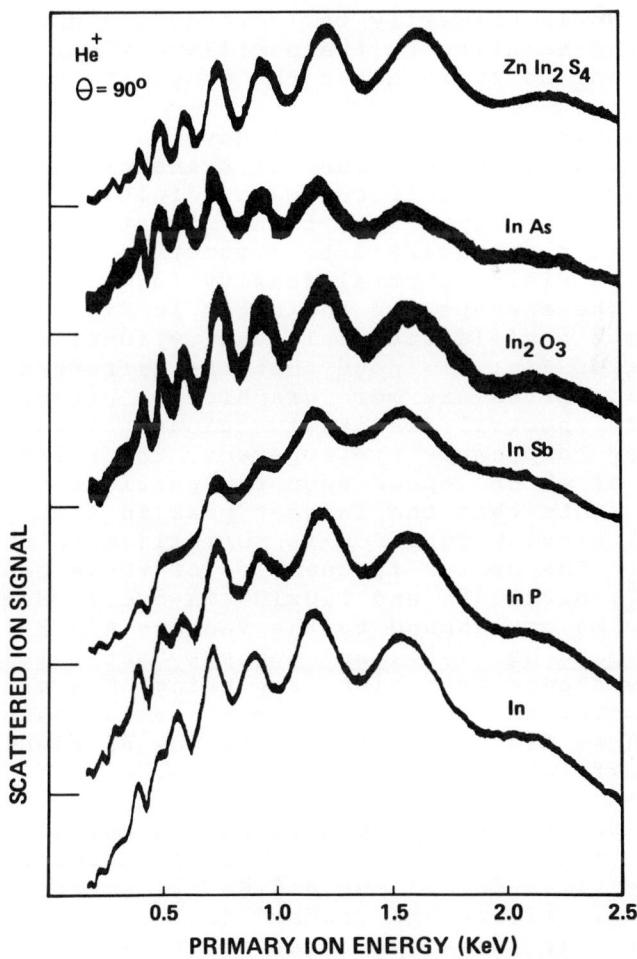

Fig. 11. *Scattered ion yield curves for ^{4}He$^+$ scattering from indium in different chemical environments. (After Christensen et al (29).)*

Fig. 11 presents the scattered ion yield curves for In in InP, InSb, InAs, In_2O_3, $ZnIn_2S_4$, and elemental In. With the exception of InSb, the oscillatory structure in the yield curves from the compounds is shifted to higher values of primary ion energy than for metallic indium. In the particular case of InSb, the scattered ion peaks for In and Sb are unresolved by the experiment, so that the yield curve for InSb is the sum of the individual yields for $^4He^+ \rightarrow In$ and $^4He^+ \rightarrow Sb$. This latter yield has been found to have only very weak oscillatory structure so that the net result of the superposition of the two yields is primarily to increase the size of the background relative to the oscillatory structure from In and possibly alter the shape of the trend.

Christensen et al (29) have analyzed the yield data of Fig. 11 with a Fourier transform technique. In essence, each yield curve was digitized, the oscillatory portions were then extracted from the background and converted to a function of 1/v, and, finally, a yield spectral density function was computed. These steps are indicated in Fig. 12. Figure 13 shows the yield spectral density functions for the data of Fig. 11; note that the differences in the yield curves are more graphically presented by the yield spectral density functions. Three distinct frequency components are apparent, the relative amplitudes of which depend upon the particular In compound. (Note that the largest peak in each yield spectral density function is normalized to unit amplitude.) The center frequencies of these peaks are 3.3×10^{15}, 6.2×10^{15}, and 1.0×10^{16} Å-cycles/sec. These frequencies correspond to the factors $\pi/2$ β in Eq. (8) or $\pi\beta$ in Eq. (10), and suggest that multiple charge exchange mechanisms are being observed. Furthermore, the apparent harmonic relationship among these frequency components may be significant in the extraction of the mechanisms.

As a check on the analysis technique, a Fourier transform analysis of total charge exchange cross sections for $Li^+ \rightarrow Na$ and $Na^+ \rightarrow Li$ scattering was performed. Melius and Goddard (30) calculated cross sections for these two cases that closely fit the measured cross sections of Daley and Perel (20). Their calculations were based on having two interfering states coupled to higher lying states. The

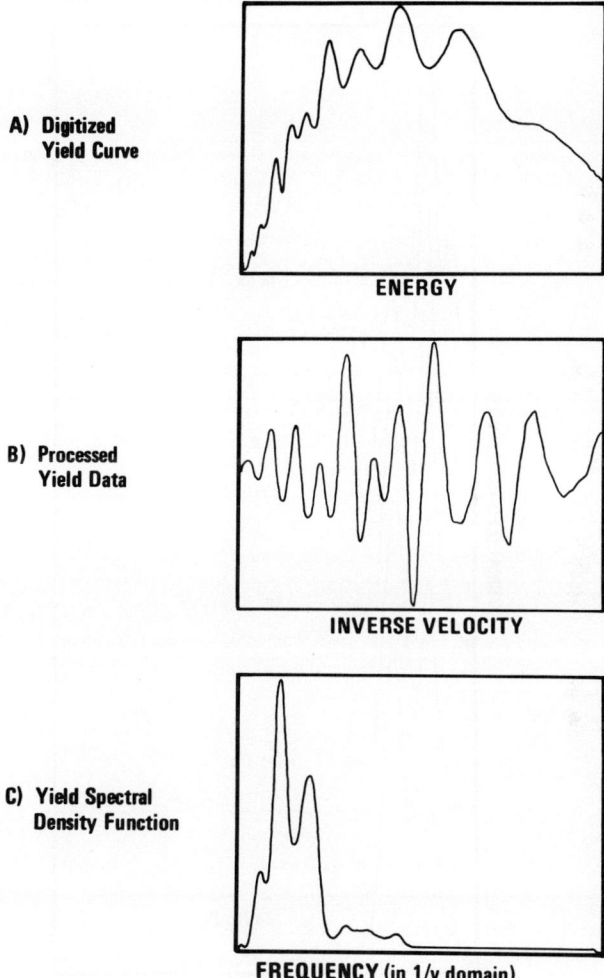

Fig. 12. Example of processed signals for $^4He^+ \to In$ (in InP), $\theta = 90°$. a) Digitized yield curve. b) Yield curve after normalization with respect to analyzer transmittance, $E^{-1/2} \propto 1/v$ abscissa transformation, and removal of slowly varying trend. c) Magnitude of the Fourier transformation of "b" after interpolation and smoothing. (After Christensen et al (29).)

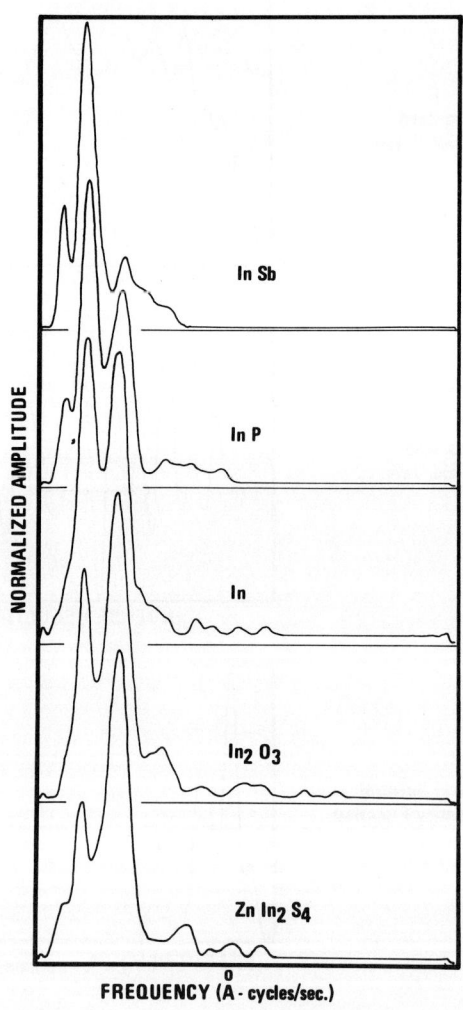

Fig. 13. Yield spectral density functions for $^4He^+$ scattering from indium in different chemical environments. Each curve is normalized such that the largest peak has unit amplitude. (After Christensen et al (29).)

oscillations resulted solely from the two interfering states, and the coupling to higher states served mainly to adjust the phase of the oscillations.

Since a single oscillatory mechanism is involved, the Fourier analysis should give only a single component in the yield spectral density function, as was obtained (Fig. 14).

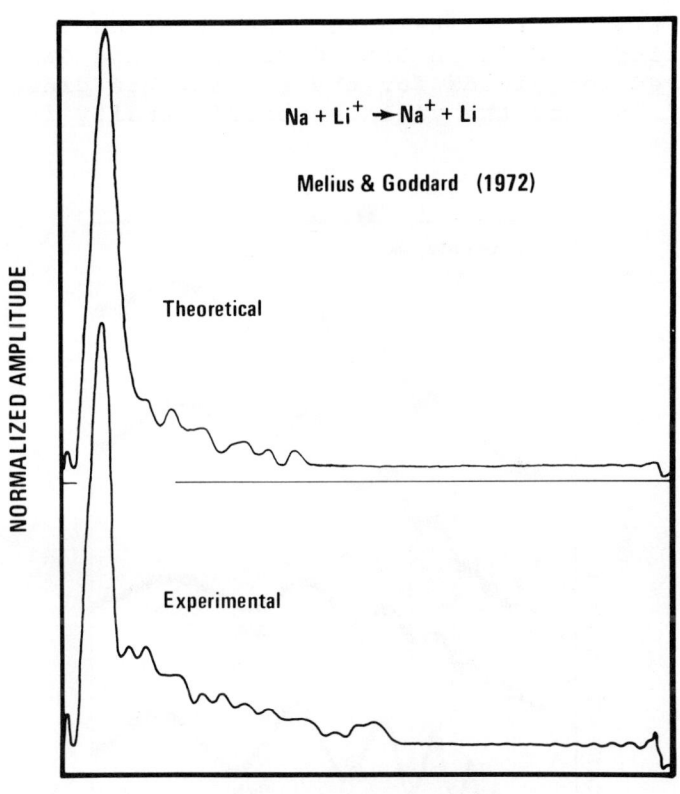

Fig. 14. *Yield spectral density functions for $Li^+ \to Na$ calculated from the theoretical results of Melius and Goddard (30) and the experimental observations of Daley and Perel (20). (After Christensen et al (29).)*

We presently speculate that the two major frequency components for $^4He^+ \to In$ may arise from two different charge transfer processes involving transfer from the 4d states of In to the He ground state. The splitting of the d states in the solid may be significant in these processes. Because the frequencies of these components do not change substan-

tially as a function of chemical environment, the
amplitude variations in Fig. 13 and the oscillatory
structure shifts with respect to primary ion energy
in Fig. 11 may both result from altered coupling to
higher lying states in the solid (30-32).

Similar results have been observed for ^{4}He$^+$
scattering from Ga in GaN, GaSb, GaP, and GaAs. The
scattered ion yields for these cases are presented
in Fig. 15, and the yield spectral density functions

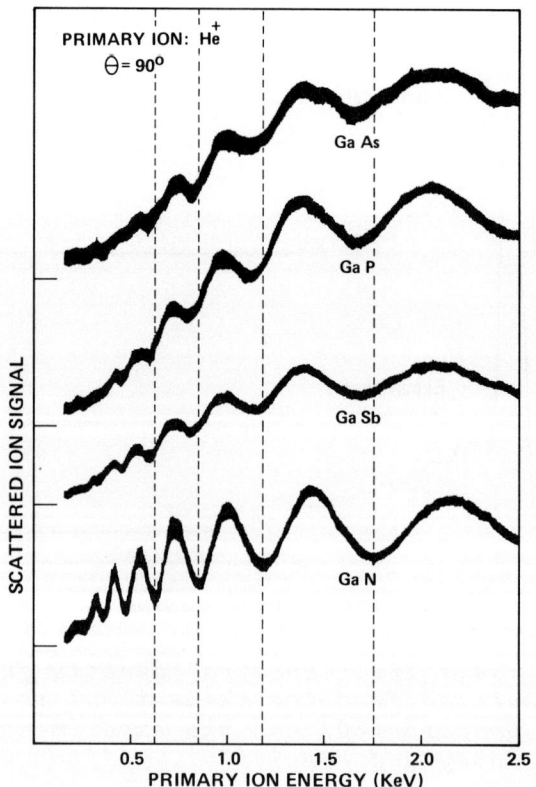

*Fig. 15. Scattered ion yield curves for ^{4}He$^+$
scattering from gallium in different chemical environments.*

in Fig. 16. Again, two major frequency components
are present with varying relative amplitudes, although the highest frequency component may well con-

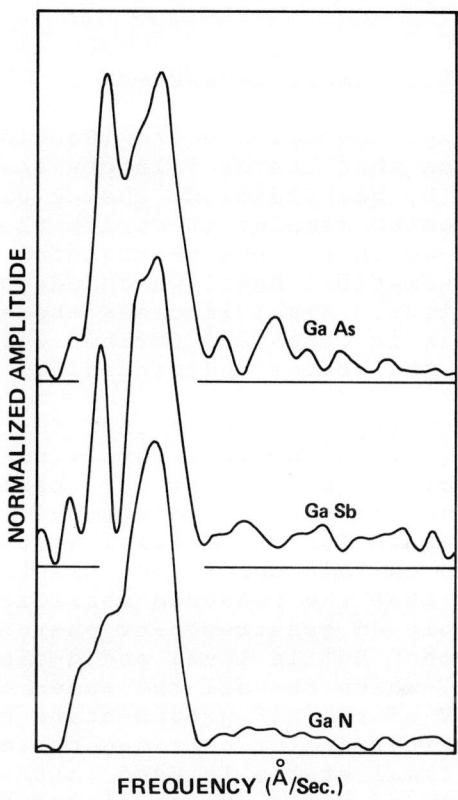

Fig. 16. Yield spectral density functions for $^4He^+$ scattering from gallium in different chemical environments. Each curve is normalized such that the largest peak has unit amplitude.

tain two subcomponents. The interpretation is similar to the indium case except that the 3d states are appropriate for Ga.

In summary, the oscillatory structure in the scattered ion yield for He^+ scattering from In and Ga has been observed to depend upon the chemical environment of the atoms. Fourier analysis of the oscillatory structure indicates two major components whose relative amplitudes depend significantly upon chemical environment. It is felt that the Fourier transform technique enhances the observation of the chemical effects and may facilitate understanding of the underlying mechanisms.

VI. CHARGE TRANSFER MECHANISMS

The experimental results (Section IV) support the conclusion that the oscillatory scattered yields are produced by quasiresonant charge exchange. However, the problem remains to verify the electronic states involved in the charge transfer process through a theoretical model which adequately explains the observations. Also, in cases where non-oscillatory structure is observed, further work is required to determine the states and mechanisms involved.

Lichten (22) has shown that for asymmetric collisions (i.e., $A^+ + B \rightarrow A + B^+$) involving charge exchange, one of the conditions for quasiresonant charge exchange is that the electronic energies be nearly degenerate for the initial state and final state. Based on this condition, Erickson and Smith (7) proposed that the measured oscillatory ion yields were the result of quasiresonant charge transfer between the vacant He^+ 1s level and d-electron levels in the solid, which for all the cases observed were within ±10 eV of the He^+ ground state energy (24.58 eV). This criterion for near-degeneracy of the initial and final states is used in the expanded classification of yield curves discussed in Section III.

Examples of the experimental support for He^+ 1s$\leftrightarrow$d interaction are presented in Figs. 6-8. As an illustration, the five adjacent elements Ag, Cd, In, Sn, and Sb, have 4d-electron binding energies (16) of 7 eV, 13 eV, 20 eV, 28 eV, and 35 eV, respectively. As the d-electron energy approaches 24.58 eV, pronounced oscillations in scattered ion yield are superimposed upon the Class I curve (Fig. 7). As the d-electron energy exceeds 24.58 eV by increasingly greater amounts, the oscillations subside. This dependence upon energy defect is consistent with quasiresonant charge exchange. Also, the oscillations are strongest for the smallest defect (Pb, $\Delta E < 1 eV$) and the onset of oscillation occurs at a lower ion velocity in accordance with theory (30, 33,34) (e.g., compare Ga and Ge with levels at ~ 22 eV and 33 eV respectively).

Additional justification of the He$^+$ 1s↔d interaction is presented by Tolk et al (17) who estimated the value of β (Eq. 9) for these states in the cases of ^{4}He$^+$→Ga (in GaP) and ^{4}He$^+$→Pb to be 19 eV Å and 22 eV Å, respectively. These compare favorably with their respective experimental values of 23.5 eV Å and 17.7 eV Å (17). Corresponding experimental values that we obtained are 26 eV Å for ^{4}He$^+$→Ga (in GaN) and 23 eV Å for ^{4}He$^+$→Pb (Fig. 17).

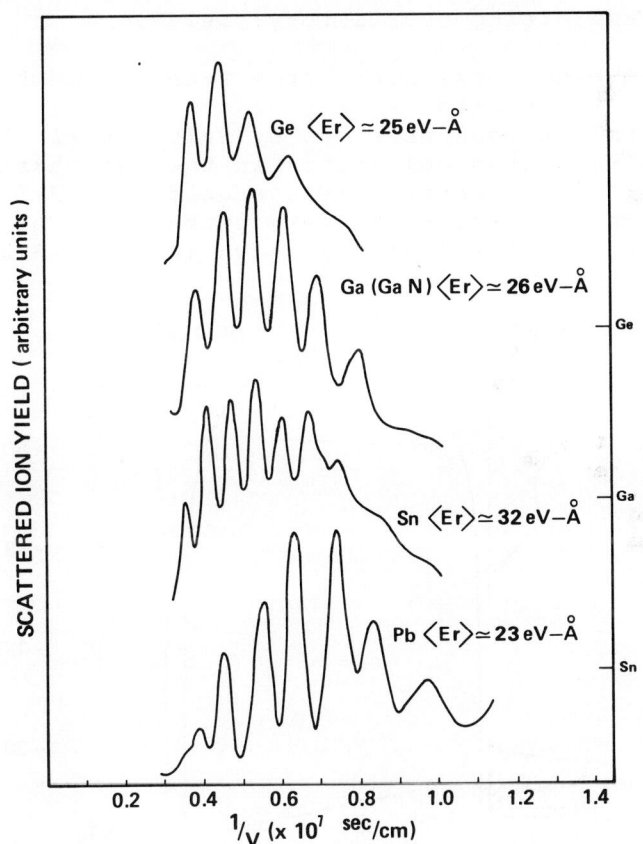

Fig. 17. Scattered ion yield curves versus reciprocal incident ion velocity indicating <Er> values computed from the 1/v differences between minima or maxima (Eq. 11).

A possible method for checking the target atom state involved would be to look for Auger electrons emitted from the solid due to the de-excitation of the target atom following transfer of a core electron. Fano and Lichten (35) predicted that electrons with discrete energies should be observed from gas phase ion-atom collisions due to Auger and autoionization processes; they have been observed by Rudd et al (36) for $Ar^+ \to Ar$ and $Ar^+ \to Cu$ (ion-solid) collisions, and more recently by Grant et al (37) for $Ar^+ \to Mg$, $Ar^+ \to Si$ (ion-solid) collisions. These experiments were not performed over a range of primary ion energies and did not use ion-target combinations exhibiting oscillatory yields.

Another possible charge transfer mechanism involves the excited states of neutral He. Such a mechanism has been observed by Salop et al (38) for $He^+ \to$ alkali atom collisions in the gas phase. Energetically, excited-state transfer in $He^+ \to$ solid collisions appears to be very probable. The lowest He excited states (15) are shown in Fig. 18 aligned

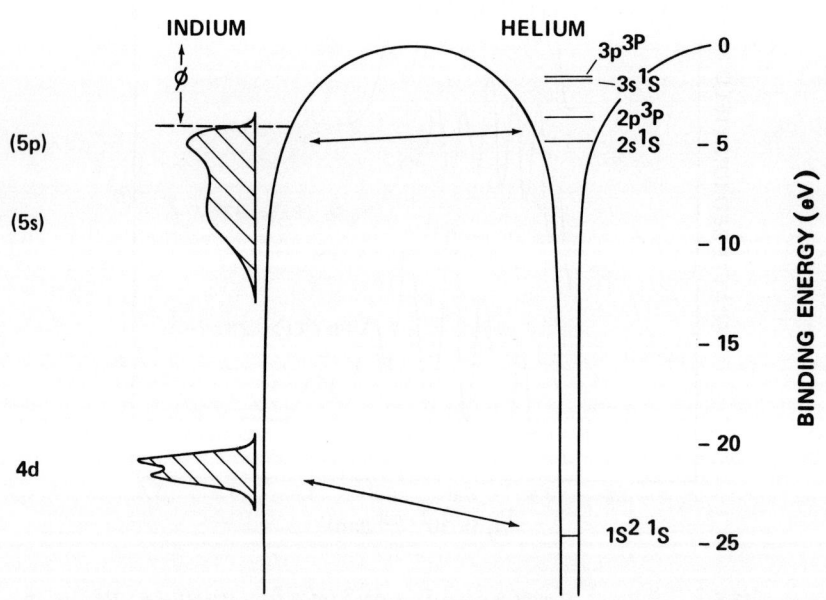

Fig. 18. Comparison of the energy levels of In metal and the ground state plus several excited states of a helium atom. Core level and excited-state charge transfer mechanisms are schematically indicated.

on an energy scale with the experimentally determined electronic density of states for In (39). The binding energies are referred to the vacuum level assuming an indium work function of 4 eV. Note that the He $2s^1S$ and $2p^3P$ states are very nearly degenerate with the upper valence band states of the In. Salop et al indicate that strong evidence exists for charge transfer to the n = 2 states, but charge transfer to only the $3p^3P$ was detected (38). Furthermore, for ion-solid collisions, Brongersma et al (41) concluded that neutralization of the incident ions occurs mainly at ion-atom separations small compared to lattice separations. Hence, atomic-like processes may be an important aspect of neutralization in ion-solid collisions.

Measurement of photon emission from the scattered helium would provide information on excited states involved in charge transfer and would help determine the mechanisms involved. For completeness, however, this would require the measurement of high energy photons which makes this experimentally difficult. Optical emission measurements would be especially useful in the cases where non-oscillatory structure is observed.

The present experimental data (Sections IV and V) strongly suggest that some aspects of the ion-solid collision might be very adequately modeled by an ion-atom collision. It might be appropriate to use perturbed atomic initial states to incorporate the effect of the crystalline environment. This approach is further supported by the similarities in the charge transfer characteristics of $H^+{\rightarrow}H$ and $H^+{\rightarrow}H_2$ collisions (24). Good theoretical results will probably require multistate molecular wavefunction calculations such as those used by Melius and Goddard (30) for $Na^+{\rightarrow}Li$ and $Li^+{\rightarrow}Na$ collisions.

VII. SUMMARY AND CONCLUSIONS

To date, the majority of surface studies utilizing low energy (0.1-3 keV) ion scattering have relied mainly upon the kinematics of the scattering process. Recent observations of structure in the energy dependence of scattered ion yield indicate that certain types of electronic interactions during the scattering process behave very much like ion-atom collisions. Oscillatory structure observed for the specific cases of He^+ scattering from Ga, Ge, As, In, Sn, Sb, Tl, Pb, and Bi is attributed to quasiresonant charge transfer between the vacant He^+ 1s state and the outer d states of these atoms. This structure is observed to be sensitive to the chemical environment of the target atom. In addition, Fourier analysis of the oscillatory structure suggests that more than one oscillatory charge transfer mechanism may be present. Excited-state charge transfer processes may also be occurring. Much additional experimental and theoretical work will be required to understand charge transfer processes and allow them to be used as a probe of surface characteristics.

VIII. ACKNOWLEDGEMENTS

It is our pleasure to acknowledge H. Helbig and P. Adelman for helpful discussions and permission to use Fig. 10 prior to publication. We also wish to acknowledge K. N. Maffitt, J. T. McKinney, J. R. Onstott, and R. H. Plovnick for critically reading the manuscript.

IX. REFERENCES

1. Smith, D. P., Surf. Sci. 25, 171 (1971).
2. Suurmeijer, E. P. Th. M., and Boers, A. L., Surf. Sci. 43, 390 (1973).
3. Buck, T. M., Methods of Surface Analysis, Ed. A. Czanderna, Elsevier, New York (1975).
4. Smith, D. P., J. Appl. Phys. 38, 340 (1967).
5. Begemann, S. H. A., and Boers, A. L., Surf. Sci. 30, 134 (1972).
6. Heiland, W., and Taglauer, E., J. Vac. Sci. Technol. 9, 620 (1972).
7. Erickson, R. L., and Smith, D. P., Phys. Rev. Letters 34, 297 (1975).
8. Ball, D. J., Buck, T. M., MacNair, D., and Wheatley, G. H., Surf. Sci. 30, 69 (1972).
9. Taglauer, E., and Heiland, W., Surf. Sci. 47, 234 (1975).
10. Robinson, M. T., Oak Ridge National Laboratory Report ORNL-3493 (1963).
11. Bingham, F. W., J. Chem. Phys. 46, 2003 (1967).

12. Hagstrum, H. D., and Becker, G. E., Phys. Rev. B8, 107 (1973).
13. Rusch, T. W., and Erickson, R. L., J. Vac. Sci. Technol. 13, 374 (1976).
14. Erickson, R. L., and Smith, D. P., U. S. Patent 3, 920, 989.
15. Moore, C. E., Atomic Energy Levels, Vol. 1 (U.S. GPO, Washington, D. C., 1971), NSRDS-NBS 35.
16. Bearden, J. A., and Burr, A. F., X-ray Wavelengths and X-ray Atomic Energy Levels (U.S. GPO Washington, D. C., 1967), NSRDS-NBS 14.
17. Tolk, N., Tully, J. C., Kraus, J., White, C. W., and Neff, S. H., Phys. Rev. Letters 36, 747 (1976).
18. Ziemba, F. P., Lockwood, G. J., Morgan, G. H., and Everhart, E., Phys. Rev. 118, 1552 (1960).
19. Perel, J., Vernon, R. H., and Daley, H. L., Phys. Rev. 138, A937 (1965).
20. Daley, H. L., and Perel, J., in Sixth International Conference on the Physics of Electronic and Atomic Collisions, Abstracts of Papers, Boston, 1969, (MIT Press, Cambridge, 1969), p. 1051.
21. Ziemba, F. P., and Everhart, E., Phys. Rev. Letters 2, 299 (1959).
22. Lichten, W., Phys. Rev. 139, A27 (1965).
23. Bates, D. R., Massey, H. S. W., and Stewart, A. L., Proc. Roy. Soc. (London) A216, 437 (1953).
24. Lockwood, G. J., and Everhart, E., Phys. Rev. 125, 567 (1962).
25. Smith, F. J., Phys. Letters 20, 271 (1966).
26. Perel, J., Daley, H. L., and Smith, F. J., Phys. Rev. A1, 1626 (1970).
27. Lockwood, G. J., Helbig, H. F., and Everhart, E., Phys. Rev. 132, 2078 (1963).
28. Helbig, H., and Adelman, P., J. Vac. Sci. Technol. 14, 468 (1977).
29. Christensen, D., Mossotti, V., Rusch, T., and Erickson, R., Chem. Phys. Letters 44, 8 (1976).
30. Melius, C. F., and Goddard, W. A. III, Phys. Rev. Letters 29, 975 (1972).

31. Bates, D. R., and Williams, D. A., Proc. Phys. Soc. 83, 425 (1964).
32. Smith, F. J., Proc. Phys. Soc. 84, 889 (1964).
33. Gurnee, E. F., and Magee, J. L., J. Chem. Phys. 26, 1237 (1957).
34. Rapp, D., and Francis, W. E., J. Chem. Phys. 37, 2631 (1962).
35. Fano, U., and Lichten, W., Phys. Rev. Letters 14, 627 (1965).
36. Rudd, M. E., Jorgenson, T. Jr., and Volz, D. J., Phys. Rev. 151, 28 (1966).
37. Grant, J. T., Hooker, M. P., Springer, R. W., and Haas, T. W., J. Vac. Sci. Technol. 12, 481 (1975).
38. Salop, A., Lorents, D. C., and Peterson, J. R., J. Chem. Phys. 54, 1187 (1971).
39. Pollak, R. A., Kowalczyk, S., Ley, L., and Shirley, D., Phys. Rev. Letters 29, 274 (1972).
40. Hagstrum, H. D., Phys. Rev. 96, 336 (1954).
41. Brongersma, H. H., Hazewindus, N., Van Nieuwland, J. M., Otten, A. M. M., and Smets, A. J., J. Vac. Sci. Technol. 13, 670 (1976).

nonadiabatic neutralization at surfaces: oscillatory ion scattering intensities

John C. Tully and Norman H. Tolk
Bell Laboratories

We have carried out an experimental and theoretical study of the neutralization mechanisms responsible for producing irregular and oscillatory structure in the energy dependence of ion scattering yields. Analysis of the angular dependence of oscillatory structure provides evidence that a specific binary ion-atom interaction dominates the process. We present a theory of ion neutralization at surfaces which incorporates both adiabatic (direct resonance and Auger) and nonadiabatic neutralization mechanisms, and which takes into account the energy, symmetry, localization and lifetime of surface electronic states. The theory reproduces the observed irregular and oscillatory ion intensities and establishes without doubt that this behavior arises from a nonadiabatic near-resonant charge exchange mechanism.

I. INTRODUCTION

The intensity, kinetic energy and angular distribution of ions scattered from surfaces contain important information about the composition and structure of the surface, about the identity and location of adsorbed species, and in some cases about the electronic properties of the surface (1-5). Utilization of this information requires an accurate and detailed understanding of ion neutralization at surfaces. In this chapter we describe experimental and theoretical studies designed to shed light on these processes.

We begin in the next section by outlining the usual adiabatic picture (6) of ion-surface neutralization, using 300-2500 eV He^+-Cu scattering as an illustration. We then discuss the importance of nonadiabatic effects, referring primarily to the irregular and oscillatory ion scattering intensities described by Rusch in the previous chapter (7).

In Sec. IV we present an analysis of the angular dependence of oscillatory ion scattering intensities, focusing on experimental measurements of He^+ - GaP scattering (8). These studies provide strong evidence that a localized ion-atom interaction dominates the collision process. This binary interaction picture provides the basis for an a priori theoretical treatment of ion-surface scattering, presented in Sec. V. The theory is applied to 300-2500 eV scattering of He^+ by Ga, Pb, Ge and Bi. The calculated scattering yields are not in quantitative agreement with experiment, but the qualitative similarities are sufficient to establish without doubt that the recently observed oscillatory ion scattering intensities are produced by a nonadiabatic near-resonant charge exchange process. The theory encompasses effects due to the energy, symmetry, and lifetime of surface electronic states, and clarifies the roles of the various neutralization processes occurring at solid surfaces.

II. ADIABATIC NEUTRALIZATION

In most low-energy (<100 eV) ion-surface collisions, neutralization is thought to occur almost exclusively by an adiabatic mechanism; i.e., neutralization is accomplished through electronically resonant processes which require no exchange of energy between electronic and nuclear motion. Hagstrum (6) has presented a very fine overview of adiabatic neutralization in the opening chapter of this book. We confine ourselves to a simplified model calculation which illustrates the major points of direct interest here and which is amenable to generalization to nonadiabatic neutralization processes described later.

Following Hagstrum (6,9) we define a neutralization probability per unit time $R_t(s)$. $R_t(s)$ encompasses all adiabatic mechanisms, including Auger and direct resonance neutralization. If the adiabatic assumption is valid, $R_t(s)$ will be independent of the past history of the collision and will depend only on the distance s of the ion from the surface, as defined in Fig. 1. We define an ion survival probability P_s by

$$P_s = |a_o(t=\infty)|^2 , \qquad (1)$$

where $a_o(t)$ is the amplitude of the wave function describing the unneutralized ion in the vicinity of the surface. The rate of change of a_o with time is then given by

$$\dot{a}_o(t) = -\frac{1}{2}\hbar^{-1}\Gamma_o(t)\,a(t) , \qquad (2)$$

where

$$\Gamma_o[s(t)] = \hbar R_t[s(t)] ; \qquad (3)$$

i.e., Γ_o is the energy width associated with the transition rate R_t. Γ_o depends on time since it depends on the distance s which changes as the ion approaches and then scatters off the surface. Thus, if the transition rate R_t is known as a function of s, the ion survival probability P_s can be computed from Eq. (1) by integrating Eq. (3) along some appropriately selected ion trajectory. The ion scattering intensity I is proportional to P_s;

$$I \propto \sigma P_s , \qquad (4)$$

where σ is the differential cross section for scattering into angle θ, defined in Fig. 1.

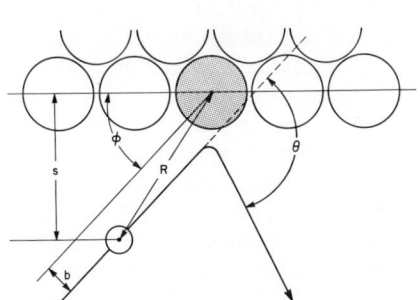

We have applied this simple model to scattering of 300-2500 eV helium ions by a copper surface. We have computed the ion trajectory by numerical integration of the classical equations of motion using the Thomas-Fermi-Moliere (10) atom-atom interaction potential for He - Cu; i.e.

Fig. 1

$$V(R) = \frac{z_1 z_2 e^2}{R}\,[\,0.35\,exp(-0.3R/R_f)$$

$$+ 0.55\,exp(-1.2R/R_f) + 0.1\,exp(-6R/R_f)\,], \quad (5)$$

where R is the radial distance between the ion and the atom, R_f is the Firsov screening length

$$R_f = \left(\frac{9}{128}\pi^2\right)^{1/2} \left(z_1^{1/2} + z_2^{1/2}\right)^{-2/3}, \qquad (6)$$

and Z_1 and Z_2 are the atomic numbers of the ion and surface atom. The differential cross section σ appearing in Eq. (4) can be computed from this potential by standard techniques (10).

We have employed the simple form suggested by Hagstrum for the neutralization rate,

$$R_t(s) = A\ exp(-as). \qquad (7)$$

Since the neutralization probability depends on the ratio A/a but is almost independent of the individual values of A and a, we have fixed the parameter a at 1.3 $Å^{-1}$ and have varied only A. For A chosen to be 2.8×10^{15} sec^{-1}, we obtain the results shown in Fig. 2. This value of A corresponds to a ratio A/a of 2.2×10^7 cm/sec, well within the range of values estimated for this quantity by other groups (6).

Figure 2 illustrates the behavior typical of an adiabatic neutralization process. The cross section σ decreases smoothly with increasing ion collision energy, Fig. 2a. The ion survival probability P_s, Fig. 2b, increases with energy as the ion spends less time in the neutralization region. The ion intensity, which is proportional to the product of σ and P_s, is therefore smoothly varying and has a maximum at some intermediate energy. The ion intensity curve computed by this simple model is compared in Fig. 2c with results of Rusch and Erickson (11). Quantitative agreement is poor. This is due primarily to the fact that the experiments report the maximum <u>height</u> of the "surface peak" corresponding to specular reflection of the He$^+$ by an individual surface atom, with no correction for background. The theory computes the <u>area</u> of the surface peak with background subtracted. Nevertheless, our primary interest here is in the overall shape of the ion intensity curves, and these are similar for experiment and theory.

III. OSCILLATORY INTENSITIES

The adiabatic neutralization model just described can produce only smooth ion intensity curves of the type illustrated in Fig. 2. However, the previous chapter by

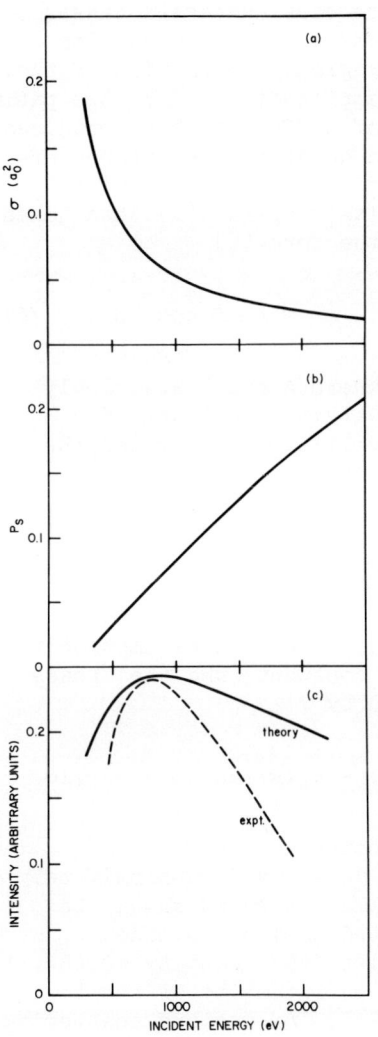

Fig. 2

Rusch (7) contains a beautiful display of the varied oscillatory and irregular ion intensity curves that are actually encountered experimentally. Behavior of this kind has now been observed in several laboratories (3,7,8, 11-14).

Oscillatory intensities have been found only in cases where the solid exhibits an electronic level whose energy with respect to the vacuum level is within a few eV of the ionization potential of the incident atomic species, e.g., 24.5 eV for He^+ scattering. For this reason it has been thought that the effect arises from a near-resonant charge exchange mechanism similar to that established in gas-phase collisions (15, 16). Conclusive evidence for the validity of this mechanism is provided by the angular distribution measurements and the theoretical calculations described in the next two sections.

A qualitative picture of the charge exchange mechanism can be obtained from Fig. 3. Because the near-resonant condition is satisfied, there are two distinct potential curves a and b with energies $E_{a,b}(R)$ corresponding at large R to an He^+ ion in the vicinity of the surface S (curve b), and a neutral He with an electron removed from the near-resonant surface state (curve a). As the approaching He^+ reaches the "interaction region" near R_m where the exchange interaction between the two states becomes comparable to their splitting, the scattering wave packet splits into two parts. Each part evolves independently along its own potential curve, $E_a(R)$ or $E_b(R)$, and develops a quantum mechanical phase proportional to $\int E_{a,b} \, dt$. As the atom

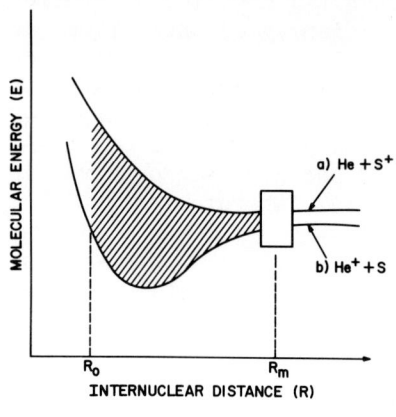

Fig. 3

recedes and again passes through the interaction region, combination of the amplitudes for the two paths of different phase produces a quantum interference effect. The ion survival $|a_o|^2$ of Eq. (1) then takes the form (8)

$$|a_o|^2 = A + B \cos^2 \delta/2 , \quad (8)$$

where A and B are slowly varying functions of collision energy and the phase δ is given by

$$\delta = \hbar^{-1} \int (E_a - E_b) dt \simeq \frac{2}{\hbar v} \langle \Delta E \Delta R \rangle . \quad (9)$$

The last term in Eq. (9) is obtained under the assumption that the collision velocity v is nearly constant, and the "phase development area" $\langle \Delta E \Delta R \rangle$ is defined by

$$\langle \Delta E \Delta R \rangle = \int_{R_o}^{R_m} [E_a(R) - E_b(R)] dR \quad (10)$$

where R_o is the classical turning point.

If the assumption of constant velocity is approximately satisfied, then ion intensity oscillation peaks should be nearly equally separated when plotted against inverse velocity, with spacing proportional to the quantity $\langle \Delta E \Delta R \rangle$. Figure 4 shows a plot of the measured intensity of the Ga surface peak for He$^+$ scattered by GaP, for various scattering angles θ. The peaks are indeed very nearly equally spaced, and the spacing changes in a regular way with θ.

IV. ANGULAR DEPENDENCE OF OSCILLATIONS

We have carried out experimental studies of the angular dependence of oscillatory structure for scattering of He$^+$ by Pb and GaP targets (8). We find, first, that the observed oscillation peak positions are almost totally independent of orientation angle ϕ defined in Fig. 1. Second, peak positions do depend on scattering angle θ, as shown for the Ga case in Fig. 4.

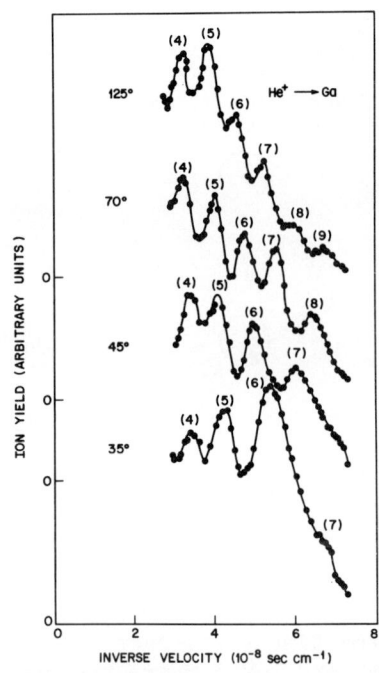

Fig. 4

These observations provide important qualitative information about the neutralization process. Consider the schematic illustration of Fig. 5. Fig. 5a is intended to illustrate a hypothetical situation in which the ion interacts with the surface as a whole, and does not see the effects of any individual atom until it is extremely close to it. The coordinate R in Fig. 3 should then be replaced by the perpendicular distance s. It is clear that, in this limit, if the orientation ϕ were changed, this would change the time that an ion moving with some fixed velocity v would remain in the phase development region. Then $v_\perp$ would be the appropriate quantity in Eq. (9), so when plotted against v the oscillation peak positions would exhibit a strong dependence on ϕ, in contradiction to our experimental observations. Furthermore, this model would predict that the peak positions would depend on scattering angle θ in a way which is inconsistent with experiment. We should mention that the experiments were performed with target samples that were not precisely flat, so ϕ may vary on a microscopic level. This would reduce somewhat the extreme sensitivity on ϕ predicted by this limiting picture, but there is absolutely no way that it could eliminate it entirely.

Experiments therefore demonstrate that the alternative limiting picture of Fig. 5b is more nearly correct; i.e., the ion feels primarily a <u>binary interaction</u> with a particular surface atom. This is consistent with the independence of peak positions on orientation ϕ, and suggests that the radial distance R and total velocity v are the correct quantities to appear in Fig. 3 and Eq. (10). Furthermore, this picture accounts for the observed dependence of He^+ - Ga scattering on scattering angle θ shown in Fig. 4.

Let us no longer invoke the constant velocity assumption and rewrite Eq. (9) as

$$\delta = \frac{2\mu}{\hbar} \int_{R_o}^{R_m} [E_a(R) - E_b(R)]\{2\mu[E - V(R)] - L^2/R^2\}^{-1/2} \, dR, \quad (11)$$

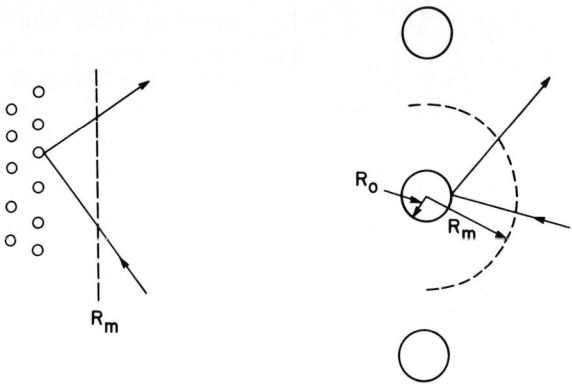

Fig. 5

where $V(R)$ is the effective ion-atom binary interaction potential, E is the collision energy, and L and μ are the angular momentum and reduced mass of the ion-atom scattering pair. L is related to the final scattering angle θ by the well-known classical deflection function,

$$\theta = \pi - 2L \int_{R_o}^{\infty} R^{-2} \{2\mu[E - V(R)] - L^2/R^2\}^{-1/2} \, dR. \quad (12)$$

Thus, for any assumed interaction potential $V(R)$ and energy difference $\Delta E = E_a - E_b$, and for any scattering angle θ, we can compute the energies E at which oscillation peaks will occur by requiring δ to be an integral multiple of 2π.

The circles in Fig. 6 are the experimental peak positions for He$^+$-Ga as obtained from Fig. 4. The solid lines were computed from Eqs. (11) and (12) using the crudest possible forms for the interactions. ΔE was assumed to be constant for $R < R_m = 2.7$ Å, and $V(R)$ was taken to be Ze^2/R. The solid lines of Fig. 6 correspond to the best-fit values of ΔE and Z, 8.8 eV and 25, respectively. The fact that this simple model is able to account satisfactorily for the observed angular dependence is strong evidence that the oscillatory behavior is determined predominantly by a specific localized ion-atom interaction.

V. NONADIABATIC NEUTRALIZATION

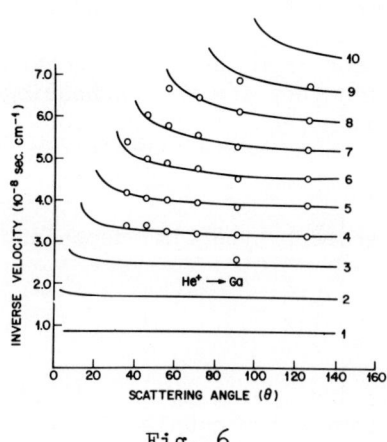

Fig. 6

The binary ion-atom interaction picture established in the last section provides a basis for extending the theory of ion neutralization at surfaces to nonadiabatic processes. We consider here the simplest situation in which there is a single near-resonant electronic level in the surface. We denote the electronic wave function describing the initial ion state by ϕ_0, and that describing a neutral atom with an electron removed from the near-resonant surface state by ϕ_1. We expand the total time-dependent electronic wave function of the system in terms of these basis functions,

$$\Phi(t) = b_0(t)\phi_0 + b_1(t)\phi_1 , \qquad (13)$$

where b_0 and b_1 are expansion coefficients. At time t=0 before the collision, $b_0 = 1$ and $b_1 = 0$.

We now define the matrix elements of the electronic Hamiltonian H_{el} with respect to the two basis wave functions ϕ_0 and ϕ_1:

$$<\phi_0| H_{el} |\phi_0> = W_0 + \frac{1}{2} i\Gamma_0$$

$$<\phi_1| H_{el} |\phi_1> = W_1 + \frac{1}{2} i\Gamma_1 \qquad (14)$$

$$<\phi_0| H_{el} |\phi_1> = W_{01} .$$

The diagonal matrix elements are taken to be complex, with the real part W_0 or W_1 corresponding to the energy of the state, and the imaginary width $i\Gamma_0$ or $i\Gamma_1$ describing dissipation. In the present context, Γ_0 describes destruction of the ion state; i.e. neutralization by processes other than near-resonant charge exchange. Therefore Γ_0 describes adiabatic neutralization, mainly Auger, and is precisely equivalent to its earlier definition in Eqs. (2) and (3). Γ_1 accounts for any processes that may destroy the localized hole created by removal of an electron from the near-resonant level. Both Auger and delocalization (hopping to neighboring sites) may contribute to Γ_1. The off-diagonal interaction W_{01} is the quantity which produces nonadiabatic transitions between states ϕ_0 and ϕ_1; i.e., which promotes near-resonant charge exchange.

The wave function $\Phi(t)$ is governed by the time-dependent Schrödinger equation,

$$i\hbar \partial \Phi(t)/\partial t = H_{el} \Phi(t) . \qquad (15)$$

Substituting Eq. (13) into Eq. (15) and operating on the left by $\langle\phi_0|$ or $\langle\phi_1|$, we obtain the following two coupled differential equations:

$$i\hbar \dot{a}_0(t) = -\frac{1}{2} i\Gamma_0(t) a_0(t)$$

$$+ W_{01}(t) a_1(t) \exp\{-\frac{i}{\hbar} \int_0^t [W_1(t') - W_0(t')] dt'\} \qquad (16)$$

$$i\hbar \dot{a}_1(t) = -\frac{1}{2} i\Gamma_1(t) a_1(t)$$

$$+ W_{01}(t) a_0(t) \exp\{-\frac{i}{\hbar} \int_0^t [W_0(t') - W_1(t')] dt'\} . \qquad (17)$$

The new expansion coefficients a_0 and a_1 are related to the old ones by

$$a_i(t) = b_i(t) \exp[\frac{i}{\hbar} \int_0^t W_i(t') dt'] . \qquad (18)$$

With this definition, the probabilities associated with each state are given simply by $|a_i(t)|^2 = |b_i(t)|^2$.

The theory is thus a direct extension of the simple adiabatic theory discussed in Sec. II. We again select some appropriate ion trajectory, e.g., that determined by a Thomas-Fermi-Moliere potential, and solve numerically the coupled equations (16) and (17) along this trajectory for the amplitudes a_0 and a_1. The ion survival probability is given, as before, by Eq. (1) and the scattered ion intensity by Eq. (4). Note that if the off-diagonal interaction W_{01}

were zero or, equivalently, if the energy splitting $W_1 - W_0$ were sufficiently large to make the phase integral in Eq. (16) oscillate rapidly, then Eq. (16) would reduce to Eq. (2), and the theory would become identical to that of Sec. II; i.e., neutralization would procede adiabatically. Nonadiabatic neutralization occurs via population of the near-resonant state ϕ_1, with the energy discrepancy being taken up by kinetic energy of the ion. This kinetic energy change of a few eV has a negligible effect on the ion trajectory for ion energies greater than 100 eV.

The theory as presented above is largely phenomenological. A rigorous derivation and analysis of the inherent approximations are given elsewhere (17). We confine ourselves here to the heuristic presentation above, and proceed now to apply the theory to 300 - 2500 eV scattering of He^+ from Ga, Pb, Ge and Bi surfaces.

We use, as before, Thomas-Fermi-Moliere potentials to define the ion trajectories. We need, in addition, the interactions defined in Eq. (14). It is here that the binary nature of the ion-surface interaction demonstrated in the previous section comes to our aid. We assume that the off-diagonal interaction W_{01} depends only on the radial distance R between the ion and surface atom. Specifically, we assume that W_{01} arises solely from an ion-atom exchange interaction (charge transfer), and approximate it by the crude expression

$$W_{01} \simeq \frac{1}{2} [W_0(\infty) + W_1(\infty)] <\phi_0|\phi_1> . \qquad (19)$$

The wave function ϕ_0 describing the vacant orbital on He^+ is taken to be a 1s hydrogenic function with a binding energy of 24.5 eV. Similarly, the wave function ϕ_1 describing the vacant surface state orbital is taken to be a hydrogenic function with the appropriate binding energy and n, ℓ and m quantum numbers, e.g. $3d_{z^2}$ for Ga. The overlap term in Eq. (19) was evaluated exactly for these hydrogenic functions.

The energy difference $W_1 - W_0$ required in Eqs. (16) and (17) was approximated by the very simple expression

$$W_1(s) - W_0(s) = W_1(\infty) - W_0(\infty) + e^2/4s , \qquad (20)$$

i.e., the energy splitting is equal to the asymptotic splitting of the two near-resonant levels adjusted by the bare image potential.

The rate of destruction $\Gamma_1/\hbar$ of the near-resonant surface level was taken to be zero for this primitive set of calculations. It is shown elsewhere (17) that calculated ion scattering intensities can be sensitive to the width Γ_1 of the surface state, at least for widths greater than about 0.2 eV. Indeed, it might very well be possible to extract information about surface state widths from ion backscattering measurements. However, in all the cases treated here (states lying 20 eV or so below the Fermi level in Ga, Pb, Ge and Bi), the widths are expected to be of order 0.1 eV or less (18), so it is a good approximation to assume these widths are zero.

The Auger rate $\Gamma_0/\hbar$ was again assumed given by Eq. (7), with the parameter a fixed at 1.3 Å^{-1} as before, and the parameter A treated as adjustable. Thus the only adjustable quantity appearing in the calculation is the Auger parameter A, and it was selected, as in Sec. II, simply to make the overall envelope of the intensity vs. energy curve have the correct general shape. The only effect of the Auger rate on the detailed structure of the intensity curves is a slight damping. All calculations were performed for $\theta = 90°$ and $\phi = 45°$.

Figure 7 shows the calculated intensities of He^+ scattered by Ga, which possesses a 3d level about 22 eV below the vacuum level. The parameter A is 5.0×10^{15} sec^{-1}. Comparison with the experimental results of Rusch and Erickson (11) is certainly not quantitative, but it is nevertheless very encouraging. The fact that the theory does reproduce the observed oscillatory behavior is conclusive proof that this effect is due to the proposed near-resonant charge exchange mechanism.

Calculated and experimental (19) intensities for He^+ - Pb (5d level 25 eV below the vacuum level with A = 3.7×10^{15} sec^{-1}) are shown in Fig. 8. Qualitative agreement is again surprisingly good, even including irregular features of the spectra. These irregular features were thought to possibly arise from the fact that the 5d level in Pb is split by about 2 eV due to spin-orbit coupling. This explanation is not correct since these features are reproduced by the present calculation which ignores this complication. Rather, the irregularities arise because the off-diagonal coupling $W_{01}(R)$ varies rapidly with R, going through zero at several places as a result of the nodal behavior of the 5d orbital. This causes the simple picture of a single interaction region in Fig. 3 to break down.

Not all calculations agree as well with experiment as He^+ - Ga and He^+ - Pb. For He^+ - Ge (3d, 33 eV, A = 2.5×10^{15} sec^{-1}), theory predicts that the oscillatory structure should be almost washed out whereas it appears clearly in the experiment (12), Fig. 9. For He^+ - Bi (5d, 30 eV,

A = 3.7×10^{15} sec^{-1}) both experiment (12) and theory show highly irregular structure, but quantitative agreement is terrible, Fig. 10.

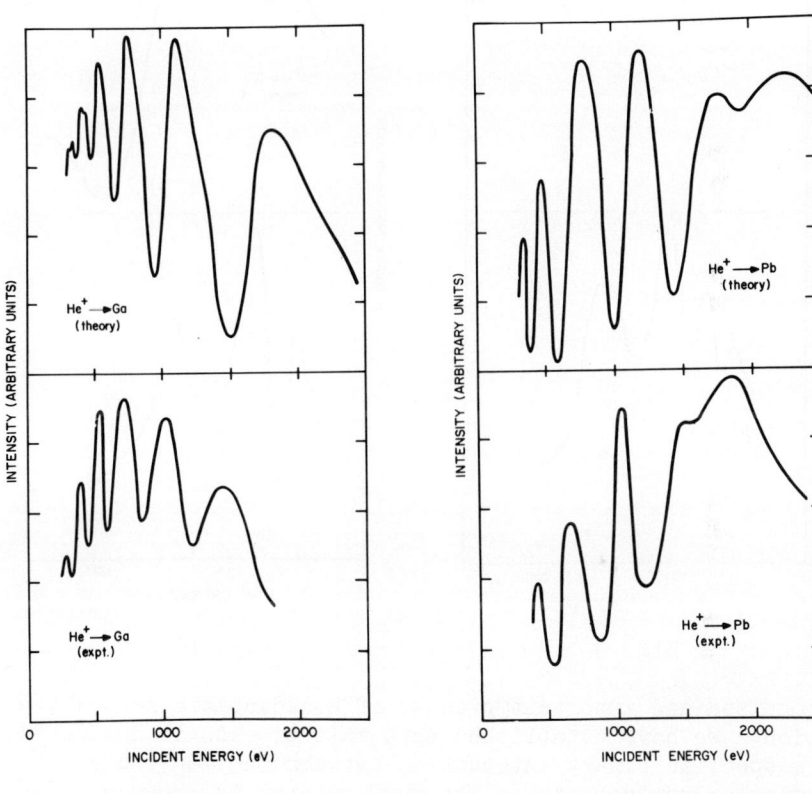

Fig. 7   Fig. 8

Considering the crude interaction potentials employed, these discrepancies are not surprising. Efforts to implement more accurate interactions based on atomic Hartree-Fock functions are in progress. It is hoped that the theory will then become reliable and predictive.

VI. DISCUSSION

The experimental and theoretical studies described above have contributed to our understanding of the physics of ion-surface collisions and, in particular, of the newly

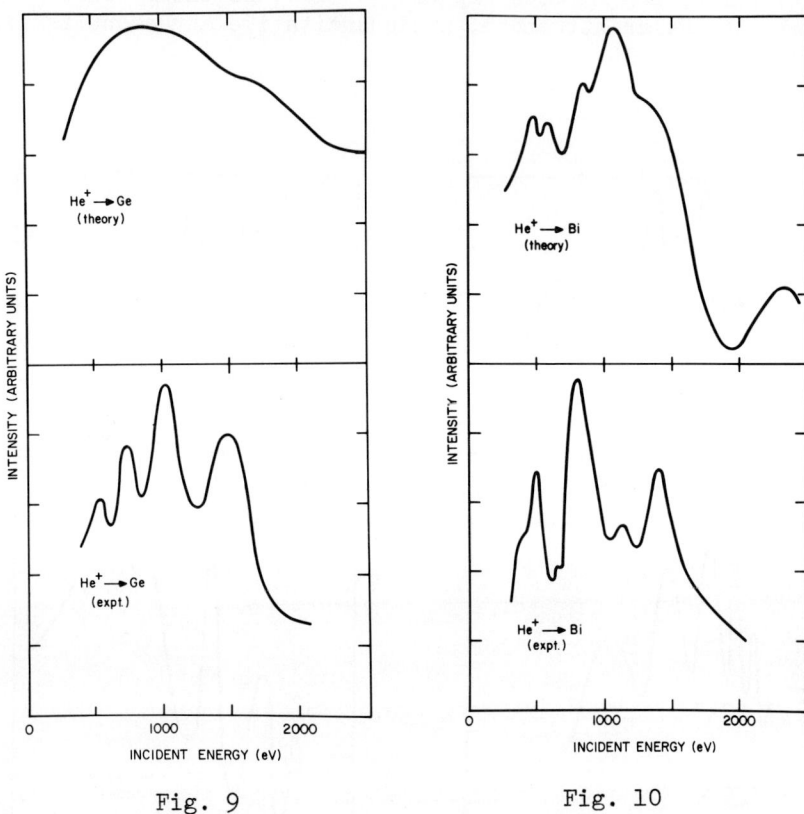

Fig. 9   Fig. 10

recognized and important process of nonadiabatic neutralization. We have established that this process is dominated by a specific binary interaction between the ion and a particular surface atom. The surface atom is, however, perturbed by the proximity of its neighboring atoms, and this perturbation is reflected in ion neutralization probabilities.

The theory described here was applied in a very primitive way. Even in this crude form, it can produce oscillatory ion scattering intensities when they are present and it can predict the dependence of ion neutralization on surface properties such as the energy, lifetime, symmetry and localization of electronic states. The theory can also be applied to neutralization into excited states and to excitation and de-excitation processes. It is our feeling that this theory can provide a framework for future experimental studies in this area, and that these studies should provide a wealth of exciting new information about the nature of solid surfaces and about inelastic ion-surface interactions.

VII. REFERENCES

1. Smith, D. P., Surface Sci. 25, 171 (1971).
2. Buck, T. M., in Methods and Phenomena: Methods of Surface Analysis, ed. by S. P. Wolsky and A. W. Czanderna (McGraw-Hill, N.Y., 1975).
3. Taglauer, E. and Heiland, W., Appl. Phys. 9, 261 (1976).
4. Heiland, W. and Taglauer, E., in Inelastic Ion-Surface Collisions, ed. by N. H. Tolk, et al. (Academic Press, N.Y., 1977).
5. Buck, T. M., in Inelastic Ion-Surface Collisions, ed. by N. H. Tolk, et al. (Academic Press, N.Y., 1977).
6. Hagstrum, H. D., in Inelastic Ion-Surface Collisions, ed. by N. H. Tolk, et al. (Academic Press, N.Y., 1977).
7. Rusch, T. W., and Erickson, R. L., in Inelastic Ion-Surface Collisions, ed. by N. H. Tolk, et al. (Academic Press, N.Y., 1977).
8. Tolk, N. H., Tully, J. C., Kraus, J., White, C. W., and Neff, S. N., Phys. Rev. Letters 36, 747 (1976).
9. Hagstrum, H. D., Phys. Rev. 96, 336 (1954).
10. Robinson, M. T., and Torrens, J. M., Phys. Rev. B9, 5008 (1974).
11. Rusch, T. W., and Erickson, R. L., J. Vac. Sci. Tech. 13, 374 (1976).
12. Erickson, R. L., and Smith, D. P., Phys. Rev. Lett. 34, 297 (1975).
13. Brongersma, H. H., and Buck, T. M., Nucl. Instrum. Methods 132, 559 (1976).
14. Helbig, H. F., and Adelman, P. J., unpublished.
15. Lichten, W., Phys. Rev. 139, A27 (1965).
16. Smith, F. T., Lectures in Theoretical Physics XIC, 95 (1969).
17. Tully, J. C., "Neutralization of Ions at Surfaces", submitted to Phys. Rev. B.
18. McGuire, E. J., Phys. Rev. A6, 1043 (1972).
19. Kraus, J. and Tolk, N. H., unpublished results.

sputtering processes: collision cascades and spikes

P. Sigmund
*Physical Laboratory II, H. C. Ørsted Institute,
DK-2100 Copenhagen Ø*

Sputtering of solids by ion bombardment is discussed mainly from a qualitative point of view. The elementary processes are described on the basis of few experimental observations and results of collision theory. Special attention is given to the mechanisms of sputtering from atomic collision cascades and spikes. Simple estimates are given for the sputtering yield, and the relation to some differential quantities is discussed. Comments on the sputtering of clusters and on the significance of crystal lattice structure and target surface on the sputtering process conclude the chapter.

I. INTRODUCTION

One may define sputtering as the erosion of solid surfaces by energetic particle bombardment. This definition stresses the radiation damage aspect of the sputtering process. In a research program that is governed by this aspect, sputtering experiments will usually be directed toward quantitative determination of surface erosion; this can be accomplished by weight-loss measurements, interferometric or microscopic observation of surface changes, etc. Historically, this aspect has been the dominating one for about a century of research in sputtering phenomena (1850-1950) during which almost all work dealt with the special case of cathode sputtering in gas discharges. Yet sputtering is a rather universal phenomenon. Whenever solid surfaces are exposed to energetic particle bombardment, surface erosion must be considered in an attempt to understand the overall behaviour of a system. This includes the significance of sputtering in the technological design of accelerators, fission and fusion reactors, and in astrophysical phenomena.

One may also define sputtering as the emission of atomic particles from surfaces under energetic particle bombardment. This definition stresses the dynamical aspects of the sputtering process, i.e., the atomic collision processes that are initiated by an impinging particle, to the extent that they may lead to sputtering. The English term "sputtering", even more than the German "Kathodenzerstäubung" comprises an indication of the microscopic nature of a sputtering event: Apparently, it is implied that matter is emitted in discrete blobs from such areas of a surface that are hit by bombarding particles. In order to study the dynamic aspects of sputtering, one may direct one's attention toward the emitted particles, i.e., the emitted atomic species, their distribution in angle and energy, and the atomic and molecular states. Such studies of the sputtering source have proved to be a useful tool in surface science, in as much as observation of the sputtered particles may allow conclusions on the composition of the sputtered surface. Identification of sputtered particles by means of optical spectrometry and mass spectrometry have been useful in this context. Even more important, sputtered particles can be deposited on a collector, and analysed by means of standard techniques.

It is by no means the purpose of these considerations to artificially divide up the field of sputtering into two more or less unrelated aspects. Indeed, great unjustice would be done to all those numerous erosion measurements that actually have contributed a major part of the present microscopic understanding of the emission process. However, it is important to notice the different nature of, on the one hand, (post-bombardment) erosion measurements that often require high doses, i.e., substantial erosion and major changes of the structure and composition of a surface, and, on the other hand, measurements of an instantaneous flux of sputtered particles, where the time interval for a measurement may be small enough to require the sputtering of less than a monolayer only.

An understanding of the *erosion* aspect of sputtering requires, by necessity, a study of the *overall behaviour* of the sputtered particles. On the other hand, when interest is directed toward specific aspects of the *emission* behaviour, one may succeed in describing such an aspect as an *isolated phenomenon*. An obvious example, as will be seen, is the comparatively rare event of emission of a highly energetic recoil atom under ion bombardment: Little knowledge about sputtering is necessary to make predictions concerning such an event. Yet, little can be said about the overall sputtering process if one has knowledge only about those energetic recoil atoms. This is a fairly typical situation. Most experimental techniques to study sputtered particles deal only with a small fraction of those particles. In some cases (e.g. in radioactive-tracer experiments) it can be ensured that the extracted information is representative of the overall behaviour of sputtered particles; in others (e.g. ionic emission, optical spectroscopy on sputtered particles) it may require great care (or may even appear impossible) to draw conclusions from a small minority of sputtered particles on the majority.

The present chapter is intended to give the reader a qualitative understanding of the processes leading to sputtering by heavy-particle, i.e., mostly ion bombardment. An attempt is made to identify those processes that determine the overall behaviour, and to point out some processes that are likely to be important in the understanding of specific phenomena. For more detailed information, the reader is referred to Behrisch's classic (1), to Andersen's recent summary (2), and to two sets of lecture notes by the present author (3,4). Moreover, a comprehensive monograph is in preparation (5).

II. OBSERVATIONS ON METALS

Numerous sputtering experiments have been performed on thick, metallic targets bombarded with ion beams in the eV-keV range at normal incidence. The erosion of the bombarded surface is measured conveniently in terms of the sputtering yield S which is the average number of sputtered target atoms per incident ion. Fig. 1 shows the sputtering yield of polycrystalline copper for krypton ions in the range of 1 keV - 1 MeV, measured by means of weight loss by a number of authors. Although the experimental data are not the most recent ones, the scatter is representative. The data have been taken at high bombardment doses, where the copper target is essentially saturated with krypton. The general

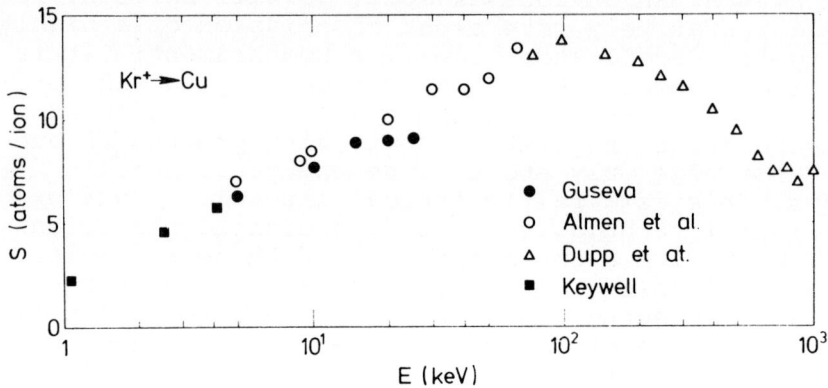

Fig. 1. Measured sputtering yields for krypton ions on polycrystalline copper at normal incidence. Compiled from /6/.

shape of the yield versus energy curve is characteristic for this class of experiments. Absolute yields range between $\sim 10^{-2}$ and $\sim 10^{+2}$ atoms/ion, dependent on ion and target, with the highest yields being obtained with the heaviest ions bombarding materials with the lowest sublimation energies, and vice versa. The position of the maximum in the S(E) curve depends primarily on the type of bombarding ion; it moves toward lower energies with decreasing ion mass. For bombardment at oblique incidence, the yield increases up to a maximum at some large angle -- normally greater than 60° -- and then drops to zero.

A qualitative description of this behaviour is found when the S(E) curve is compared with the nuclear stopping power, $(dE/dx)_n$ for the Kr^+-Cu system (Fig. 2). This quantity, calculated by Lindhard et al. (7) from classical scattering theory on the basis of the Thomas-Fermi model of interacting atoms, determines the rate (energy loss per travelled path length) at which a moving ion dissipates its energy among the target atoms in elastic (or quasi-elastic) collisions. The term "nuclear stopping" is used to specify that kinetic energy is transferred primarily to the center-of-mass of a target atom, i.e., the nucleus, while electronic excitation processes are treated separately, and make up the electronic stopping power, $(dE/dx)_e$. The approximate proportio-

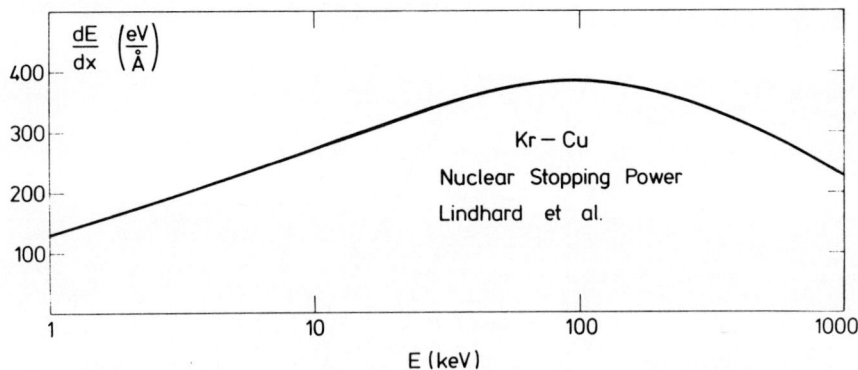

Fig. 2. Calculated nuclear stopping power (7) for krypton penetrating copper.

nality of the two curves in Fig. 1 and Fig. 2 is striking, and by no means unique for the Kr^+-Cu system. The simplest way to convert this observation into an approximate formula for the sputtering yield (which is a dimensionless quantity) appears to be as follows,

$$S = \frac{x_1}{W_o} \left(\frac{dE}{dx}\right)_n \quad (1)$$

where x_1 is a length and W_o an energy. Apart from a geometry factor, x_1 is the effective thickness of the sputtered layer, and W_o a typical energy of sputtered atoms. Figs. 1 and 2 indicate that this ratio is of the order of

$$W_o/x_1 \sim 40 \; eV/Å. \quad (2)$$

Fig. 3 shows an energy spectrum of sputtered gold atoms, measured by means of time-of-flight spectrometry in a particular direction of the sputtered beam (8). The spectrum shows a fairly typical behaviour (copper behaves similarly (8)), i.e., a pronounced maximum below 10 eV and a rapid decrease -- approximately as the inverse square of the energy of a sputtered atom -- at higher energies. It is obvious that the overwhelming majority of sputtered atoms have quite small energies, less than $\sim$ 20 eV.

Following Fig. 3, we insert $W_o \sim 10$ eV as a ty-

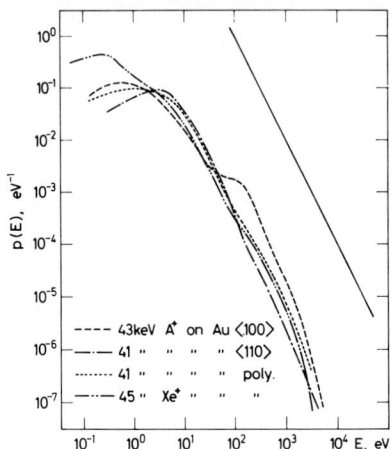

Fig. 3. Measured energy spectra of gold atoms sputtered at right angles to incoming argon or xenon beam (8). Courtesy of the authors and Taylor & Francis, Ltd.

pical value in Eq. (2), and find $x_1 \sim 0.25$ Å.

Within this simple picture, the observed increase in sputtering yield with increasing angle of incidence is caused by the increased energy dissipation within a certain depth from the surface. This would suggest an inverse-cosine dependence on the angle of incidence.

The present, simplified picture, which is implicit in many early theories of sputtering, has its quite definite limitations, which will be discussed below. The guidelines for the present order-of-magnitude estimates have been given by experimental results. No specific assumptions have actually been made about the sputtering process. In particular, the quantity x_1 has been introduced by a dimensional argument, and its significance is not clear at this point.

III. BASIC PROCESSES

Much of the early discussion of the sputtering process was dominated by the question of whether sputtering was a collision or an evaporation process (1). Ingenious experiments have been performed to give an answer to this question. Perhaps the question was posed the wrong way. It appears that the significance of atomic collision processes in sputtering has not seriously been questioned ever since Stark's early work (9), and that thermodynamical theories have been developed not so much because of being more accurate than collision theories, but because of a seemingly greater chance of arriving at practicable results (10). At present, collision theory has arrived at a reasonable level of understanding, and it is rather the knowledge of the thermodynamic aspects of the sputtering process that is lacking behind.

I should like to make some comments on the above question, but before doing so, I should like to put it into a somewhat broader perspective.

We note first that sputtering is a specific type of radiation damage, the latter term being understood as radiation-induced displacement of atoms from their equilibrium sites. Thus, an intimate connection must exist between the mechanism of radiation damage in any given system, and the mechanism of sputtering in the same system. In other words, if a sample is exposed to radiation under conditions where defects are produced in the bulk, chances are high that the surface will be sputtered as an accompanying effect. Mechanisms of radiation damage are manifold, dependent on the material, and have been studied extensively (11). Therefore, one expects a manifold of sputtering processes on different materials.

It is well known (11) that both nuclear and electronic stopping may cause radiation damage, the former by direct displacement of atoms in elastic collisions, the latter by means of indirect processes that are specific to different target materials. For example, in alkali halides, radiationless decay of excitons has been invoked as a mechanism of producing color centers (12), and in many insulators, fission products leave damage tracks that have been identified clearly as being caused by the ionizing action of those projectiles (13). Sputtering measurements on such systems have been quite scarce. Pronounced ionization sputtering effects have been demonstrated on alkali halides by UV and low-energy electron irradiation (14), i.e., a type of irradiation with negligible nuclear stopping. In case of heavy-ion bombardment, where electronic and nuclear stopping compete, the sputtering behaviour becomes complex, once nuclear and electronic stopping mechanisms act simultaneously (15). While these processes appear interesting and important for a wide variety of insulators and ionic crystals, little is known about the detailed sputtering mechanism. Systematic experiments aiming at a separation of the different mechanisms can make use, e.g., of the rather different energy dependencies of electronic and nuclear stopping (Fig. 4). Note, in parti-

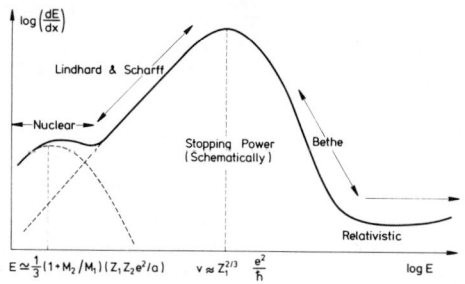

Fig. 4. Schematic energy dependence of nuclear and electronic stopping power.

cular, that in case of ionization sputtering the sputtering yield may be presumed to increase until much higher energies than in case of sputtering by nuclear collisions (16).

In a metal, the ionizing action of a bombarding ion serves mainly to heat the target. While experiments normally take proper care to ensure that the overall heating effect of the beam is small enough not to make the target evaporate, one might still expect evaporation from "hot spots", i.e., small regions around the points of impact where energy is dumped in electronic collisions, and converted into heat. In case of competing electronic and nuclear stopping, such evaporation, if present at all, must be far exceeded by a similar effect of nuclear stopping, since nuclear energy is deposited in a much smaller region in space. A possible effect at high energies, where electronic stopping dominates by 2-3 orders of magnitude, does not appear to be corroberated by experimental data (17).

We now return to the mechanism of sputtering by means of elastic collisions, or nuclear stopping. It was mentioned in sect. 2 that this mechanism dominates in ion-bombarded metals. Whatever the detailed mechanism, sputtering by nuclear collisions takes place when sufficient energy, ≳ 10 eV, is transferred to target atoms. Thus, this effect is much less sensitive to specific target properties than various ionization sputtering mechanisms. In particular, no drastic differences have so far been pointed out in the general behaviour of sputtering yields between metals and semiconductors.

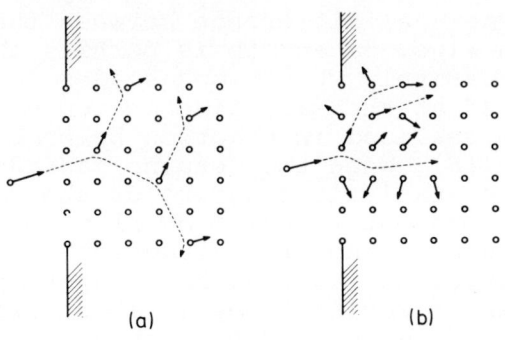

Fig. 5a. Linear collision cascade. The structure is preserved, and a small fraction of the atoms is in motion.

Fig. 5b. Dense cascade (spike). The structure is destroyed locally. All atoms within the spike volume are in motion.

Fig. 5 shows two different views of the way how energy may be dissipated by an energetic particle in a solid via elastic collisions. Fig. 5a shows a *collision cascade*, and Fig. 5b a *spike*.*
In both cases, energy is ultimately shared by a great number of atoms. The difference between the two situations lies in the fact that while only a small fraction of all atoms within a certain cascade volume are in motion in Fig. 5a, essentially all atoms within a certain spike volume move in Fig. 5b. The picture of a cascade applies when there is a long mean free path between significant collision

---

* The term "spike" has had a long history, and has covered over a great variety of concepts over the years. In this chapter, the term will be used only to characterize the situation sketched in Fig. 5b.

events, such that the energy is spread out over a large volume. Energy continues to dissipate until some sort of equilibrium is reached. But in a genuine collision cascade, equilibrium is reached at such small energies per atom where sputtering does not take place. Conversely, in a spike, energy is dissipated at a high rate, within a small volume, and shared among the atoms within the spike volume at a level that is relevant for sputtering, i.e., a few eV or more.

The practical difference between the two situations in Figs. 5a and 5b is perhaps characterized most illustratively when two cascades (or spikes) are generated at the same time on top of each other. This can be realized in practice by bombardment with a molecule (18). When two genuine cascades are superimposed, the statistical nature of the collision events will prevent a strict overlap in the two cascades. Because of the small fraction of atoms moving in either case, the two cascades superimpose so the total number of moving atoms is about twice as large as in either cascade, and the energy distribution essentially unchanged. Obviously, the system is *linear*. When two spikes are superimposed, on the other hand, energy will be dissipated essentially within a single spike volume, i.e., the total number of atoms in motion will increase only insignificantly. Consequently, the available energy per atom must double approximately. Obviously, such a system is *nonlinear*. In a linear system, one expects the sputtering yield to be proportional to the number of impinging particles, regardless of their correlation in time and space. In a nonlinear system, such a simple relationship is not expected: The sputtering yield of a molecule is expected to differ from the sum of the sputtering yields of the constituents of the molecule.

Having arrived at this point, the reader may already have realized that the distinction between a collision cascade and a spike is a matter of the minimum energy, an atom must have in order to be called "in motion". If this minimum energy is sufficiently high, the number of "moving" atoms will always be small enough that the cascade concept applies. Conversely, any cascade is like a spike if the minimum energy is sufficiently small. Thus, with regard to sputtering, the cascade or spike aspects dominate depending on whether the mean energy ultimately dissipated per each atom within the slowing-down region is smaller or greater, respectively, than the minimum energy U_0 for a sputtering event, i.e.,

$$\frac{\nu(E)}{N\Omega} \quad \begin{array}{l} <U_0: Collision\ Cascade \\ >U_0: Spike \end{array} \quad (3)$$

Here, $\nu(E)$ is that part of the impinging energy E that is *not* lost to ionization during the slowing-down process (19), Ω the slowing-down volume, and N the number density of target atoms. U_0 is generally smaller than the quantity W_0 introduced in eq. (1).

Even a very crude estimate of the quantities entering Eq. (3) on the basis of existing knowledge of ion ranges leads to the result (20) that the available spectrum of ion/target combinations as well as ion energies covers the whole range between pronounced cascade and pronounced spike behaviour. Roughly, spikes are generated by heavy ions in heavy targets at not too high ion energies, and vice versa for cascades. With regard to intermediate situations, the ion type is the most crucial factor, and the energy the least. A beautiful illustration of these features is shown in Fig. 6 where sputtering yields

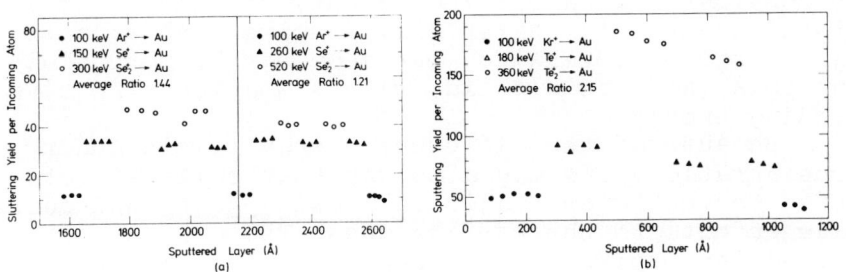

Fig. 6. Comparison of sputtering yield per incoming atom for bombardment of gold by atomic and molecular ions at the same velocity (18). Courtesy of the authors and the American Institute of Physics.

due to molecular-ion bombardment are compared with those for atomic ions at the same velocity (18).

Approximate linearity is observed for light-ion bombardment, while a drastic *enhancement* is observed for heavy-ion bombardment.

No measurements of energy spectra have yet become known that compare molecular- and atomic-ion bombardment. The present considerations imply that no difference is observed in the spectra at higher energies, $\sim$ 100 eV or more, while a significant enhancement in the molecular case, corresponding to the observed increase in yield, must be present around the maximum in the spectrum. This is consistent with the observation that the *total* sputtered energy shows a linear behaviour (18). Therefore, even in case of pronounced spike behaviour, one cannot expect a Maxwellian tail (1) in the spectrum of sputtered particles.

This suggests an alternative formulation of the old question of whether sputtering is a collision or an evaporation process: There is no doubt that at least part of the sputtered atoms are sputtered as a result of linear atomic collision cascades. What are the conditions under which a significant -- or the dominating -- portion of low-energy sputtered particles originate in spike processes? A partial answer has been outlined in Eq. (3). Once U_0 is given, standard penetration theory (3) can provide at least a qualitative criterion. A delicate question is, however, whether U_0 within this context can be considered a material constant, or whether it is itself a variable. After all, a spike with an average energy of several eV per atom may be more similar to a gas than a solid, with a smaller effective binding energy.

Because of the different energy levels under consideration, one may also say that spike effects refer to the *later* stages of slowing-down, and cascade effects to the *earlier* ones. The concept of cascade and spike effects separated in *time* (rather than *space* (21)) originates in earlier work of Thompson and Nelson (22).

IV. FLUX OF RECOIL ATOMS IN A COLLISION CASCADE

Fig. 7 shows Monte-Carlo-simulated orbits of keV protons slowing down in aluminium (23). The length of the tracks shows rather little fluctuation from event to event, but the high frequency of elastic scattering events at large angles causes a con-

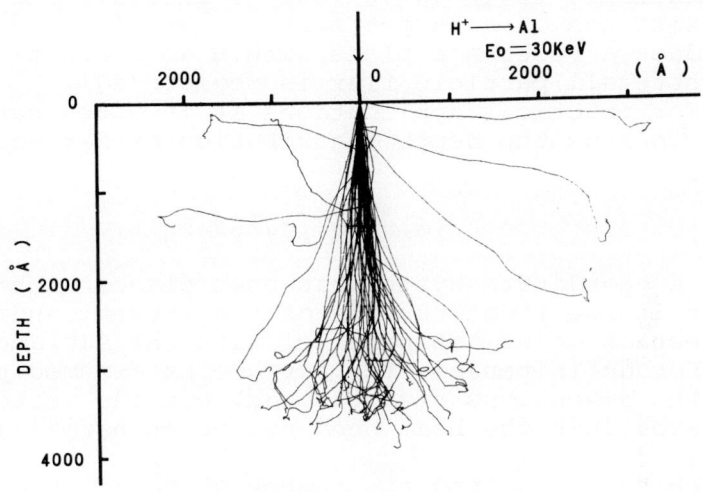

Fig. 7. Trajectories of 30 keV protons slowing down randomly in aluminium (23). Courtesy of the authors and Akademie Verlag, Berlin.

siderable spreading of the initially well-collimated beam. We note that whenever a proton is deflected by a measurable angle, the recoiling nucleus will carry away some energy to create a cascade or spike. Allowing for further dissipation of energy by elastic collisions of recoil atoms, we can define a deposited-energy distribution $F_D(\vec{r},\vec{v})$ as the average density of energy deposited in nuclear motion at a vector distance $\vec{r}$ from the origin of the primary ion with initial velocity $\vec{v}$. The average is taken over many slowing-down events. Energy is considered as "deposited" when dissipation among atoms has proceeded so far that the energy per atom is much smaller than the initial energy, and the range small compared to the overall dimensions of a cascade. (With regard to spikes, special problems arise, cf. below). It follows from the definition that

$$\int F_D(\vec{r},\vec{v})d^3r = \nu(E) \qquad (4)$$

where $\nu(E)$ is the function introduced in Eq. (3), i.e., allowance has been made for energy to be lost in electronic excitation processes. Some energy may also be lost through the target surface. This effect may or may not be included in Eq. (4). (In actual calculations, it is convenient to initially ignore the existence of a target surface, and instead to introduce a reference plane within an infinite target where all particle flux is registered).

For practical applications it is often sufficient to know the depth distribution of deposited energy,

$$F_D(x,\vec{v}) = \int dy \int dz \, F_D(\vec{r},\vec{v}) \,, \qquad (5)$$

where a coordinate system has been placed with the origin in the front surface of the target, and the x-direction being antiparallel with the surface normal. For definiteness, the surface is assumed planar over the dimensions of a cascade, but the vector $\vec{v}$ indicates that the beam need not be at normal incidence.

We now look into the number of recoil atoms in a cascade. Let the initial energy be E, and count all atoms that recoil initially with an energy greater than some value E_0 which is assumed $<<E$, but above the spike limit given by Eq. (3). The total number of these atoms cannot exceed $\nu(E)/E_0$. Since energy is also spent in atoms with smaller recoil energies, one actually obtains, as an average over many cascades (24), a number

$$N(E,E_0) \underset{\sim}{\sim} B \cdot \frac{\nu(E)}{E_0} \quad for \ E>>E_0 \qquad (6)$$

with

$$B \underset{\sim}{\sim} 6/\pi^2 \,; \qquad (6a)$$

Eq. (6) is rather insensitive to the specific assumptions concerning the slowing-down process, except that no significant amount of ordered motion along the main crystal lattice directions should take place, and that E_0 must be outside the spike region.

There is a tendency for the overall velocity distribution of recoil atoms to be isotropic. This follows both from the scattering of the ion beam (Fig. 7), and the fact that low-energy recoil atoms predominantly belong to second and higher generations. It is one of the main results of the transport theory of sputtering (6) that the isotropy statement is exact in the asymptotic limit of $E>>E_0$ (Eq. (6)).

Let us now consider a constant flux of ψ[ions/sec]. Then we obtain a stationary flux of recoil atoms. In the average, each recoil atom with an initial energy above E_O will spend a time $dt_O = dE_O/|dE_O/dt|$ in the interval $[E_O, E_O+dE_O]$. Therefore, at any time, we have an expected number of

$$\psi N(E, E_O) \frac{d\Omega_O}{4\pi} dt_O = \psi B \frac{\nu(E)}{E_O} \frac{d\Omega_O}{4\pi} \frac{dE_O}{(dE_O/dt)} \quad (7)$$

atoms moving with energy (E_O, dE_O) in a direction specified by the solid angle $d\Omega_O$.

According to Eqs. (4) and (5), the quantity $\nu(E)$ is distributed in depth like $F_D(x, \vec{v}) dx$. Since E enters into Eq. (7) *only* through $\nu(E)$, it is plausible to assume that the particle flux is also distributed in depth like $F_D(x, \vec{v}) dx$. This is another result of the transport theory (6) that is exact in the asymptotic limit $E \gg E_O$.

The number of atoms passing the plane $x=0$ in a time interval dt with a velocity $(\vec{v}_O, d^3 v_O)$ is found in a layer $dx = |v_{ox}| dt$, i.e.

$$\psi B \frac{F_D(0, \vec{v}) |v_{ox}| dt}{E_O} \frac{d\Omega_O}{4\pi} \frac{dE_O}{(dE_O/dt)} \quad ;$$

The flux per incident ion is found by dividing by ψdt; inserting Eq. (6a), and replacing

$$\frac{dE_O}{dt} = v_O \frac{dE_O}{dx} , \quad (8)$$

we obtain the flux per incident ion,

$$S(\vec{v}, E_O, \cos\theta_O) dE_O d\Omega_O =$$
$$\frac{3}{2\pi^3} F_D(0, \vec{v}) \cdot \frac{dE_O}{E_O |dE_O/dx|} \cos\theta_O d\Omega_O , \quad (9)$$

where

$$\cos\theta_O = |v_{ox}|/v_O \quad (9a)$$

is the directional cosine of $\vec{v}_O$ with the surface normal.

Eq. (9) ignores the existence of a target surface. One major effect of such a surface is an effective binding force, or a probability function

$$P(E_o, \theta_o) \tag{10}$$

for an atom to be ejected when arriving at the plane x=0 with energy E_O at an angle θ_O. If we ignore, for a moment, the effect of the surface on the development of the individual collision cascades, we can integrate Eq. (9), and obtain the sputtering yield

$$S = \int S(\vec{v}, E_o, \cos\theta_o) P(E_o, \theta_o) dE_o d\Omega_o = \Lambda F_D(o, \vec{v}) \tag{11}$$

with

$$\Lambda = \frac{3}{2\pi^3} \int \frac{dE_o}{E_o |dE_o/dx|} \int \cos\theta_o d\Omega_o P(E_o, \theta_o) . \tag{11a}$$

In order to compare Eq. (11) with Eq. (1) we write

$$F_D(o, \vec{v}) = \alpha (dE/dx)_n \tag{12}$$

where α is a dimensionless quantity dependent on $\vec{v}$.

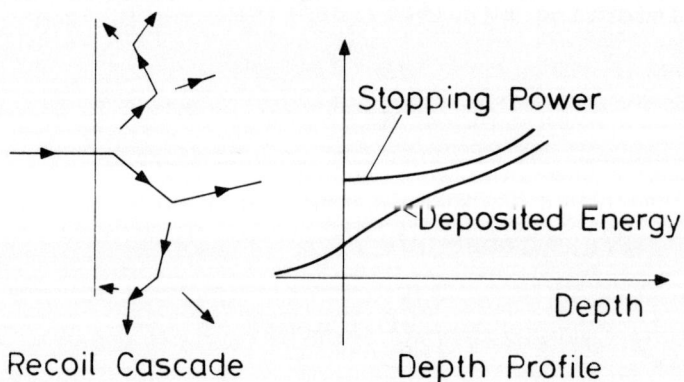

Fig. 8. *Schematic illustration of the difference between stopping power and deposited energy. The stopping medium is considered infinite, and energy dissipation starts at depth zero.*

If the deposited energy were identical with the stopping power versus depth, we would have $\alpha = (\cos\theta)^{-1}$, θ being the angle between the incoming beam and the negative surface normal. Fig. 8 illustrates that because of energy transport by recoil atoms (making α smaller) and scattering, in particular backscattering of ions (making α larger), α may differ substantially from this expression, both in magnitude and dependence on θ. As it turns out (6), for $\theta = 0$, α depends primarily on the target/ion mass ratio M_2/M_1. Calculated and experimental values of α for $\theta = 0$ are shown in Fig. 9.

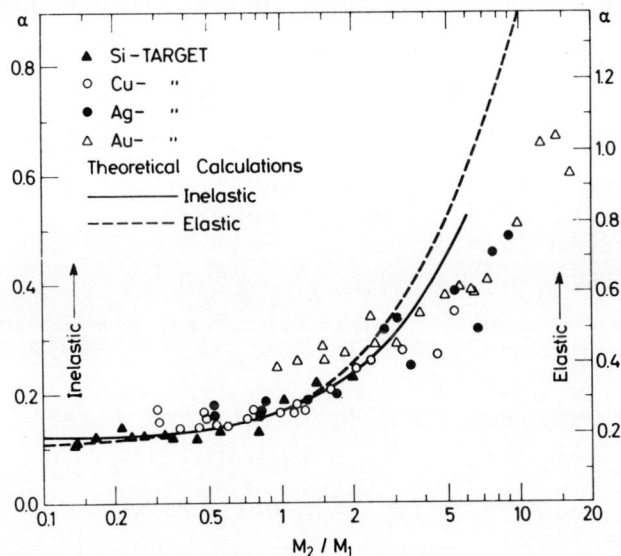

Fig. 9. Summary of sputtering yields for 45 keV ions incident on several targets. The ordinate is a factor α extracted from the measured yield S, by means of Eqs. (11) and (12). The 4 values of the constant Λ for the 4 different targets have been adjusted to the theoretically predicted curves of α versus target/ion mass ratio M_2/M_1. From ref. (18), to which the reader is referred for details. Courtesy of the authors and the American Institute of Physics.

It is obvious that the factor Λ in Eq. (11a) depends on the target material and the state of the surface, but not on the ion. The proper way to compare Eq. (11) with Eq. (1) is, therefore, to add a factor α in eq. (1), and to make the identification

$$\frac{x_1}{W_o} = \Lambda \; ; \tag{13}$$

The depth x_o from which atoms are effectively sputtered is estimated by means of the following simple argument. Assume x_o small -- since x_1 is small -- and take all recoil atoms within the layer x_o in an energy interval (E_o, dE_o) and a solid angle $d\Omega_o$,

$$\frac{6}{\pi^2} \frac{F_D(0,\vec{v})}{E_o^2} dE_o \frac{d\Omega_o}{4\pi} x_o \; ,$$

and ignore all energy loss and scattering between the point of origin and the surface. Then, the sputtering yield is

$$S \sim \frac{3x_o}{2\pi^3} F_D(o,\vec{v}) \int \frac{dE_o d\Omega_o}{E_o^2} P(E_o, \theta_o)$$

or, by comparison with Eqs. (11) and (11a)

$$x_o = \frac{\int \frac{dE_o}{E_o |dE_o/dx|} \int d\Omega_o \cos\theta_o P(E_o, \theta_o)}{\int \frac{dE_o}{E_o^2} \int d\Omega_o P(E_o, \theta_o)} \tag{13a}$$

Obviously, x_o is of the order of the range of an atom with energy $E_o \sim W_o$. If the same argument is carried out for the differential rather than the total sputtering yield, the reasonable result follows that the higher the energy of a sputtered particle, the higher the depth range from which it is likely to originate. This is consistent with the heavy dominance of low-energy recoil atoms in the spectrum.

V. SPUTTERING FROM COLLISION CASCADES. DISCUSSION

In order to arrive at numerical expressions, one needs to make some assumptions concerning the stopping power dE_o/dx and the ejection probability $P(E_o, \theta_o)$. In ref. /6/, the following expressions were used,

$$dE_o/dx = N C_o E_o , \qquad (14a)$$

$$P(E_o, \theta_o) = \theta(E_o - U_o/\cos^2\theta_o) \qquad (14b)$$

where θ is a step function, U_o a planar surface potential, taken to be equal to the experimental sublimation energy, and C_o a constant,

$$C_o = \frac{\pi}{2}\lambda_o a^2 ; \quad \lambda_o = 24 ; \quad a = 0.219 \text{ Å} ; \qquad (14c)$$

These three relations appear feasible but by no means accurate; even their general features may be open to some doubt. Eq. (14a) approximates the scattering on an exponential repulsive interaction potential between the target atoms, and Eq. (14b) assumes a planar potential barrier as in electron emission.

From Eqs. (11a) and (14) we obtain

$$\Lambda = \frac{3}{4\pi^2} \frac{1}{NC_o U_o} \qquad (15a)$$

and Eq. (13a) yields $x_o = 3/(4NC_o)$. It is seen that x_o is of the order of 5 Å (6). The fact that x_o exceeds x_1 by more than an order of magnitude shows that only a few per cent of the atoms within x_o are sputtered, i.e., the linear theory is self-consistent.

Fig. 10 shows the sputtering yield of krypton on copper calculated in this way (6). Although, the agreement with experimental data is quite satisfactory, this should not be taken as evidence for similar accuracy of the input quantities, Eqs. (14,a-c). It is implied, however, that the theory as outlined up till now provides a suitable reference standard. This has been quite well corroborated (2,6). Rather than going into the details of this comparison, and to possible improvements and extensions of the theory (25), I should like to discuss some of its general features, and the consequences for various types of sputtering experiments.

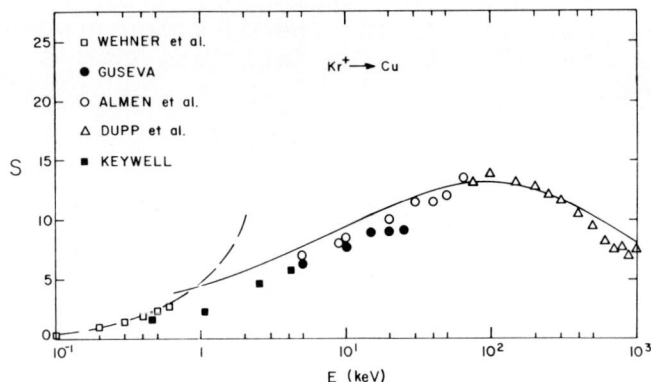

Fig. 10. Calculated and measured sputtering yields for krypton ions on polycrystalline copper. For details cf. ref. (6). Courtesy of the American Institute of Physics.

Let us first go back to Eqs. (9) and (11). It is seen that in the limit $E \gg E_0$, the properties of the ion beam (species, energy, angle of incidence) only enter through the factor $F_D(0,\vec{v})$, which in turn does not contain the energy E_0 nor direction θ_0 of the sputtered particles. In other words, the angular and energy distribution of sputtered particles is independent of the type, energy, and direction of the incident beam, in the asymptotic limit $E \gg E_0$. Conversely, the angular and energy dependence of the sputtering yield is determined by the factor $F_D(0,\vec{v})$ which is dependent on the target only through the target *mass* M_2, but not through the surface barrier $P(E_0,\theta_0)$. In other words, targets with similar mass, but very different sublimation energy (e.g. gold and tantalum) should show similar behaviour of the sputtering yield except for its absolute magnitude. Deviations from this asymptotic behaviour are expected to be of the relative order of (26) $(E_0/E)^{\frac{1}{2}}$ and, therefore, noticeable in the yield for E in the lower keV range, and in the spectrum, for all E, at high values of E_0.

As follows from the experimental energy spectrum (cf. Fig. 3) as well as the approximate

E_o^{-2} distribution (Eqs. (11a) and (14a)), the total number of sputtered particles is made up by atoms in the energy range up to 10-20 eV, i.e., a few times U_O. It is crucial, therefore, for the present estimate of the absolute sputtering yield to be valid, that the cascade picture holds true right down to U_O, i.e., that the upper inequality in Eq. (3) be fulfilled. In cases where this is not true, however, the cascade picture is still valid at somewhat higher values of E_O, certainly at $E_O \gtrsim 100$ eV.

Let us, alternatively, look at the emitted amount of *energy* per incident ion (more precisely, energy emitted via sputtered atoms),

$$\Delta E = \int dE_o \int d\Omega_o E_o S(\vec{v}, E_o, \theta_o) , \qquad (16)$$

with S being specified in Eq. (9). This quantity is determined rather more by the high-energy part of the spectrum than the yield. Ignoring surface binding and, somewhat incorrectly, assuming Eq. (9) to be valid even at high values of E_O, we obtain

$$\Delta E \sim \frac{3}{2\pi^2} F_D(0,\vec{v}) \int_0^{(E_o)_{max}} \frac{dE_o}{|dE_o/dx|} =$$

$$= \text{const } R((E_o)_{max}) F_D(0,\vec{v}) \qquad (17)$$

where $R(E_O)$ is the average path length of a recoil atom of energy E_O. Eq. (17) is a rather crude estimate of the emitted energy,* but it shows an important qualitative result: At higher (keV) energies, the stopping power dE_o/dx is rather slowly varying with energy (3,4), and therefore, the energy range $0...(E_O)_{max}$ tends to contribute rather uniformly to the total emitted energy. This means that for 100 keV Kr^+-Cu, the *emitted energy* is almost all contained

---

*A superior estimate is (27)

$$\Delta E = \int_{-\infty}^{0} dx \ F_D(x,\vec{v}) \ ;$$

this expression also contains the energy of backscattered ions, which is not included in Eq. (17).

in sputtered atoms with keV energies, while the *sputtering yield* is almost exclusively due to atoms with energies up to $\sim$ 20 eV. This is a rather drastic effect of the skewness of the energy spectrum.

There may, however, be expected even more drastic effects in the *excitation spectrum*. Let us assume some probability $Q_S(E_0,\theta_0)$ for a sputtered atom to leave the surface in some specific state of excitation or ionization. Then, the total number of sputtered atoms in this excited state will be

$$N_s = \int dE_o \int d\Omega_o Q_s(E_o,\theta_o) S(\vec{v},E_o,\theta_o) \qquad (18)$$

The relation between N_S and the spectrum of sputtered atoms depends on the behaviour of Q_S as a function of energy and angle. If Q_S varies slowly with energy, especially at low energies around the surface barrier $E_0 \approx U_0$, then N_S is proportional with S, and measurements of N_S may provide information about sputtered atoms. More frequent is the situation of a more or less steeply rising excitation function; in case Q_S rises linearly with E_0, the above argument indicates that there is *essentially no overlap* between the groups of atoms that contribute to S and N_S, respectively. *In particular, the atoms that contribute to N_S originate from much greater depths than those that make up the sputtering yield*. Considerations of this type can be applied to other excitation functions, and are of considerable importance with regard to the depth resolution of sputter-analysis.

It seems fair to assume that excitation functions Q_S most often increase with increasing energy E_0 in the eV range, possibly with some threshold at an energy higher than U_0. In cases where this is true, the conclusion is allowed that the collision cascade picture may be appropriate to describe the emission of excited particles even though the sputtering yield may originate predominantly in spike effects.

VI. SPUTTERING FROM SPIKES. QUALITATIVE ESTIMATES

Within the present concept of a spike, the collision cascade picture is valid down to recoil energies of the order of $E_0 \sim \nu(E)/N\Omega$ (cf. Eq. (3)), where spike effects take over. This immediately shows that spike effects will not show up significantly in those phenomena that hinge on the upper parts of the spectrum of sputtered atoms, such as the total emitted energy, and many phenomena involving excited atoms.

Let us, then, look at those situations where spike effects are to be expected. According to the above estimate, the most pronounced spike effects occur for small cascade volume Ω. As a qualitative estimate, Ω will be of the order of

$$\Omega \sim \pi X(E) \, (X_r(T_m))^2 \quad , \tag{19}$$

where $X(E)$ is the mean penetration depth of the ion, T_m the maximum energy of a recoil atom, and X_r the penetration depth of a recoil atom. X and X_R behave approximately as powers, $X, X_R \propto E^\alpha$, with α ranging from (3,4) ½ to 1.

Since $\nu(E)$ increases as $\propto E$, or more slowly, one concludes that ν/Ω is proportional to some *negative* power of E. Thus, the range of recoil energies E_0 where spike effects are significant *increases* with *decreasing* ion energy for a given ion-target combination. Moreover, the penetration depth at fixed energy decreases with increasing ion mass. Thus, pronounced spike effects are to be looked for in experiments with very heavy ions at not too high energies. Numbers can be found in a recent note (20).

Let us next look at the general behaviour of a spike. Obviously, once the dissipation of energy has proceeded to a stage where essentially all atoms within the volume Ω have received some amount of energy of the order of $\nu/N\Omega$, energy can be lost only by interaction with the region outside the original cascade volume, i.e., it is the *gradient* of the deposited-energy distribution that determines the rate at which energy is dissipated. Within the spike, atoms undergo collisions in which energy may be lost or gained, just about as in a gas under thermal equilibrium. Therefore, the nuclear stopping power is essentially zero during the duration of a spike.

Let us now estimate the spectrum of atoms sputtered from a spike. The principles of the derivation are the same as those in sect. 4, but the details are adjusted to the rather different physical situation.

First, the number of atoms participating in a spike is primarily the number of atoms in the spike volume, i.e., $N\Omega$. In a stationary state, with a flux of ψ [ions/s], the energy distribution of recoil atoms is given by

$$\psi\tau \; N\Omega \; f(E_o,\Theta)dE_o \quad , \qquad (20)$$

where τ is the time constant for decay of a spike, and $f(E_o,\Theta)$ a quasi-equilibrium energy distribution (normalized to 1) in a spike with an average energy Θ per atom,

$$\Theta \sim \nu(E)/N\Omega \qquad (21)$$

Proceeding as in sect. 4, we obtain the sputtering yield

$$S = \frac{1}{4\pi} \frac{\tau}{\Theta} F_D(o,\vec{v}) \int d\Omega_o dE_o f(E_o,\Theta) v_o |\cos\theta_o| P(E_o,\theta_o) \qquad (22)$$

or, by means of Eq. (11),

$$\Lambda = \frac{\tau}{4\pi\Theta} \int dE_o v_o f(E_o,\Theta) \int d\Omega_o |\cos\theta_o| P(E_o,\theta_o) \qquad (22a)$$

In analysing Eqs. (22) or (22a), one should remember that these equations are valid only in a genuine spike situation, i.e., when Θ is greater than U_o (Eq. (3)), and τ greater than the duration of a collision cascade (20), $\sim 10^{-13}$ sec.

In order to get numerical estimates, we assume the (feasible) ejection probability Eq. (14b) to be valid; in addition, we make the (much less feasible) assumption of a Maxwellian distribution $f(E_o,\Theta) =$ const. $E_o^{\frac{1}{2}}\exp(-\frac{3E_o}{2\Theta})$. Then, eq. (22a) reads

$$\Lambda = \frac{\tau}{\sqrt{3\pi M_2 \Theta}} \exp(-3U_o/2\Theta) \qquad (23)$$

The reader may object to the straight application of a Maxwellian in Eq. (22a). As a matter of fact, since only order-of-magnitude estimates of τ and Θ are available (20), the question of the energy distribution is secondary at the present stage.

The least obvious parameter occurring in Eq. (23) is the time constant τ. An estimate presented in ref. (20) was based on the assumption of energy transport in a spike being described by kinetic gas theory. While the absolute magnitude of τ as found in that manner is rather uncertain, the dependence on ion and target mass as well as ion energy is considered representative. In general, τ *increases* with *increasing* spike volume, i.e., it shows a behaviour opposite to that of Θ. The dominating influence comes from the rapidly decreasing rate of energy dissipation when the volume increases.

It follows from the general behaviour of τ and Θ that Λ, according to Eq. (23), decreases rapidly with E at low ion energy E because of the factor $\tau/\Theta^{\frac{1}{2}}$. Conversely, at high ion energy, Λ drops off rapidly because of the factor $\exp(-3U_0/2\Theta)$.

Fig. 11 shows an example of a sputtering yield curve (Au$^+$-Au) where spike effects are dominating within a certain energy range (28). One finds that Eq. (23) contains essential features of a sputtering

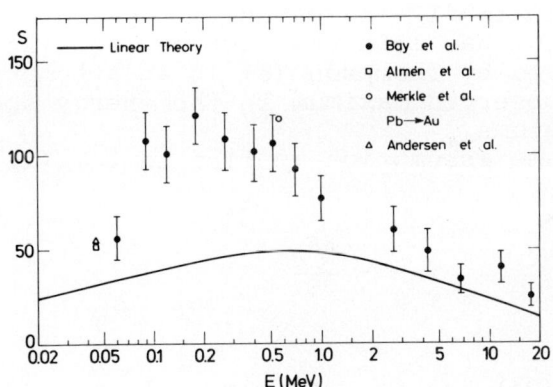

Fig. 11. Experimental sputtering yield for gold ions on gold at normal incidence (28). The theoretical curve shows the result predicted from linear collision cascade theory (6), Eqs. (11) and (15a). Courtesy of the authors and North Holland Publishing Company.

yield formula in the spike region. First of all, Λ depends on energy with a pronounced maximum at some value that depends on U_0. Also the maximum sputtering yield itself depends sensitively on U_0. Rough numerical estimates can be done by means of the expressions for τ and Θ derived in ref. (20). However, neither Eq. (23) itself nor the numbers given in ref. (20) seam accurate enough to justify a detailed comparison with experimental results yet.

In addition to the pronounced difference in the energy dependence of the sputtering yield, another characteristic difference between linear and nonlinear sputtering is expected in the energy spectrum of sputtered particles. Provided that the planar potential barrier is valid, the energy spectrum of sputtered atoms has to be corrected for the energy loss as well as the refraction effect due to the barrier. In the isotropic limit, linear cascade theory and Eqs. (14a), (14b) lead to a spectrum

$$\propto \frac{E_1}{(E_1+U_o)^3} dE_1 \cos\theta_1 d\Omega_1 \quad , \tag{24}$$

where the index "1" indicates energy and direction after an atom has left the surface. Eq. (24) was first derived by Thompson (8) in an attempt to explain the observed maximum in the energy spectrum of sputtered atoms.

The same assumption applied to Eq. (22) yields a spectrum

$$\propto E_1 e^{-\frac{3E_1}{2\Theta}} dE_1 \cos\theta_1 d\Omega_1 \tag{25}$$

While Eq. (24) has its maximum at $E_1 = U_o/2$, Eq. (25) has a maximum at $E_1 = 2\Theta/3$, i.e., at a position that depends sensitively on the bombarding ion as well as its energy.

VII. CLUSTER EMISSION

A matter of particular interest is the sputtering of clusters of two or more atoms, an effect that has been known for a long time (29), but that is still rather little understood.

Most observations of sputtered clusters are based on mass spectrometry of charged sputtered particles, i.e., metallic molecular ions as well as organic molecules (30), the latter occurring as impurities from pump oil etc. or after deliberate adsorption. I should here like to make a few qualitative comments on sputtered metallic clusters.

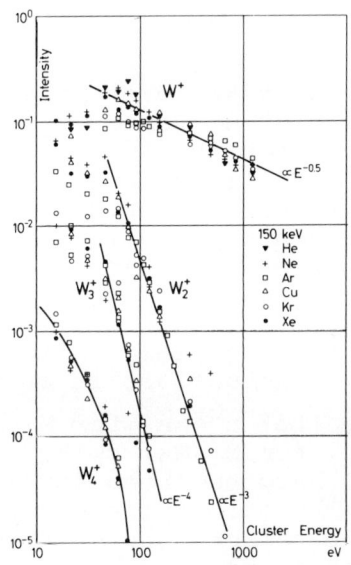

Fig. 12. Energy spectrum of positive atomic and molecular ions sputtered from tungsten (31). Courtesy of the author and Gordon and Breach Publishers.

Fig. 12. shows an energy spectrum of metallic molecular ions sputtered from tungsten (31). It is seen that molecular ions constitute a measurable fraction of the sputtered ions. Since the ion yield is small compared with the sputtering yield, this does not necessarily imply that neutral molecular particles constitute a significant fraction of all sputtered particles. The possibility of this being the case must, however, be taken into account. Measurements of Gerhard et al. (32) indicate indeed that the fraction of clusters may be as high as several percent of the sputtered particles.

The main mechanism invoked so far (33,34) assumes atoms to be sputtered independently from the surface and to cluster in front of the surface. Attempts to simulate the sputtering of molecules on a computer have been partially successful (35).

It appears likely that the analytical theories (33,34) underestimate the clustering effect. Obviously, clustering occurs as a close correlation in real and velocity space of two (or more) sputtered atoms, and thus is proportional to the square (or a higher power) of the number n of sputtered atoms in an individual event. We have

$$\overline{n} = S \qquad (26)$$

when an average is taken over many bombarding particles. The yield of diatomic particles is then proportional to n^2 which is close to S^2 only in case of small fluctuation. All evidence, however, points in the direction of the sputtering yield being a heavily fluctuating quantity (36), in particular in the case of $M_1 < M_2$ where ions are scattered appreciably. Fig. 7 indicates that the few trajectories that end close to the target surface may contribute a major part of the sputtering yield. Although the detailed theory of such fluctuations is an intricate task, one may presume that $\overline{n^2} >> \overline{n}^2$ for $M_1 < M_2$.

Conversely, for $M_1 > M_2$, spike effects may become important which have not been considered in refs. (33) and (34). The main spike effect is expected to originate in the rather more uniform velocity distribution of moving atoms in the target, which should favour the very emission of molecules. It appears that the computer simulations (35) have not been done under conditions that favour spike effects.

It appears reasonable to assume that whatever the detailed mechanism of cluster ejection, the effect will be most pronounced in the low-energy end of the spectrum, i.e., around the maximum (cf. Fig. 12). This complicates the possible investigation of spike effects by means of energy spectra, in as far as time-of-flight spectrometry is concerned (8).

VIII. CRYSTAL LATTICE EFFECTS

When sputtering experiments are performed on single crystals the most spectacular effects observed are the anisotropy of the sputtering yield with respect to the direction of the incoming beam (37), and a more or less pronounced deviation of the angular distribution of sputtered particles from the rather uniform cosine distribution (38), Eq. (24) or (25). As a rule of thumb the sputtering yield goes through a relative minimum when the beam is close to a channeling direction (39), and the distribution of sputtered atoms shows maxima (spots) around closely packed directions. The two phenomena are correlated only in as much as they express a general anisotropy of the whole slowing-down process, and not just of the initial and final stages.

Historically, crystal lattice effects had been considered essential in the interpretation of sputtering experiments on *any* material with a regular structure, i.e., also polycrystals, because the angular dependencies observed on single crystals were so obviously more complicated than what was expected for an isotropic medium. Two observations made in the late 1960's showed that despite the apparent complexity, the assumption of an isotropic medium was justified to describe the overall behaviour of the sputtering yield at least for a polycrystal. With regard to the incident beam it became evident (39) that directions of low sputtering yield alternate with angular regions of exceptionally high yield; averaged over the solid angle, a considerable degree of compensation was observed, in much the same way as in conventional channeling experiments (40). With regard to the ejected-atom distribution, careful scans (41,42) showed that even under conditions where pronounced spot patterns were produced the spots contributed only a minor fraction to the total sputtering yield. Thus, irrespective of the detailed interpretation of anisotropy effects, their significance in the overall behaviour of sputtering on polycrystals is sufficiently small to be considered as one of several secondary effects.

Because of this, and in view of the increased technological significance of various aspects of sputtering, the investigation of single crystal sputtering has become less intense in recent years.

IX. THE SURFACE

There is no doubt that the target surface plays an important role in the sputtering process. Experimental techniques have developed to a stage where sputtering experiments *can* be performed under controlled surface conditions. Similarly, a number of aspects in the theory could be treated to a considerable amount of accuracy at the present stage. It is feasible that the importance of sputtering in surface physics, in particular surface analysis, warrants a major effort in this direction. The task is complicated, however, when it comes to details. Here are some of the effects to be considered:

  i) Deficiencies of existing cascade theory with regard to the intersection with an ideal target surface.
  ii) Establishing a valid expression for the ejection probability of an atom from a well-defined but real surface.
  iii) Finding an accurate expression for the propagation of atoms and/or energy at the 1-10 eV level in a solid.
  iv) Consideration of real surface structures at the microscopic level, including differences between bulk and surface lattice spacings as well as surface defects and microstructure.
  v) Effects associated with impurities.

The last point alone splits up into a complex of problems that has been, and will be, the main topic of international conferences for a while. The first point has been solved in principle (25,43), and does not require too much extra effort in order to be settled. The second point has been analyzed in some detail for ideal surfaces (44). Little effort has been invested in the third and fourth point.

The reader may have noticed that, with the exception of the first point, the above problems require solutions that are more or less specific to the individual material under consideration. Such loss of universality is supposedly more than compensated by a gain in accuracy, but also warrants a multiplication of necessary effort.

Acknowledgements

Thanks are due to N.H. Tolk for his patient persuasion and to H.H. Andersen for comments on the manuscript.

REFERENCES

1. Behrisch, R., Ergeb.exakt.Naturw. 35, 295 (1964).
2. Andersen, H.H., in "Physics of Ionized Gases" (V. Vujnović, Ed.), p. 361. Institute of Physics, Univ. Zagreb, Yugoslavia, 1974.
3. Sigmund, P., Rev.Roum.Phys. 17, 1079 (1972).
4. Sigmund, P., in "Physics of Ionized Gases 1972", (M. Kurepa, Ed.), p. 137. Institute of Physics, Univ. Belgrade, Yugoslavia, 1972.
5. Behrisch, R., Editor,"Sputtering by Ion Bombardment". Springer Verlag, in preparation.
6. Sigmund, P., Phys.Rev. 184, 383 (1969).
7. Lindhard, J., Nielsen, V., and Scharff, M., Mat.Fys.Medd., Dan.Vid.Selsk. 36, no. 10 (1968).
8. Thompson, M.W., Farmery, B.W., and Newson, P.A., Phil.Mag. 18, 361, 377, 415 (1968).
9. Stark, J., Z.Elektrochemie 14, 752 (1908); ibid. 15, 509 (1909).
10. Hippel v., A., Ann.Phys.(Leipzig) 80, 1043 (1927).
11. Dupuy, C.H.S., Editor,"Radiation Damage Processes in Materials", Noordhoff, Leyden (1975).
12. Pooley, D., in ref. 11, p. 809, 325.
13. Fleischer, R.L., Price, P.B., and Walker, R.M., "Nuclear Tracks in Solids, Principles and Applications". Univ.Calif.Press (1975).
14. Townsend, P.D., and Elliott, D.J., in "Atomic Collision Phenomena in Solids" (D.J. Palmer et al., Eds.), 328. North Holland Publ. Co., Amsterdam, Netherlands, 1970.
15. Biersack, J.P., and Santner, E., Nucl.Inst.Meth. 132, 229 (1976).
16. Haff, P.K., Appl.Phys.Lett. 29, 473 (1976).
17. Robinson, J.E., and Thompson, D.A., Phys.Rev.Lett. 33, 1569 (1974).
18. Andersen, H.H., and Bay, H., J.Appl.Phys. 45, 953 (1974); ibid. 46, 2416 (1975).
19. Lindhard, J., Nielsen, V., Scharff, M., and Thomsen, P.V., Mat.Fys.Medd., Dan.Vid.Selsk. 33, no. 10 (1963).
20. Sigmund, P., Appl.Phys.Lett. 25, 169 (1974); ibid. 27, 52 (1975).
21. Brinkman, J.A., J.Appl.Phys. 25, 961 (1954).
22. Thompson, M.W., and Nelson, R.S., Phil.Mag. 7, 2015 (1962).
23. Ishitani, T., Murata, K., and Shimizu, R., Jap.J.Appl.Phys. 10, 1464 (1971);ibid. 11,125(1972).

24. Sigmund, P., Appl.Phys.Lett. 14, 114 (1969).
25. Sigmund, P., Proc. Conf. on Atomic Collisions in Solids, Kiev (1974).
26. Littmark, U., and Sigmund, P., J.Phys.D 8, 241 (1975).
27. Sigmund, P., Can.J.Phys. 46, 731 (1968).
28. Bay, H.L., Andersen, H.H., Hofer, W.O., and Nielsen, O., Nucl.Inst.Meth. 132, 301 (1976).
29. Honig, R.E., J.Appl.Phys. 29, 549 (1958).
30. Benninghoven, A., Jaspers, D., and Sichtermann, W., Appl.Phys. 11, 35 (1976).
31. Staudenmeier, G., Rad.Eff. 13, 87 (1972).
32. Gerhard, W., and Oechsner, H., Z.Phys. B22, 91 (1975).
33. Können, G.P., Tip, A., and de Vries, A.E., Rad. Eff. 21, 269 (1974); ibid. 26, 23 (1975).
34. Gerhard, W., Z.Phys. B22, 31 (1975).
35. Jackson, D.P., Can.J.Phys. 53, 1513 (1975); Harrison, D.E., and Delaplein, C.B., J.Appl. Phys. 47, 2252 (1976).
36. Westmoreland, J.E., and Sigmund, P., Rad.Eff. 6, 187 (1970).
37. Rol, P.K., Fluit, J.M., Viehböck, F.P., and de Jong, M.,"Proc. Int. Conf. Ionization Phenomena in Gases", Vol. 4, p. 257. North Holland Publ. Co., Amsterdam, 1960.
38. Wehner, G.K., Phys.Rev. 102, 690 (1956).
39. Onderdelinden, D., Can.J.Phys. 46, 739 (1968).
40. Lindhard, J., Mat.Fys.Medd., Dan.Vid.Selsk. 35, no. 14 (1965).
41. Schulz, F., Thesis, Univ. München (1967).
42. Olson, N.T., and Smith, H.P., Phys.Rev. 157, 241 (1967); J.Appl.Phys. 39, 3579, 4849 (1968).
43. Littmark, U., and Maderlechner, G., in "Physics of Ionized Gases 1976, Contributions", p. 139. J. Stefan Institute, Ljubljana (1976).
44. Jackson, D.P., Rad.Eff. 18, 185 (1973).

secondary ion production due to ion-surface bombardment

Klaus Wittmaack
Gesellschaft für Strahlen- und Umweltforschung mbH
8042 Neuherberg, Germany

In this paper secondary ion production due to ion-surface bombardment is reviewed. Special emphasis is given to a discussion of experimental data which can be used to check the validity of certain emission models. For clean metals bombarded with rare gas ions in the keV energy range the dominant processes leading to the emission of positive ions are autoionization and (for light elements) kinetic ionization, the latter phenomenon involving 1s or 2p hole production. Analysis of secondary electron spectra indicates a high probability for interaction between Auger electrons and sputtered particles. Accordingly ionization due to Auger electron impact is likely to be important in some cases. Secondary ion emission in the presence of electropositive and/or electronegative elements is discussed in some detail. The enhancement in the positive ion yield due to the presence of oxygen supports the bond breaking model. Different from photoemission studies the experimental results cannot be explained on the basis of band structure considerations. The yield of negative metal ions is found to be strongly related to the electron affinity. The yield of clusters, both homonuclear and heteronuclear, exhibits a strong dependence upon the bond strength. Bombardment induced changes in surface structure and stochiometry are shown to be of pronounced importance. The role of recoil implantation in producing saturation with oxygen is emphasized. The possibility of achieving substrate-independent secondary ion emission pattern from oxygen saturated surfaces is pointed out. Finally it is demonstrated that the current state of the theories of secondary ion emission is unsatisfactory.

I. INTRODUCTION

The phenomenon of secondary ion emission due to primary ion impact on solid surfaces has been known for many years. Detailed studies on this effect, however, have been carried out only for about ten years. Early reviews on secondary ion emission by Fogel (1) and Carter and Colligon (2) indicate that only a few uncorrelated experimental data were available in 1967/68. A theoretical basis was missing completely.

Growing interest in secondary ion emission originated from the potential use of this effect in analytical applications. Accordingly improvements in instrumentation have been much more spectacular than progress in understanding of the basic physical processes leading to secondary ion emission.

At the early stage of development in the field of Secondary Ion Mass Spectrometry (SIMS) people were fascinated by the microanalytical capabilities of the method. Sophisticated instrumentation such as the ion microscope of Castaing and Slodzian (3) and the scanning ion microprobe of Liebl (4) opened a wide field of applications. High cost of these magnetic type mass spectrometers, however, prevented rapid spread. Renewed popularity of SIMS is due to the use of much less expensive quadrupole mass filters. The first instruments of this type were designed for large area analysis (5-7). Very recently a quadrupole-equipped UHV scanning ion microprobe was developed by Wittmaack (8,9).

Due to its high sensitivity SIMS has rapidly become a widely used technique for surface and bulk analysis and for in-depth profiling. This is documented by a variety of review articles on instrumentation (10-19) and applications (10-13, 16-25). Whereas these reports (10-25) repeatedly cover similar or even the same aspects and data no attempt has been made as yet to deduce an overall picture of the basic physical phenomena involved. Recent review articles on secondary ion emission models (22,26,27) cover only selected aspects of this field.

In this study we have tried to collect experimental and theoretical data which are of importance for a better understanding of the secondary ion emission process. Due to the limited space available in this book it has not been possible to cover all experimental phenomena. For example, channeling effects and anisotropic ion emission can be touched only very briefly. Moreover the emission of adsorbates, either organic or inorganic, is outside the scope of this paper. Instead we have concentrated on the emission of ions characterizing the bulk of the sample. The strong variation of the secondary ion yields throughout the periodic table of elements is discussed as well as the pronounced influence of the chemical activity of the primary ion. Particular interest, however, has been

devoted to experimental data which show the dependence of the
secondary ion yield upon the primary ion energy for bombardment of clean surfaces with noble gases in the keV energy
region. Comparison with results on ion-induced Auger electron
emission seems to provide an interpretation of previously unexplained phenomena. In the field of chemically enhanced emission due to the presence of electronegative elements such as
oxygen the importance of bombardment induced changes in surface composition is pointed out. It will be shown that SIMS
may allow quantitative analysis of the bulk composition if
selective sputtering and recoil implantation of oxygen are
controlled adequately.

II. BASIC CONSIDERATIONS

In this section we will briefly discuss basic phenomenological equations as well as problems envisaged in the determination of quantitative result.

Secondary ion production may be discussed by separating
the process somewhat arbitrarily into two parts (i) energy
dissipation and sputtering and (ii) ionization and escape from
neutralization. As far as the phenomenological formalism is
concerned we will proceed along this line. The more accurate
description given later on will show that this simplified picture does not account for details of the ion production mechanism.

The effect of sputtering is fairly well understood. Following the ideas outlined by Sigmund (28) the sputtering process can be described as follows. An energetic primary ion
hitting the surface of a solid looses its energy in a series
of elastic and inelastic collisions with target atoms (and
electrons). Struck target atoms initiate a series of collisions
whereby energy and momentum is dissipated within a certain
volume in the solid, the overall event being termed "collision
cascade". In general the target surface will intersect the
cascade, which in other words means that energy and momentum
has been transferred to surface atoms. A surface atom which has
received enough backward directed momentum to overcome the
potential barrier at the surface (which is taken to be equal to
the surface binding energy) will leave the solid, i.e. it will
be sputtered. The process is shown schematically in Fig. 1,
which also illustrates that details of the cascade (sub)structure may depend upon specimen material, i.e. upon atomic number
and number density of the target as well as upon atomic number
and impact energy E_o of the primary ion.

The total sputtering yield S represents an average over
a large number of slowing-down events with eventually pronounced fluctuations of the sputtering yield of individual

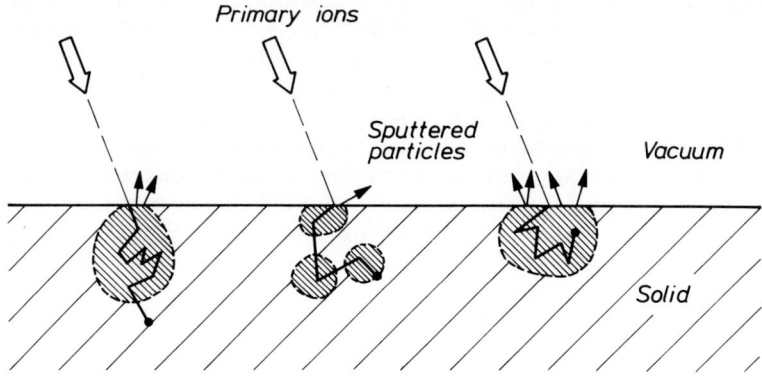

Fig. 1. Schematic illustration of ion penetration, development of collision cascades and sputtering.

ions /29/. S [atoms/ion] is defined as the number of target atoms sputtered per incoming primary ion.

The sputtered particles exhibit an energy spectrum $s(E)$ such that

$$S = \int_0^{E_m} s(E)dE, \qquad (1)$$

where E_m is the maximum energy of the sputtered particles.

Similarly the total yield of positive (negative) secondary ions, $S^{+(-)}$ [ions/ion] is defined as the number of secondary ions emitted per incoming primary ion. For simplicity we will discuss only the positive branch of the secondary ion species. The following arguments, however, hold for negative ions as well.

The differential secondary ion yield $s^+(E)dE$ is related to the differential sputtering yield $s(E)dE$ by

$$s^+(E) = \alpha^+(E)s(E), \qquad (2)$$

where $\alpha^+(E)$ is the spectral ionization probability which depends upon the particle energy (30,31). The total secondary ion yield is

$$S^+ = \int_0^{E_m} \alpha^+(E)s(E)dE. \qquad (3)$$

From Eq. (3) we see immediately that a mean ionization probability defined by $\langle\alpha^+\rangle = S^+/S$ cannot be used in a quantitative discussion of ionization phenomena. Details of the secondary ion production mechanism can only be understood if $\alpha^+(E)$ is known.

Out of the quantities defined above the total sputtering yield can be determined with little difficulty. This is due to the fact that one has to determine merely the amount of target material lost due to bombardment with a measured number of primary ions. Evaluation of S^+ and/or $s^+(E)$, on the other hand, requires much more effort because the emitted secondary ions have to be mass analyzed and transported to the detector. The secondary ion intensity I^+ [counts/s] recorded during bombardment of a pure sample is given by

$$I^+ = i_o \gamma \beta(M^+) T(E_1,R) \int_{E'}^{E''} s^+(E) dE, \qquad (4)$$

where $E' = E_1 - \Delta E/2$ and $E'' = E_1 + \Delta E/2$. i_o [ions/s] is the primary ion flux, γ is the isotope abundance, $\beta(M^+)$ is the detector efficiency which depends upon the mass and the velocity of the ion (32,33) and $T(E_1,R)$ is the transmission of the ion optical system which depends upon position E_1 and width ΔE of the energy window as well as upon the mass resolution R of the mass spectrometer. Although Eq. (4) represents only a simplified picture of the complete process of secondary ion detection (among others the angular dependence of ion emission has not been taken into account) it indicates that many experimental parameters have to be determined before quantitative data about the emission process can be obtained.

Unfortunately, in most of the experiments reported in the literature these problems are disregarded and quite often "absolute" data have been reported without specifying important quantities such as position and width of the energy band pass. Such data can be regarded only as qualitative or semi-quantitative at the most. To make this obvious we consider just one aspect of Eq. (4). Since the energy distribution of secondary ions, $s^+(E)$, exhibits a peak at or below about lo eV and a more or less steep tail towards higher energies (Sect. III. B.1) it is clear that experimental integration over a small energy window as in Eq. (4) does by no means reflect the total secondary ion yield. The correct way of determining the total yield would be a measurement of the energy and the angular distribution of secondary ions.

In view of the fact that reliable absolute yield values are practically non-exisiting, as one might also conclude from a compilation of literature data (22), we will discuss aspects of the total secondary ion yield only briefly.

To ease interpretation of the results shown below, the instrumentation used by the respective authors is briefly described in the appendix.

III. EXPERIMENTAL RESULTS

A. General Observations

There are at least three characteristic features of secondary ion mass spectra which tend to confuse scientists who are not familiar with the SIMS technique.

(i) The spectra of both positive and negative secondary ions are usually very complex, exhibiting not only singly and multiply charged atomic ions but also cluster ions of nearly all possible combinations of target atoms. Even multiply charged clusters have been observed, e.g. $(^{29}Si^{30}Si)^{2+}$ /34/. There is also preliminary evidence for the existence of doubly negative secondary ions such as S^{2-} (P. Williams and C.A. Evans, private communication).

(ii) The secondary ion intensities, in particular those of the monatomic species are very sensitive to the presence of either electropositive or electronegative atoms at the target surface. On the other hand, these yield enhancing elements show up with high intensity in SIMS surface analysis of previously air exposed samples bombarded with noble gas ions. The intensity of these impurity peaks (in particular Na^+, K^+, O^-, F^-, S^- and Cl^-) quite often exceeds the intensity of the main constituents of the target. The (surface) impurities disappear only after prolonged sputtering.

(iii) The secondary ion yields of monatomic species emitted from rare gas ion bombarded (sputter cleaned) pure specimens vary strongly but nonmonotonically with the atomic number of the target element, Z_2. The Z_2-variations are quite different for positive and negative ions, the positive ion yield usually exceeding the negative ion yield by orders of magnitude (35).

These three features will become more obvious from the data given below.

B. Secondary Ion Emission from Clean Elements (Noble Gas Ion Bombardment)

1. Mass Spectra and Energy Distributions

In one of the first systematic investigations of secondary ion emission Honig (36) found that in addition to monatomic ions dimers and trimers were emitted from 200-400 eV krypton and xenon bombarded silver. These experiments have been extended later on to a study of the cluster size distribution of semiconductors and insulators such as germanium, graphite and

diamond (37). An interesting result was that elements with a
low ionization potential produce mainly positively charged
atoms and molecules, whereas elements with a high electron
affinity exhibit mostly negative ions. Moreover Honig found
alternating intensities in the mass spectrum of C_n^- clusters,
with local maxima for even-numbered species (37). Finally
Honig demonstrated that sputtering and sublimation of graphite
results in very similar negative ion cluster patterns (37).

Following the early investigations of Honig (36,37),
cluster ion emission from clean surfaces has been studied by
various authors (34,38-60). An example is given in Fig. 2

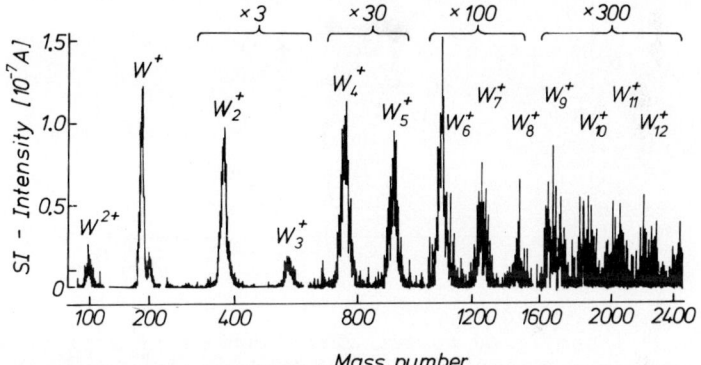

Fig. 2. Positive secondary ion emission spectrum of polycrystalline tungsten bombarded with 150 keV xenon ions. Secondary ion energy 30 eV (42,44). Spectrometer type 1.

which shows the mass spectrum of tungsten bombarded with 15o
keV xenon ions, as reported by Staudenmaier (42,44). Clusters
containing up to 12 tungsten atoms can be identified in Fig. 2.
However, there is no real cut-off at the high mass end of the
cluster distribution. This result was supported by additional
experiments in which Staudenmaier improved the detection sensitivity by secondary ion acceleration (44). Clusters containing up to 34 tungsten atoms could be detected, the cluster
intensity I_n^+ decreasing with increasing number of constituents,
n, as $n^{-5.5}$, for n > 10 (44).

Energy spectra of cluster ions are shown in Figs. 3 and 4.
It is obvious that the distributions become the narrower the
larger the cluster, a result also reported by other authors
(38,43,49,51,59-61). Note that the width (FWHM) of the Al_6^+
distribution in Fig. 4, corrected for the spectrometer resolution, amounts to only about 1 eV (60). A pronounced effect of
cluster size is also observed in the high energy tails where
the intensity drops off the more rapidly the larger the number

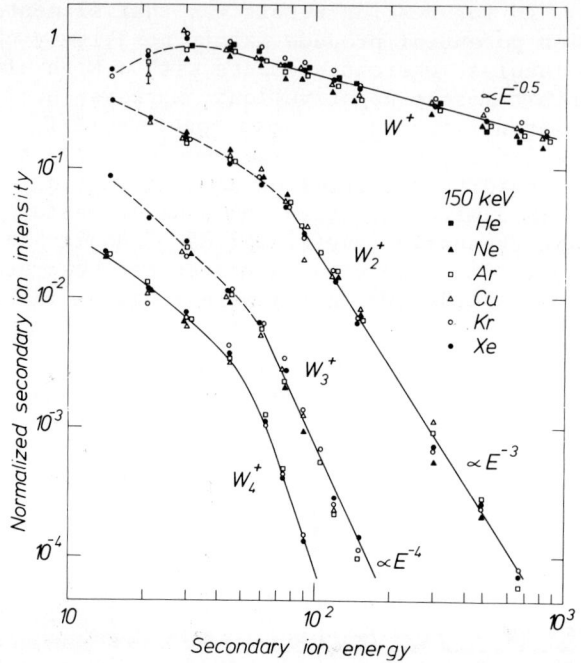

Fig. 3. Normalized energy distributions of atomic ions and cluster ions sputtered from tungsten with various 150 keV primary ions (44) (see also (42)). Spectrometer type 1.

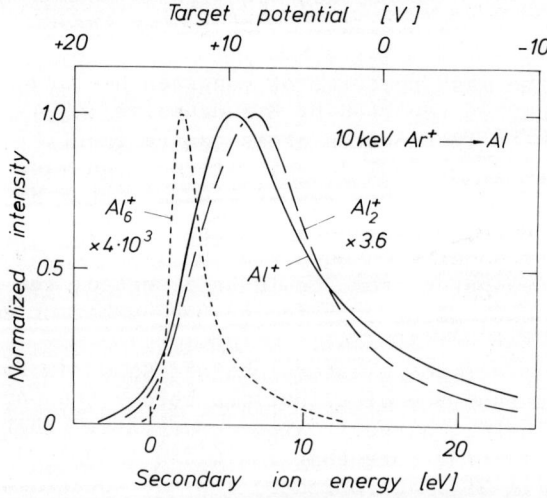

Fig. 4. Normalized energy distributions of atomic ions and cluster ions sputtered from aluminium (uncorrected)(60). Spectrometer type 2.

of constituents. These results can be interpreted on the basis of stability considerations (42,44,62,63). The effect can be used in quantitative analysis of atomic ions to discriminate against eventually interfering molecular ions (49,64).

Another interesting result of Fig. 3 is that the shape of the energy distributions of atomic ions and cluster ions does not depend upon projectile mass (42,44). This effect has been observed also for an Al-Mg alloy (49). It seems to indicate that the energy spectrum in the collision cascade is not sensitive to the amount of energy deposited into atomic processes.

The influence of the sputtering yield on the intensity of atomic ions and cluster ions has also been investigated. For 150 keV bombardment of polycrystalline tungsten Staudenmaier (44) found the normalized W^+ intensity, $I(W^+)/i_o$, to be proportional to the calculated sputtering yield. Accordingly one can assume that the ionization mechanism does not change with projectile mass. The normalized intensity of clusters, on the other hand decreased with decreasing projectile mass, i.e. with decreasing sputtering yield (44). This is in agreement with results reported by Gerhard and Oechsner who demonstrated that the yield of neutral dimers sputtered from polycrystalline metals by 1 keV argon ions is roughly proportional to the square of the sputtering yield, $I(M_2) \propto S^2$ (65).

Related effects have been observed by Herzog et al. (49) who studied the influence of the projectile mass on the population of the cluster ion spectra of aluminium bombarded with 12 keV rare gas ions. It was found that the intensity distributions of Al_n^+ clusters drop off the more rapidly the smaller the projectile mass. For example, the relative intensity $I(Al_4^+)/I(Al^+)$ was 10^{-2}, 10^{-3} and $< 10^{-5}$ for xenon, argon and helium bombardment, respectively (49).

Intensity distributions I_n^{p+} for silicon bombarded with noble gas ions have been reported by various authors (46,49, 55,57). Some results are compiled in Fig. 5. The influence of projectile mass on the relative cluster intensity I_n^+ is obvious again. Moreover we see that experimental parameters such as bombardment energy, angle of incidence and/or energy band pass of the mass spectrometer affect the cluster size distribution. The influence of the projectile energy on the intensity of multiply charged ions, which can also be deduced from Fig. 5, will be described more thoroughly in Sect. B.2.

An important effect noticed by Storp (55) should be mentioned. The cluster size distribution is the same for vacuum annealed, crystalline silicon and bombarded (amorphous) silicon, respectively.

In Figs. 2-5 the monatomic secondary ions exhibited the largest intensity of all species. This is not always the case. Wittmaack and Staudenmaier (60) have demonstrated recently that in many cases (positive) diatomic ions show much larger

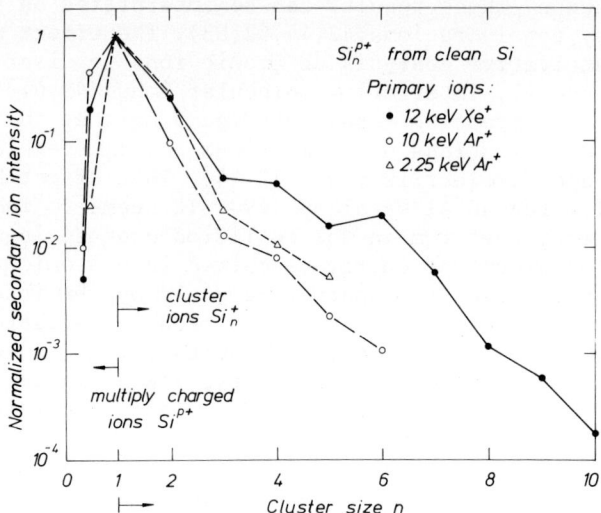

Fig. 5. Intensity distributions of singly and multiply charged ions and cluster ions sputtered from silicon under various conditions. $\circ(46)$, $\bullet(49)$, $\triangle(55,57)$. *Instrument type 3 (46,49) and 4 (55,57), respectively.*

peak intensities than monatomic ions. An example is presented in Fig. 6. Intensity ratios $I_2^+/I^+>1$ have been observed for transition elements such vanadium, zirconium, niobium, molybdenum, tantalum and tungsten (60). Since only 15 elements were investigated it is much likely that there are even more metals with $I_2/I^+>1$. It is important to note, however, that reliable intensity ratios can be detected only under ultrahigh vacuum conditions (60). Presence of impurities such as oxygen usually enhances the monatomic yield much more pronounced than the diatomic yield (58) (see Sect. C.2.). Consequently poor vacuum conditions will result in $I_2^+/I^+<1$.

The energy spectra shown in Figs. 4 and 6 provide important information about the low energy part of the secondary ion distributions. At energies above about 10 eV, however, these curves do not represent true distributions because they have not been corrected for ion optical effects which cause an energy dependent ion discrimination.

For mass spectrometers in which acceleration fields are applied to improve secondary ion transport the correction factor is usually proportional to the ion energy, i.e.

$$I_{corr} = IE/E_c \quad \text{for} \quad E \geq E_c \quad (5a)$$

and

$$I_{corr} = I \quad \text{for} \quad E \leq E_c \quad (5b)$$

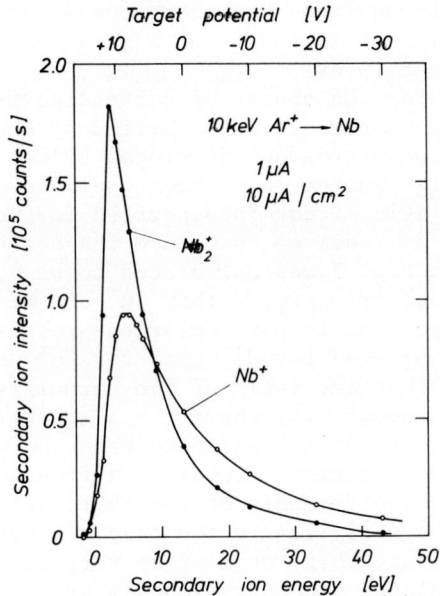

Fig. 6. Energy distributions of monatomic and diatomic secondary ions sputtered from niobium (uncorrected) (59). Instrument type 2.

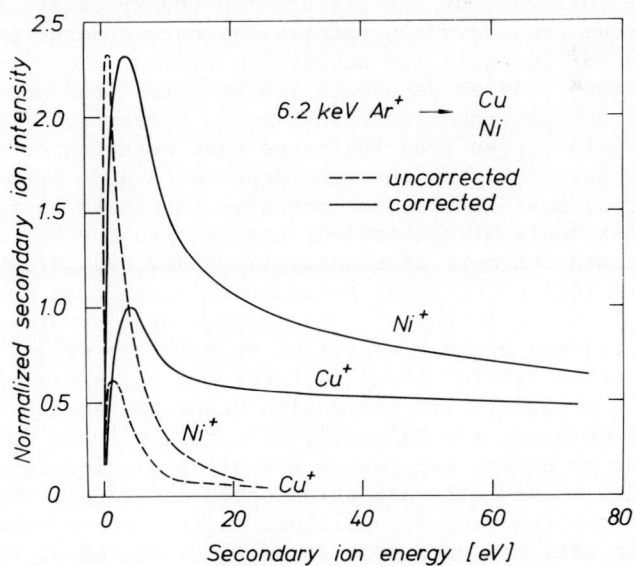

Fig. 7. Uncorrected and corrected energy distributions of secondary ions sputtered from nickel and copper (52). Instrument type 5.

In the quadrupole-equipped secondary ion mass spectrometer (8,9) used for the measurements of Figs. 4 and 6 the critical energy E_c is about 10 eV.

An example for the amount of correction required is given in Fig. 7. The measurements were carried by Blaise and Slodzian (52) using the ion microscope developed by Castaing and Slodzian (3). High energy resolution could be obtained by means of the electrostatic mirror incorporated in the system. The critical energy E_c required to deduce the corrected energy distributions in Fig. 7 was calculated to be 0.7 eV (52).

It is obvious from Fig. 7 that Cu^+ and Ni^+ exhibit different energy spectra. In particular the relative peak height is much more pronounced for Ni^+ than for Cu^+. A quantitative analysis shows that the ratio of the normalized peak heights $r(Ni^+)/r(Cu^+)$ is about 2.4, where $r = I^+(peak)/ I^+(100\ eV)$. A behaviour similar to nickel has been observed also for the other ferromagnetic elements cobalt and iron, whereas the remaining transition elements of the third series showed energy spectra very similar to the case of copper. These differences demonstrate that one must be very careful when trying to compare "absolute" secondary ion yields of different elements.

2. Dependence of the Secondary Ion Yield upon the Primary Ion Energy

The influence of the primary ion energy on the yield of the various secondary ion species has been studied only by a few authors. In early low energy bombardment experiments Bradley (66) studied secondary ion emission from molybdenum, tantalum and platinum near threshold. Comparison with sputtering yield curves (68) indicates that emission of ions and neutrals does not show the same dependence upon primary ion energy and, most likely, not the same threshold energy.

Experiments at bombardment energies in the keV range have been carried out more frequently (40,58-60,67). Although Hennequin (40) pointed out several years ago that not only the secondary ion yield but also the degree of ionization increases with increasing argon energy (for example for Cu^+) this effect has attracted little attention later on. A roughly linear increase in secondary ion yield with argon ion energy (2-8 keV) has been observed for Mg^+, Si^+, Ti^+, Fe^+, Ni^+ and Cu^+, whereas there was no energy dependence for Al^+ (40). The latter result may be due to an influence of adsorbed and recoil implanted oxygen since the vacuum conditions in the set-up used by Hennequin were not very good ($\geq 2\times10^{-7}$ Torr 69). This problem will be discussed in more detail in Sect. III. C.2.

Brochard and Slodzian (67) have studied the emission of Al^+, Al^{2+} and Al^{3+} from pure aluminium (and copper-aluminium alloys) bombarded with 1 to 6.2 keV argon ions. At low primary ion energies their set-up did not allow a correct measurement of the primary current density and the secondary ion extraction efficiency. Accordingly only intensity ratios I^{2+}/I^+ and I^{3+}/I^{2+} were presented as a function of the primary ion energy. I^{2+}/I^+ was found to increase by three orders of magnitude between 1 and 2 keV. Between 2 and 6.2 keV the additional increase amounted to only 50%. For I^{3+}/I^{2+} an increase of only about 50% was observed between 1.5 and 6.2 keV (obviously Al^{3+} was not detectable below 1.5 keV). Brochard and Slodzian concluded from their results that 2p hole formation in aluminium becomes increasingly probable at argon energies above 1 keV (67). Since production of 2p holes is a necessary requirement for the emission of multiply charged ions from clean aluminium the measurement of Al^{2+} and Al^{3+} allows a detection of this inner shell excitation process.

Information about the relative importance of either symmetric or unsymmetric collisions has been achieved from measurements of the intensity of Al^{2+} ions emitted from copper - aluminium alloys. It was found that $I^{2+}(Al)$ is roughly proportional to the square of the aluminium concentration (67). From this result one can conclude that 2p vacancies are preferably produced in symmestric aluminium - aluminium collisions and not by direct collisions between primary argon ions and target atoms.

Studies on the variation of the secondary ion yield with increasing primary ion energy have also been carried out by Wittmaack and coworkers (58-60). Results for argon bombarded silicon are shown in Fig. 8. Different from the measurements by Brochard and Slodzian (67) the current density could be controlled adequately. One can see that at constant beam current and current density the intensity of all ion species (except SiO^+) emitted from silicon increases with increasing primary ion energy (58). The most pronounced effect is observed for Si^{2+} and Si^{3+}. As in the case of aluminium the intensity of these ions can be assumed to reflect the probability for the production of 2p holes in silicon. Note that similar to aluminium the intensity ratio I^{3+}/I^{2+} for silicon is constant throughout the range of impact energies investigated and that I^{2+}/I^+ becomes constant above 8 keV. Below 5 keV I^+ tends to approach a value independent of the primary ion energy. This indicates that the emission of singly charged ions from silicon does not necessarily require 2p hole formation.

In a discussion of the emission of multiply charged ions from aluminium and silicon one important experimental detail has to be taken into consideration. Whereas the results of Ref. (67) were obtained from an integral measurement over a large part of the secondary ion spectrum the data shown in Fig. 8 where recorded at secondary ion energies of about 5 to 10 eV (58-60). Secondary ions of such a low energy are likely to be emitted

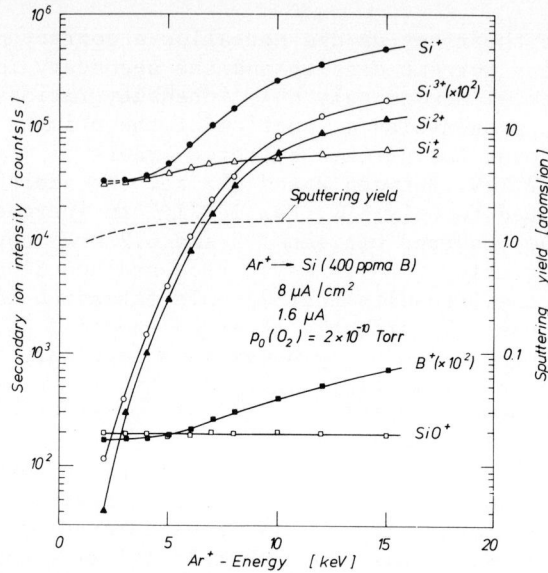

Fig. 8. Energy dependence of the secondary ion intensity of various ions emitted from argon bombarded silicon (58,59). Instrument type 2.

from the random cascade and not directly from a close binary collision between the projectile and a target atom (or between target atoms). This idea will be of importance in a quantitative interpretation of secondary ion formation (Sect. V.).

With regard to Fig. 8 we like to point out that not only I^{2+} and I^{3+} but also I^+ increases much more rapidly with increasing primary ion energy than the sputtering yield. Since the energy spectra of both sputtered particles and secondary ions vary only negligibly in the range of primary ion energies discussed, the ionization probability $\alpha^+(E)$ must be a function of the bombardment energy, E_0, i.e. we have $\alpha^+ = \alpha^+(E_0,E)$. Adequate theories of secondary ion emission must take into account such a dependence upon both the impact energy and the emission energy.

Finally we mention that there is only a very small effect of the bombardment energy on the secondary ion intensity of dimers (58,59). The increase in I_2^+ exceeds the increase in sputtering yield (28,70) only slightly (Fig. 8). A variation of the form $I_2^+ \propto S^2$ (65) is in agreement with the experimental data. However, in view of the small change in sputtering yield covered in Fig. 8 these results should not be taken as a proof of such a model of dimer emission.

Comparison of I_2^+ with I^{p+} indicates that dimer ionization is not due to 2p hole production. This fact may be explained by assuming that energetically the requirements for dimer ionization (and escape from neutralization) can be fulfilled much more easily than the conditions for cluster stability.

The influence of impact energy on secondary ion emission has been investigated for 13 other elements (60). In all cases I^+ increased more rapidly with increasing primary ion energy than the sputtering yield. However, the effect was most pronounced for silicon. At least in part this is due to the fact that the experiments were restricted to bombardment energies between 2 and 15 keV. For different projectiles and at other energies strong intensity variations may show up not only for silicon. For example, pronounced effects occur for argon bombarded aluminium between 1 and 2 keV (67), as discussed above.

3. Z_2-Variation if the Secondary Ion Yield

Pronounced differences in the secondary ion yield of the elements bombarded with inert gases have been observed frequently (18,20,25,35,40,41,52,60,71-76). The first thorough study was carried out by Beske (72). Although the general trend of his data has been reproduced several times it is clear now (58,60) that his measurements must have been strongly affected by the presence of oxygen (residual gas pressure 2 to 3 x 10^{-6} Torr (72)).

To demonstrate the Z_2-variation of the secondary ion emission from clean surfaces we have compiled data which were obtained under more adequate vacuum conditions (Fig. 9). The

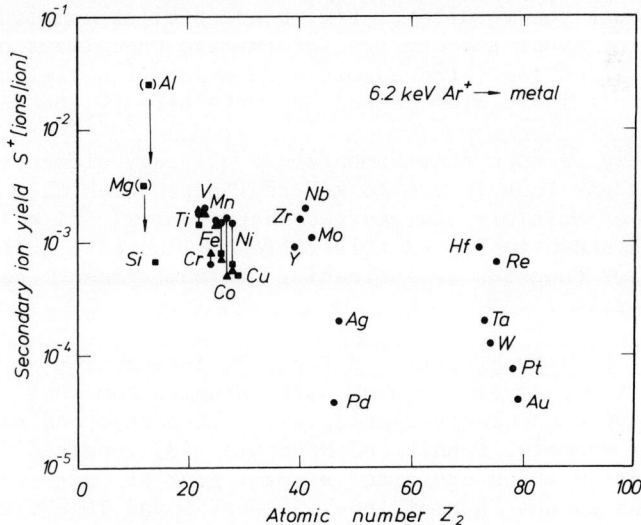

Fig. 9. Secondary ion yields of various metals. ■ absolute yields (40), ▲● relative yields (52) (see text). Instruments type 5 (52) and type 6 (40,69).

results shown are due to Hennequin (40) and Blaise and Slodzian /52/. The relative yields of Ref. (52) (circles and triangles) have been converted somewhat rigorously to absolute yields by using the copper yield of Ref. (40) for calibration (squares). Squares refer to integral measurements 40 whereas circles and triangles represent secondary ion intensities taken at the low energy peak or at 100 eV respectively (52).

Hennequin considered the vacuum and bombardment conditions in his studies (residual pressure $\geq 2 \times 10^{-7}$ Torr, current density 1 to 10 $\mu A/mm^2$ (69)) as sufficient to assure that there was only negligible influence of the residual gas on the secondary ion emission (69). Studies at elevated oxygen pressure, however, showed a yield enhancement of less than a factor of 10 for Al^+ and Mg^+ (71). It is known that an increase in yield by more than two orders of magnitude should be observed when starting with a clean aluminium surface (67,77). Accordingly the secondary ion yield for Al^+ in Fig. 9 is least one order of magnitude too large. The data for the other metals can be expected to be less strongly affected (but in general still too large).

Keeping this in mind we realize that the secondary ion yields of typical metals differ by about two orders of magnitude. The variation in ionization probability is even larger because many elements with a low secondary ion yield exhibit a high sputtering yield, for example palladium, silver and gold (72).

Although the results of Fig. 9 and similar compilations presented by other authors are informative one always has to realize that not only the absolute secondary ion yields but also the relative yields depend upon the primary ion energy (60).

Finally we note that bombardment of heavy elements($Z_2>20$) with inert gas ions in the lo keV region produces only small fractions of multiply charged secondary ions, $I^{2+}/I \leq 1\%$ (60, 72,78). As shown in Fig. 2 this ratio may increase to more than 10% for tungsten by increasing the bombardment energy to 150 keV (42,44).

4. Current Density Effects and Particle Interaction in Vacuum

In a few secondary ion emission studies somewhat curious effects have been observed which were often regarded as artefacts. For example, Dennis and McDonald (43) reported Cu^+ energy spectra which exhibited a small peak at "zero" energy in addition to the "normal" peak between 5 and 10 eV. Recent results from the same group (79) again showed a low energy hump in the Cu^+ spectrum, now at about 5 eV. In neither case this peak has been discussed. Note that the results reported in Ref. (43,79) were obtained with a quadrupole-equipped mass spectrometer. Such systems are known to introduce problems because of the strong energy dependence of the quadrupole transmission below 10 eV (60,80).

Detailed investigations of the "pre-peak" intensity in energy spectra have been carried out by different group (45, 78,80). These studies had in common that acceleration fields were applied between target and mass spectrometer to improve the secondary ion transport. Ionization phenomena occurring in vacuum at some distance from the target thus resulted in a shift to lower or even "negative" energies. The effects have been found to depend strongly upon the primary ion current density (45,80). For example, the peak intensity in the secondary ion spectrum of Au^+ was found to increase by more than a factor of 10 due to raising the current density in a 5 μA argon ion beam (80). Simultaneously the peak shifted to lower energies by about 1 eV. In addition a pronounced low energy tail was observed which extended to "negative" energies (80). A quantitative discussion of the effect indicated that ionization in vacuum at distances of the order of the focussed beam diameter (~100 μm) could in fact occur via interaction of sputtered metastable particles (80). Since details of the processes are not well understood presently, ionization due to interaction with the primary ion beam cannot be excluded. Note that processes in vacuum show up the more clearly the samller the surface ionization probability (such as in case of gold).

In this section we should mention another astonishing effect which is obviously induced by ion implantation. The spectra of argon bombarded aluminium show strong emission of argon clusters Ar_n^+ and mixed clusters $Al_m Ar_n^+$ (34,49). High intensities of such clusters have been observed only for this particular projectile-target combination. The reason for the occurrence of these polyatomic ions is not clear. However, build-up of argon bubbles in aluminium can be assumed to be a necessary requirement for the formation of Ar_n^+ and $Al_n Ar_m^+$. More detailed studies might provide interesting information about the stability of inert gas clusters.

5. Channelling Effects and Anisotropic Secondary Ion Emission

As pointed out in the introduction channelling effects and anisotropic secondary ion emission can be discussed only briefly. Most of the relevant studies were carried out to investigate projectile-target interaction or sputtering rather than secondary ion emission.

Bukhanov et al. (81) investigated anisotropic secondary ion emission from copper and differences between the energy spectra for emission in either random or low index directions. MacDonald and coworkers determined focussing energies (43) as well as critical angles for channelling and relative anisotropic sputtering yields (82) from secondary ion emission data. Channelling has also been studied by Bernheim (83) and Laurent and Slodzian (84), the latter authors including temperature effects. Staudenmaier (48) investigated anisotropic emission of monatomic and polyatomic secondary ions emitted from tungsten. All the above experiments provide only little information about the relevant ionization processes.

C. Secondary Ion Emission in the Presence of Electropositive or Electronegative Elements

1. Yield Enhancement Due to the Presence of Electropositive Elements

In one of the early secondary ion emission experiments Krohn (85) reported high yields (> 1%) of negative metal ions ($Cu_n^-, Ag_n^-, Au_n^-, Al_n^-, Sn_n^-$ and Ni_n^-) from targets bombarded by a beam of positive cesium ions (1.1 keV). A further increase in yield by a factor of 10 could be achieved by directing an auxiliary beam of neutral cesium at a copper target. Most astonishing the yield of Al_n^- and Sn_n^- clusters was larger than (or at least equal to) the monatomic yield for n > 4, the dimer yield exceeding the monomer yield by more than an order of magnitude (85). The high stability of Al_2^- has been confirmed (and interpreted) recently by Leleyter and Joyes (53) for 10 keV argon bombarded aluminium.

Following the work of Krohn (85) (whose set-up seems to have been the first quadrupole-equipped secondary ion mass spectrometer) Hortig et al. (86,87) used cesium-coating to achieve high yields of negative ions from inert gas bombarded surfaces.

Andersen (88) measured relative intensities of negative ions emitted from elements bombarded with 9.5 keV cesium. The yields of negative ions were found to be strongly correlated with the electron affinities of the neutral atoms (88). Comparison of Cu⁻ emission from either xenon or cesium bombarded copper demonstrated that the observed pronounced difference in negative ion yield is related to the surface chemistry produced by electropositive ion bombardment and not to the projectile mass. Without work function measurements Andersen claimed the yields of negative ions "to be proportional to an exponential function involving the difference between the work function of the cesium bombarded surface and the electron affinity of the element" (88). Since the actual surface concentration of cesium will depend upon the steady state conditions achieved during continuous implantation and sputtering (89) quantitative relations should only be given after carefully controlling the relevant parameters.

Negative ion yields from cesium bombarded copper have also been investigated by Abdullayeva et al. (90). Other authors have used positive ion emission from alkali ion bombarded molybdenum (91), aluminium and copper (92) to study angular and energy distributions of sputtered particles.

Detailed quantitative studies of secondary ion emission from alkali ion bombarded surface are not available.

2. Yield Enhancement Due to the Presence of Electronegative Elements

The influence of electronegative elements on the secondary ion yield was discovered in a comparison of ion emission from pure elements and the respective oxides (3,). Later on it was shown that enhanced emission of positive ions can be obtained not only from oxides but also from metal targets sputtered at elevated oxygen pressure (58,67,71,77,93-98). Moreover bombardment with oxygen ions has been found to strongly enhance the ion yield (70,94,98-101). The latter effect could be demonstrated to result from ion implantation of oxygen (70, 100,101). Yield enhancement due to oxygen adsorption without simultaneous ion bombardment has been studied by Benninghoven and coworkers (20,21,25,55,74,102-108).

Oxygen adsorption has equally been assumed by many authors to cause the enhancement under steady state bombardment conditions (77,96). The possibility of recoil implantation of adsorbed oxygen has been considered by Hennequin (109). Other authors have also discussed this effect, both in connection with secondary ion emission (70) and photoemission studies (110,111). Experimental evidence for recoil implantation of absorbed oxygen has been obtained recently by Rutherford backscattering depth analysis of silicon samples bombarded with xenon at elevated oxygen pressures (112).

In secondary ion emission experiments recoil implantation causes transients in the intensity of oxygen sensitive ions after variations of the oxygen pressure. Examples are shown in Fig. 10a. The experiments were carried out as follows. At a base pressure of 10^{-8} Torr a silicon sample was first sputter cleaned with a large area lo keV argon ion beam. Then the beam was interrupted and oxygen was introduced to constant pressures around 10^{-5} Torr. After turning the beam on again the SiO$^+$ intensity increased rapidly from a very low initial value which was independent of the time of oxygen exposure before bombardment (10 s $\leq$ t $\leq$ 10 min). The shape of the I(SiO$^+$) transients and the bombardment fluence required to achieve equilibrium were found to depend strongly upon oxygen pressure (Fig. 10a). This result indicates that the amount of (temporarily adsorbed) oxygen available at the surface determines the time (fluence) dependence of the transients. Comparison with ion collection curves (89) supports the idea that the transients reflect the build-up of the oxygen surface concentrations induced by recoil implantation.

The depth distributions of oxygen can be studied after

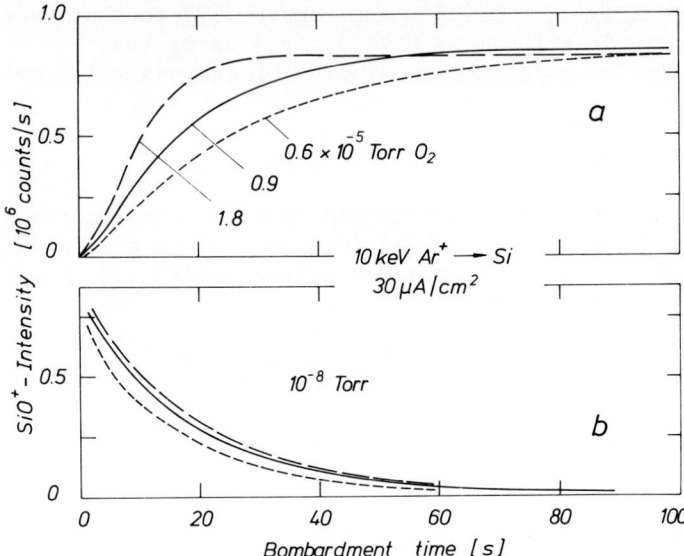

Fig. 1o.(a) Transients of the SiO^+ intensity observed during bombardment of silicon at elevated oxygen partial pressures. b) Depth profiles of recoil implanted oxygen, showing the steady state distributions produced in (a). (K. Wittmaack, unpublished results). Instrument type 2.

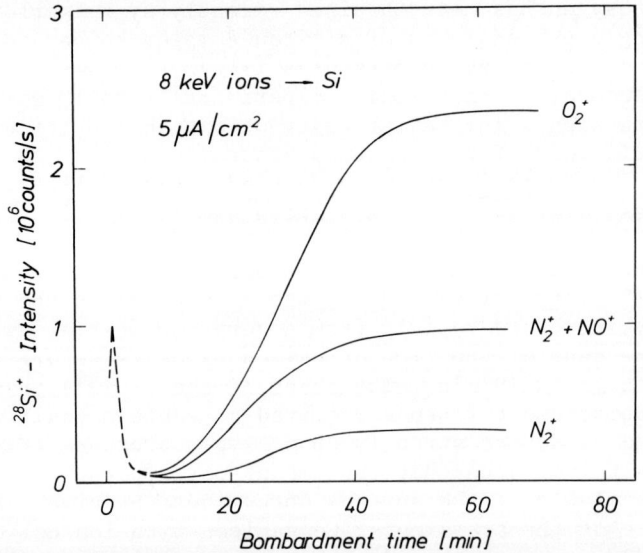

Fig. 11. Implantation induced variation of the Si^+ intensity emitted from as-supplied silicon upon bombardment with different primary ions (70). Instrument type 2.

closing the oxygen leak. Results of depth profiling studies
are shown in Fig. 10a. As expected (113-115) the recoil implantation distributions are peaked at the surface. Using a
sputtering yield of 1.5 atoms/ion (70) the 10% concentration
level of the distributions is found at a distance of 20-30 Å
from the surface.

Differences between recoil implantation and direct implantation become obvious from a comparison of the build-up
of the oxygen surface concentration. Fig. 11 shows the time
(fluence) dependence of the Si^+ intensity emitted from a silicon crystal covered with a natural oxide (70). (It will be
shown below that Si^+ can be used to monitor the oxygen surface
concentration semi-quantitatively.) Neglecting the peaking due
to sputtering of the surface oxide we see that the implantation
induced increase in intensity (i.e. the build-up of the oxygen
surface concentration) is delayed considerably as compared to
recoil implantation (Fig. 10a). This is due to differences in
the range distributions. Whereas implantation distributions
are Gaussian to first order recoil implantation distributions
exhibit a peak at the surface followed by a roughly exponential
tail. Note that bombardment with nitrogen or oxygen-nitrogen
mixtures also enhances the Si^+ intensity (Fig. 11) (70).

Knowledge of the recoil implantation effect eases interpretation of many of the following results. Whereas Fig. 10
demonstrated oxygen induced effects only at relatively high
pressures more detailed measurements by Maul and Wittmaack (58,
97) showed that yield enhancement can be observed already in
the 10^{-9} Torr range. An example is presented in Fig. 12. Boron
doped silicon was bombarded with 4 keV argon ions at various
oxygen partial pressures, the ultimate residual pressure being
8×10^{-10} Torr (oxygen partial pressure 2×10^{-10} Torr) (58,
97). The secondary ion intensities recorded represent steady
state values of transients similar to those in Fig. 10a.

As one can see the monatomic ions Si^+ and B^+ are very
sensitive to oxygen, the silicon intensity starting to increase
at pressures as low as 10^{-8} Torr (Fig. 12). For both Si^+ and
B^+ the increase in intensity amounts to more than three orders
of magnitude. Note that Si^+ and B^+ very nicely follow the SiO^+
variation. This indicates that the yield of monatomic ions is
already enhanced by the presence of spurious oxygen. In fact,
if one assumes the SiO^+ intensity to be a measure of the oxygen
surface concentration one finds the yield enhancement for Si^+
and B^+ to be roughly proportional to the oxygen concentration
(except for the saturation region). This important result indicates that changes in the band structure of the sample cannot
be responsible for the enhancement effect at low oxygen concentrations. Although one must expect the local density of
states to change due to the presence of oxygen it is a completely unfounded speculation to assume that these local

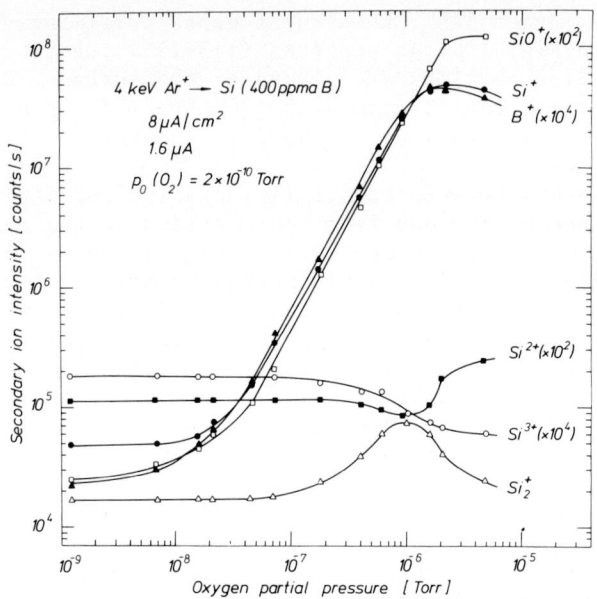

Fig. 12. Steady state intensity of various secondary ions emitted from argon bombarded silicon as a function of the oxygen partial pressure in the target chamber (97). For similar results at 10 keV see Ref. (58). Instrument type 2.

changes resemble the differences between the band structure of silicon and silicon dioxide.

We like to emphasize this point of view already at this stage because oxygen enhanced photon emission has been discussed by many authors within the frame of the band structure model (79,110,111,116-119). In the following we will recall experimental results which demonstrate that enhancement in secondary ion emission is due to breaking of molecular bonds (24,71). By this we mean that the ionization probability of sputtered atoms will be enhanced according to the ionic character of the molecular bond between oxygen and target atoms.

The bond breaking model is supported by some data shown in Fig. 12. As one can see $I^{2+}(Si)$ increases at high oxygen partial pressures after having passed through a flat minimum. On the other hand, $I^{3+}(Si)$ decreases monotonically for $p(O_2) > 10^{-7}$ Torr. This decrease might reflect the reduced probability for 2p hole production (27). Quantitative estimates, however, cannot be given since the silicon sputtering yield is also reduced in that range of oxygen pressures.

Definite evidence for reduced 2p hole production at elevated oxygen pressure comes from measurements of ion induced Auger electron emission. Hennequin (109) observed that the Auger electron peak from 10 keV argon bombarded magnesium and aluminium disappears completely at high oxygen pressures, a finding which compares well with the absence of Auger peaks in the secondary electron spectrum of argon bombarded oxides MgO and Al_2O_3 (109). These experimental results clearly demonstrate that introduction of oxygen changes the mechanism of excitation. Models of ion and photon emission in which this is not taken into account (119) must be viewed with reservation.

On the basis of these facts we can conclude that the increase in $I^{2+}(Si)$ in Fig. 12 must be due to the onset of a new ionization mechanism. Interpretation in terms of bond breaking seems to be adequate as can be seen from the following arguments. $I^{2+}(Si)$ increases only after large amounts of oxygen have been introduced. The high concentration of oxygen can be estimated from the fact that the saturation intensity of Si^+ and SiO^+ is roughly equal to the intensity emitted from bulk SiO_2 (58). Therefore $I^{2+}(Si)$ enhancement requires the formation of an oxide-like structure. It has been shown above that this structure is produced as a result of recoil implantation of adsorbed oxygen. Oxygen adsorption without recoil implantation would not generate the required oxide thickness. Only at high oxygen concentration each silicon atom is surrounded by enough oxygen atoms to make Si^{2+} production via bond breaking effective.

The decrease in $I(Si^{3+})$ on the other hand, may indicate that the charge of the emitted ion is related to the chemical valence, i.e. Si^{2+} should be favoured in case of silicon. Further evidence for such a picture comes from Fig. 13 which

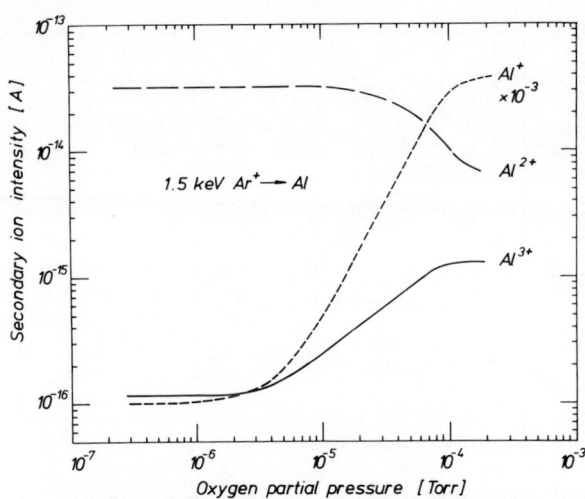

Fig. 13. Steady state intensity of singly and multiply charged ions emitted from argon bombarded aluminium as a function of the oxygen partial pressure in the target region /67/. Instrument type 7.

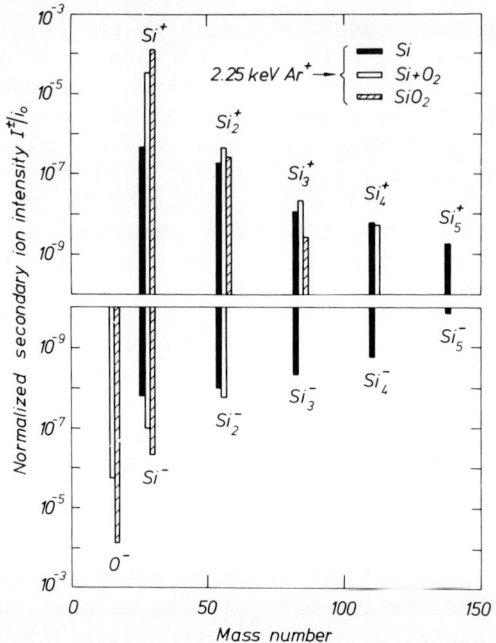

Fig. 14. Positive and negative spectrum of silicon clusters emitted from samples with different oxygen surface concentration. The intensity of O^- is shown for comparison. Compiled from Ref. (55,57). Instrument type 8.

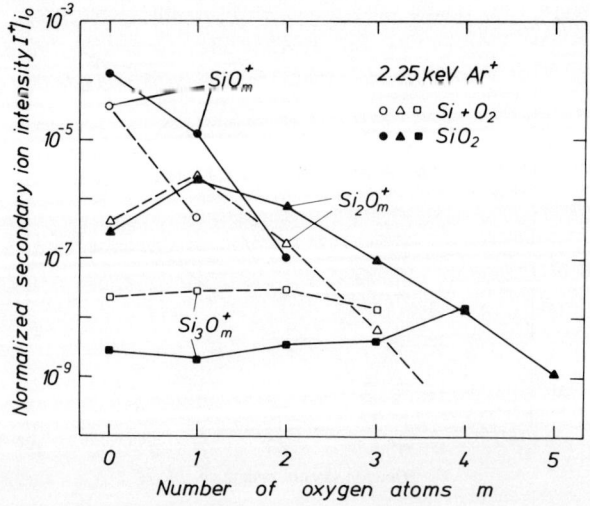

Fig. 15. Intensity of $Si_nO_m^+$ clusters emitted from silicon samples with different oxygen surface concentration. Compiled from Ref. (55,57). Instrument type 8.

presents results obtained by Brochard and Slodzian (67). As one can see $I(Al^{3+})$ increases by more than an order of magnitude due to introduction of oxygen, whereas $I(Al^{2+})$ decreases. Since aluminium is characterized by the valence three the increase in $I(Al^{3+})$ is not surprising.

We like to emphasize that for multiply charged ions the enhancement effect is pronounced only if the bombardment energy is low enough because only in that case the counteraction due to reduced 2p hole production is negligible. At higher energies the "kinetic" production of multiply charged ions increases rapidly (see Fig. 8) so that the "chemical" effect is masked completely. At 10 keV, for example, $I(Si^{2+})$ decreases at elevated oxygen pressure 58. Similarly $I(Al^{3+})$ increases by only about 40% at 6.2 keV (67). Note that only the 6.2 keV results have been cited by Kelly and Kerkdijk (110,111) to support the band structure model, whereas the 1.5 keV results and other data in favour of bond breaking have not been mentioned.

It is clear that the effectiveness of bond breaking in producing singly charged monatomic secondary ions cannot be deduced simply from Figs. 12 and 13. Additional experimental evidence would be desirable. Results in favour of this model have been obtained by Storp (55,57) who measured positive and negative secondary ion emission from clean and oxygen covered silicon and from silicon dioxide bombarded with 2.25 keV argon. Data compiled from his work are shown in Figs. 14 and 15. (Note that the mass spectra of Ref. (55) have been reproduced incorrectly in Ref. (21).)

In a comparison with the results discussed above one has to take into the fact that Storp (55) used the "static" SIMS method (21,25). This simply means that very low beam current densities were applied such that the average target thickness sputtered while taking a complete spectrum was much smaller than a monolayer. Accordingly the spectrum of silicon saturated with adsorbed oxygen was only slightly affected by recoil implantation of oxygen.

To ease comparison the data in Fig. 14 are restricted to the mass lines of oxygen-free silicon clusters $Si_n^{\pm}$ and O^-. With respect to the above discussion the most important mass lines are Si^+ and O^-. As one can see the intensities of $I(Si^+)$ and $I(O^-)$ are nearly identical in case of a SiO_2 target. Such a close one by one correspondence would be quite surprising if the ions were formed independently. Accordingly we conclude that bond breaking is mainly responsible for the Si^+ yield enhancement.

Several other important results of Fig. 14 are worth being discussed. First we note that clean silicon saturated with adsorbed oxygen $(Si + O_2)$ does not show the same cluster distribution as SiO_2. This indicates that surface saturation with oxygen does not result in an oxide sufficiently thick to yield a SiO_2 specific secondary ion emission pattern. This interpretation is supported by a compilation of the cluster spectra emitted from either SiO_2 or $Si + O_2$ (Fig. 15). As one can see the relative intensity of $Si_nO_m^+$ clusters is very sensitive to the oxygen content of the target. Whereas silicon rich clusters decrease in intensity with increasing oxygen content of the sample, oxygen rich clusters (and Si^+) increase in intensity. This can be understood easily on the basis of statistical arguments (46). The higher the oxygen concentration the smaller the probability that two or more silicon atoms are found at neighbouring sites in the solid. In fact the intensity distribution of Si_n^+ clusters drops off very rapidly for SiO_2 samples (46,55,57) (see Fig. 14).

These arguments have been recalled very recently by Morgan and Werner (120) who demonstrated that the secondary ion spectrum of oxygen bombarded iron is substantially changed at elevated oxygen pressure. This effect has been mentioned first by Prager (121). The variations are very similar to those shown in Fig. 15. They can be understood if one assumes that the saturation oxygen concentration produced by implantation at low base pressures is too small to generate an oxide like structure. Selective oxygen sputtering may also play a role. Again a high surface concentration of oxygen can be brought about by recoil implantation because the respective depth distribution is peaked at the surface.

Returning to Fig. 14 we note two other interesting phenomena. (i) The yield of negative silicon clusters is quite large. In some cases (Si_3^-, Si_4^-) it nearly equals the yield of positive clusters. (ii) The presence of oxygen enhances the emission not only of positive ions but also of negative ions. At first sight this experimental result may be somewhat surprising. If we assume, however, that negative ion emission is controlled by the electron affinity of the sputtered particle the effect can be explained easily. According to a recently published review (122) the electron affinities of silicon and oxygen are EA(Si) = 1.39 eV and EA(O) = 1.47 eV. On the basis of these data one would even expect $I(Si^-)$ to be comparable to $I(O^-)$. The fact that O^- emission from SiO_2 exceeds Si^- emission by more than two orders of magnitude (Fig. 14) may again support the bond breaking model. It should be mentioned that pronounced emission of negative monatomic ions from oxygen covered surfaces has also been observed for copper (107) (EA(Cu) = 1.23 eV (122)).

To complete the presentation of experimental results on oxygen enhanced secondary ion emission, available data for various elements are compiled in Fig. 16. "Absolute" secondary ion yields for 3 keV argon bombardment of clean metals were taken from the work of Hennequin (full squares) (40) and Benninghoven and coworkers (full circles and triangles) (21,25,

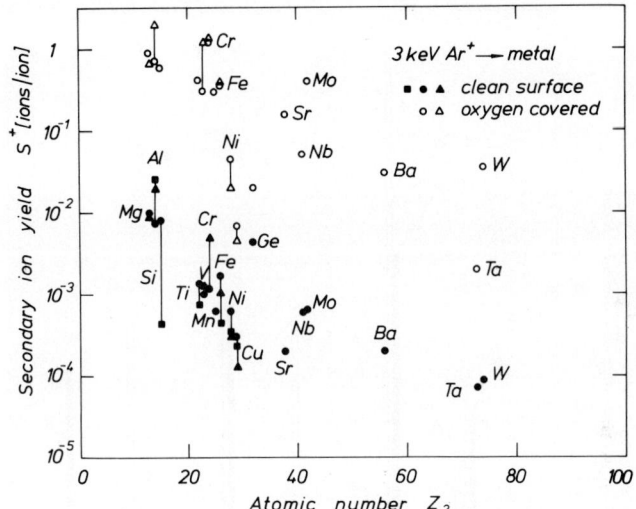

Fig. 16. Comparison of the secondary ion yield of clean and oxygen covered metals, sputtered with 3 keV argon ions. ■ (40), ▲△ (21,74), ● ○ (25,104-108). Instruments type 6 (40) and type 8 (104).

74,104-108). Note that not only the results of different groups show pronounced discrepancies in either way but also the results reported by the same group differ by up to a factor of four (Cr). An explanation for these discrepancies has not been given in the respective compilations (21,25).

The effect of oxygen surface coverage on the secondary ion yield S^+ can be estimated from a comparison of open and closed circles or triangles. The enhancement factor covers a wide range, from about 5 for germanium to 400 for tungsten. As far as the maximum enhancement is concerned these figures can be used only as lower limits, the reason again being that monolayer oxygen coverage does usually not provide an oxide-like target structure. For example, the enhancement factor for Si/SiO_2 should be about 300 (Fig. 15) instead of 70 (Fig. 16).

Differences in the enhancement factor observed under different experimental conditions (factor 1000 in Fig. 12) may be due in part to changes in the energy spectrum. Slodzian and Hennequin (71) already noted that the oxygen induced increase in secondary ion yield is most pronounced at low ion energies. This effect has been claimed to support the band structure model (119). However, possible changes in the energy distribution of sputtered particles have not been taken into account. In fact, presently available comparative results on energy distributions of neutrals and ions do not seem to allow a quanti-

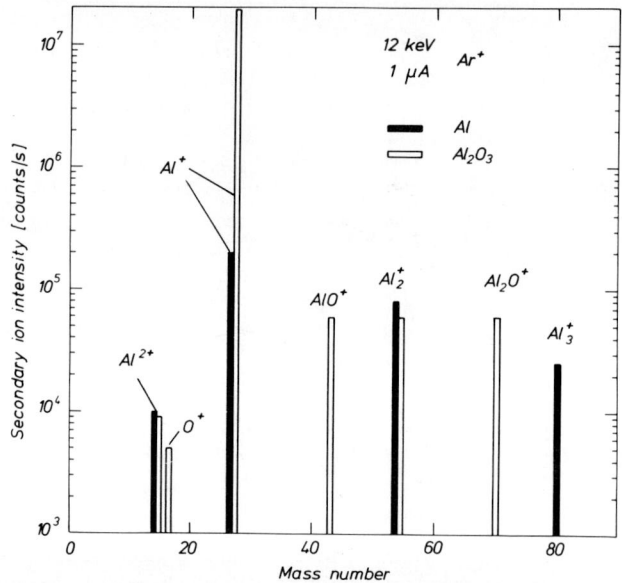

Fig. 17. Comparison of the mass spectra of Al and Al_2O_3, (K. Wittmaack, unpublished results). Instrument type 2.

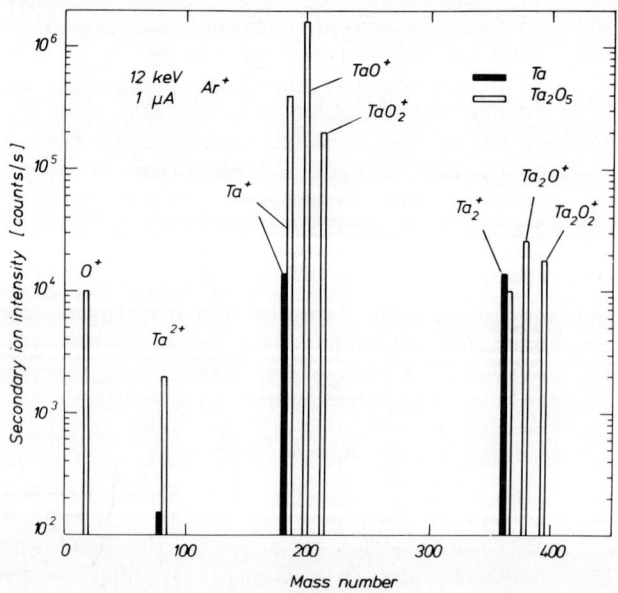

Fig. 18. Comparison of the mass spectra of Ta and Ta_2O_5, (K. Wittmaack, unpublished results). Instrument type 2.

tative discussion.

Up to now we have mainly discussed the effect of oxygen on the intensity of monatomic ions. This was partly due to the fact that Si^+ was the most intense mass line in the spectrum of oxygen rich silicon. Similarly Al^+ is the most prominent mass line in the spectrum of Al_2O_3 (Fig. 17). Many oxides, however, show a completely different spectrum. Ta_2O_2 is an example (Fig. 18). In this case TaO^+ exhibits by far the largest intensity. Moreover TaO_2^+ is comparable in intensity to Ta^+. Even larger relative intensities of these oxide ions have been observed for 3 keV argon bombardment of oxygen covered tantalum (105). Other metals showing intensity ratios $I(MO^+)/I(M^+) > 1$ are titanium, vanadium, niobium, molybdenum and tungsten (25).

Similarly large intensity ratios $I(MO^+)/I(M^+)$ have been observed when sputtering the respective (transition) elements from alloys (98,123). This agreement indicates that the pattern of ions such as M^+, MO^+ and MO_2^+ becomes largely independent of the target composition if the near surface region is saturated with oxygen. It would be of considerable importance for quantitative analysis if experimental results could support this speculation.

3. Comparison with Secondary Ion Emission from Alkali Halides

In this section we briefly discuss secondary ion emission from alkali halides. These compounds are of interest because they represent examples for which the ionic character of the bonding is out of question.

Estel et al. (124) have reported on secondary ion emission from various alkali halides bombarded with 1.3 keV argon ions. In all cases the spectra are characterized by strong lines of monatomic positive cations and negative anions of nearly equal intensity, e.g. Li^+ and F^- emitted from LiF. Moreover cluster ions are observed with rapidly decreasing intensity. These clusters comprise one or more neutral cation-anion complexes together with one additional cation or anion, e.g. $(LiF)_nLi^+$ and $(LiF)_nF^-$. Charged cation-anion complexes are either not observed or appear with small intensity, e.g. $(NaCl)^{\pm}$ (124).

From these results one can conclude that the ionic character of the bonding in the solid is in fact reproduced in the secondary ion spectrum.

In view of the characteristics of cluster ion emission from alkali halides 'irregularities' in the cluster intensities of ions emitted from SiO_2 (55) can be explained easily. $(SiO_2)_nSi^+$ and $(SiO_2)_nO^-$ are prominent lines in the sense that $I(Si_2O_3^+) > I(SiO_2^+)$, $(Si_3O_4^+) > (Si_2O_4^+)$, $I(SiO_3^-) \simeq I(SiO_2^-)$ and $I(Si_2O_5^-) > I(Si_2O_4^-)$. $(SiO)_nSi^+$ and $(SiO)_nO^-$ are intense only for n = 1. This indicates that for large enough clusters the target composition and the ionic character of the bonding are reflected in the secondary ion spectrum.

D. Comparison with Other Ion Impact Phenomena: Ion-Induced Emission of Photons, X-rays and Auger Electrons

1. Photon Emission

The complexity of the secondary ion emission pattern described above makes it highly desirable to study ion-induced processes by related techniques. Fortunately some investigations on the emission of ion-excited photons, x-rays and Auger electrons have been published very recently which seem provide useful information for a better understanding of the secondary ion emission process.

Photon emission originating from sputtered particles has been investigated by various authors (78,111,112,117-119,125-142). The most important results can be summarized as follows:

(i) Photon lines of particles emitted from clean targets exhibit pronounced Doppler broadening. Line shape analysis indicates that the mean energy of light-emitting particles is in the keV range (116,134,136,142). This differs completely from secondary ion emission where the energy distributions are peaked below about 10 eV and mean energies are only of the order of 100 eV (143).

(ii) The photon emission intensity is very sensitive to oxygen (and water vapour), the increase in intensity amounting to orders of magnitude in the extreme case (116,117,119,135,136). The enhancement effect is accompanied by a pronounced narrowing of the line width indicating that under these conditions photon emissions originates mainly from slow particles (116,136). This is explained by the fact that resonance deexcitation processes become improbable in oxide-type materials due to the existence of a forbidden gap (30,110,111,116-119,136).

(iii) A variety of metals exhibits continuum radiation in addition to line emission (79,139-141). Interesting enough the most intense continuum radiation is emitted from transition metals with incomplete d-shells, i.e. from metals which show strong secondary ion emission of dimers M_2^+ and MO^+ (see Sect. C). There is experimental evidence, however, that continuum radiation originates from neutral particles (141).

(iv) At low impact energies photon emission exhibits a mostly undefined threshold below about 100 eV followed by a gradual rise with increasing energy (117,128). Comparison with low energy secondary ion emission studies (66) indicates that although the threshold energies may be similar the secondary ion yield increases much more rapidly with increasing energy than the photon yield. This is obviously due to the fact that photon emission is governed by a higher probability for radiationless deexcitation.

From these results it is clear that there is by no means a one-by-one correspondence between photon emission and secondary ion emission.

2. X-Ray Emission

Different from photoemission only a few x-ray measurements are available which might be of use in a discussion of secondary ion emission. This is due to the fact that studies on x-ray emission have been carried out mainly at energies above about 1oo keV. Investigations on soft x-ray emission at impact energies below 1oo keV are rare.

Saris and Onderdelinden (144) measured the cross section for Ar L-shell x-ray emission in Ar-Ar collisions in the gas phase. The results shown in Fig. 19 are presented in this context because they quantitatively reflect the energy dependence of the cross section for L-shell vacancy (2p hole) production. With a fluorescence yield $\omega = 10^{-3}$ (144) the cross section for vacancy production is found to exceed 10^{-17} cm^2 at energies above about 13 keV. According to Fig. 19 one would expect no measurable 2p hole production at impact energies below 8 keV. This may change markedly if argon is embedded in a solid because in that case the steady state excitation of the projectile penetrating the solid plays an important role (145). Note that the Z_1-dependence of Ar L-shell vacancy production exhibits pronounced oscillations (146). For example, with Al$^+$ projectiles the threshold for this process is as high as 50 keV

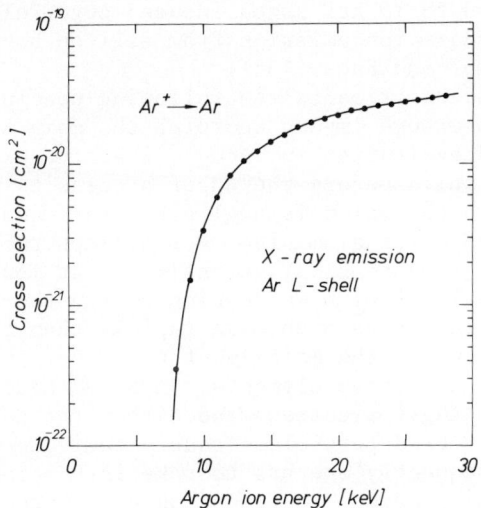

Fig. 19. Energy dependence for Ar L-shell x-ray emission excited in Ar-Ar collisions in the gas phase (144).

(146). At bombardment energies usually applied in secondary ion emission experiments (~ 10 keV) Ar L-shell production in argon-aluminium collisions is impossible.

3. Auger Electron Emission

Ion-induced Auger electron emission was first studied to check the model of kinetic secondary ion production. Hennequin and coworkers (109,147) demonstrated that 10 keV rare gas ion bombardment of beryllium, magnesium, aluminium and silicon results in very well defined Auger peaks in the spectrum of secondary electrons whereas there was no such peak for titanium, nickel and copper targets. Occurrence of Auger peaks with light element targets was attributed to $1s$ (beryllium) or 2p hole production whereas absence of Auger peaks was explained by effective shielding of inner shells by 3d electrons (109). The first experiments have been repeated later on with improved resolution (148) and were extended to alkali metals (149,150,151). Argon Auger electron emission has been studied in collision with elements from potassium (Z_2 = 18) through chromium (Z_2 = 24) (152). The respective cross sections for Auger electron production where measured at impact energies below 15 keV.

Measurements on ion-excited Auger electron emission at energies around 50 keV have been published repeatedly by Benazeth, Viel and coworkers (153-156). At these energies Auger electron emission from transition elements has been identified clearly (155). A very informative comparison of low-energy ($\leq$ 3 keV) electron and argon ion-excited Auger spectra of magnesium, aluminium and silicon has been reported by Grant et al. (157). 3 to 15 keV argon induced Auger electron emission and secondary ion emission from silicon have been investigated by Kempf and Kaus (158).

From these experiments the following conclusions can be drawn. At high enough impact energies the main Auger intensity results from deexcitation in vacuum, i.e. the excited particles have left the solid before the Auger process takes place. Auger deexcitation in the solid is negligible for magnesium and of little importance for aluminium. For silicon there is a major bulk-like peak (due to $L_{2,3}VV$ transition) at low bombardment energies (3 keV) but with increasing energy the atomic-like peak ($L_{2,3}MM$) increases much more rapidly than the bulk-like peak. In addition to the main Auger peaks one observes energy loss peaks for all three elements. This indicates that interaction between Auger electrons and either the solid (plasmon losses) or sputtered particles (quasi-atomic losses) is highly probable. Consequently one has to take into account secondary ion production by Auger electrons. Such a process has not been considered before. We will show, however, that a considerable amount of secondary ions may be produced that way (Sect. V.).

IV. THEORIES AND MODELS OF SECONDARY ION EMISSION

A. Surface Effect Models

The complexity of the experimental data summarized in the preceding section clearly indicates that a unified theory of secondary ion emission will hardly be obtainable. Nevertheless some authors have tried to deduce formulas for the ionization probability which contain only material constants such as the work function and the ionization potential.

Schroeer et al. (159) proposed the use of adiabatic perturbation theory and suggested a formula for the (positive) ionization probability rather than performing detailed calculations. The final equation contains measurable quantities such as the ionization potential, the velocity of the sputtered atom, the work function of the surface and two adjustable parameters. These parameters were determined by inserting mean velocities of sputtered neutrals and mean ionization probabilities determined by other authors under markedly different conditions. Neglecting the problems introduced due to inadequate quality of the experimental data used, the evaluation technique must be viewed as unfounded because the mean velocity of sputtered particles is usually very much different from the mean velocity of secondary ions. Moreover both characteristic velocities depend strongly upon experimental parameters such as projectile mass, bombardment energy and angle of incidence (143,160). In view of these facts the best fit obtained by Schroeer et al. (159) in determining the adjustable parameters can only be taken as a demonstration for the power of a double logarithmic plot. Note that experimental data which did not fit reasonably into Fig. 2 of Ref. (159) were omitted, e.g. the results for carbon and aluminium.

Due to the fact that other existing equations for the ionization probability are even less well founded the formula suggested by Schroeer et al. (159) has been applied by some authors in trying to fit their experimental data into a universal plot (23,60). The spread in the adjustable parameters thus determined is large enough to support the impression that the formula for the ionization probability given in Ref. (159) is not applicable. Consequently extensions of that expression (161,162) cannot provide an improved picture of the process of secondary ion formation.

Perturbation theory has also been used by Šroubek (163) to calculate ionization probabilities in an approximate way. Calculated and experimental results were found to differ by orders of magnitude.

Very recently Cini (164) presented a nonperturbative many electron theory of atom resonance ionization and ion resonance neutralization at a metal surface. Auger processes and auto-

ionization have not been taken into account. It was shown that
the qualitative behaviour of the ionization probability as a
function of the particle velocity depends upon the detailed
shape of the virtual level. The present form of the theoretical
results reported by Cini (164) does not allow a direct comparison with available experimental data.

More detailed theoretical investigations of (positive)
secondary ion production have been carried out by French groups.
The production of secondary ions of transition elements sputtered from pure metals or dilute alloys has been studied by Blaise
and Slodzian (38, 165-168). These authors propose that secondary ion production is mainly due to autoionization processes.
The autoionizing states are populated as a result of the perturbation experienced by the outer shell electrons of the atom
while it crosses the metal-vacuum interface. Thereby mulielectronic excitation becomes possible.

The transitions in the case of copper are shown schematically in Fig. 20. The mean energy E_e of the atomic levels 4s,
5s lies below the Fermi level, i.e. $E_e-\Phi>0$. Consequently, the
autoionizing state $3d^94s5s(^4D)$, which is 0.1 eV above the ionization energy, is equivalent in energy to a state occupied
by two electrons of the conduction band of the metal. The ionization probability α^+ of an atom leaving the metal is thus
proportional to the occupation probability $P(E_e-\Phi)$ which can
be calculated from the density of states.

On the basis of this picture Blaise and Slodzian have
determined relative ionization probabilities of 3d transition
elements (165,166). Differences in the energy distributions

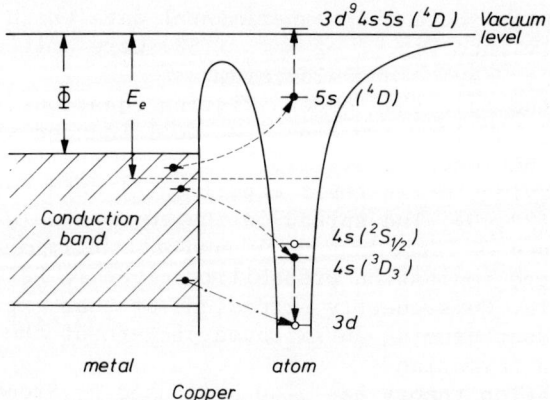

*Fig. 2o. Electronic structures of copper atom and copper
metal. Dashed lines indicate possible transitions leading to
a population of the autoionizing state $3d^94s5s(^4D)$. The dashed-dotted line represents a transition leading to deexcitaion
(165,167).*

(see Fig. 7) were attributed to differences in the survial probability of autoionizing states (166). Although absolute data have not been presented the first semiquantitative results on pure metals and alloys (167,168) are quite promising.

B. Kinetic Model

The so-called kinetic secondary ion emission has been discussed by Joyes (27,169-171). This model is based upon the electron promotion model developed by Fano and Lichten (172, 173) for the interpretation of inelastic atomic collisions in gases. As the internuclear distance between two colliding atoms becomes smaller and smaller, the atomic energy levels become molecular levels with energies depending upon the distance of approach. This is shown semiquantitatively in Fig. 21 for Al-Al collisions. Only those levels are indicated which might be of importance at bombardment energies of the order of 10 keV.

As one can see the 2p levels give rise to four molecular levels. Out of these the 4fσ level is promoted quickly and its energy crosses the conduction band at an internuclear distance of about 0.7Å(27,170). At this point the 4fσ level becomes a virtual bound state which is broadened due to interaction with the continuum. When the upper states of this level exceed the Fermi level the promoted electron can escape from the virtual bound state. As a result one of the colliding aluminium atoms is left with a 2p hole.

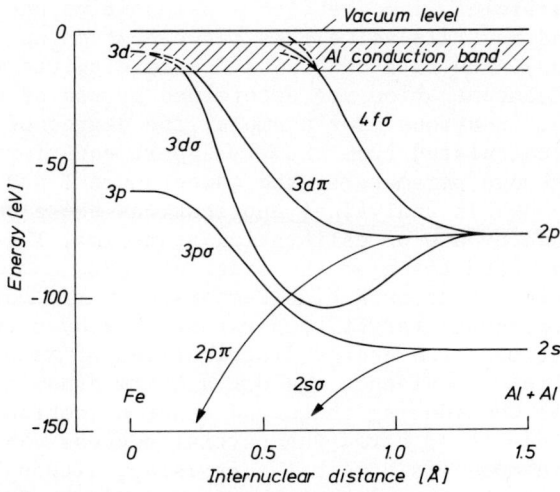

Fig. 21. Schematic representation of electron promotion in Al-Al collision. Interpolated from diagrams for Ne-Ne and Ar-Ar collisions (173).

Joyes has discussed this promotion model with special reference to differences between pure aluminium and aluminium oxide (27). He showed that the probablility for 2p Al-hole production is almost zero in Al-O collisions. This result agrees very well with the experimental finding that there are no Auger electrons in the spectrum of secondary electrons emitted from Al_2O_3 (109).

In a quantitative analysis of kinetic secondary ion emission one has to consider both the probability for 2p hole production and the lifetime of this vacancy. At bombardment energies usually applied in secondary ion emission studies 2p hole production is most probable in magnesium, aluminium and silicon. 1s holes can be produced with much smaller probability in lithium and beryllium. The lifetime of 2p holes can be deduced from calculated Auger transition rates (174,175). Up to now, however, these data have not been applied to a quantitative description of kinetic secondary ion emission.

C. Thermodynamic Models

1. Equilibrium Model

A completely different model has been described by Andersen and Hinthorne (123). These authors tried to develop an analytical method for quantitative analysis of alloys and compounds. They started with the assumption that the sputtering region resembles a dense plasma in local thermal equilibrium. This equilibrium is supposed to be established between all sputtered particles, i.e. between ions(positive and negative), electrons and neutrals as well as between atoms and molecules (dissociation). The rate constants determining the respective equilibrium concentration are determined by use of well-known thermodynamic equations. For example, the degree of ionization of atoms is calculated from the Saha-Eggert equation which contains two unknown parameters, the temperature T and the electron density n_e. In analytical applications these quantities are determined by use of calibration standards. The parameter T is usually found to be of the order of $10^4 K$.

It is clear that such high temperatures have nothing to do with the macroscopic target temperature. Moreover there is no question that measured energy distributions of secondary ions provide sufficient evidence against a thermodynamic model of ion emission. Considering these and other objections (which will not be discussed here) the partial success (98,123) of the thermodynamic model is somewhat surprising, although difficulties introduced by the high yield of oxide ions of the transition elements (Sect. III. C.2) could not be accounted for by the model (123).

Considering the partial success of the thermodynamic model

one should reverse the point of view. Since there is experimental evidence that the intensities of secondary ions emitted from oxygen saturated surfaces are governed by a Boltzmann factor, i.e. $I^+ \propto \exp(-E_I/kT)$ (98) the question arises whether this is physically plausible (E_I is the ionization energy of the atom and k the Boltzmann constant). It has been proposed (98) that the characteristic energy kT may be related to the energy density (energy per atom) in the collision cascade. In fact the calculated energy densities (176) are of the same order of magnitude as the measured kT values. At present it is not clear whether this rough agreement is accidental or not. Anyway it will be of interest to investigate in more detail the possible influence of the energy density of the cascade on the secondary ion emission process.

2. Nonequilibrium Model

Equations developed for the description of nonequilibrium surface ionization have been used by Jurela (76) to calculate mean temperatures of the sputtering centres from experimentally determined ionization probabilities α^+ and α^-. The agreement between $T(\alpha^+)$ and $T(\alpha^-)$ was considered to support the applicability of the respective equations. A more detailed analysis of the model shows, however, that strong objections should be made against such an interpretation. It is known, for example, that the work function (contained as one parameter in the model) is not an adequate quantity in calculations of the secondary ion yield (96). In fact Jurela determined 'realistic' temperatures (76) although the experimental conditions were such that the measured ionization probabilities cannot be considered to represent ion emission from clean surfaces.

D. Ion Neutralization

In the preceding sections we have mainly considered the excitation processes which may result in ionization. It is known, however, that non-radiative deexcitation and neutralization processes take place with very high transition rates as long as the atom or ion is close to the surface (177). The probability $P(\infty,v)$ for a particle to escape from deexcitation is usually written as (30)

$$P(\infty,v) = \exp(-A/av), \qquad (6)$$

where A is the transition rate at the surface, a is a critical distance and v the particle velocity normal ot the surface.

The ratio A/a can be estimated if we make the simplifying assumptions that the excitation probability is independent of the particle velocity. The energy distribution of sputtered particles is taken to be of the form (161)

$$s(E) \propto E/(E+E_b)^3, \qquad (7)$$

where E_b is the surface binding energy. We restrict the discussion to excited atoms emitted normal to the surface. The energy distribution of particles escaping deexcitation is then given by

$$s^+(E) = s(E)\exp(-b/\sqrt{E}), \qquad (8)$$

where $b = 0.7 \times 10^{-6}$(cm/s)$\sqrt{m}A/a$ and m and E are particle mass [u] and energy [eV].

Eqs. (7) and (8) have been plotted in Fig. 22 for E_b=5eV and various values of b. One can see that the shape of $s^+(E)$ is strongly affected by the magnitude of b, i.e. by the ratio A/a. Fig. 22 thus allows to estimate A/a even if the energy distribution of sputtered neutrals is not known in much detail. Peak positions of $s^+(E)$ between 5 and 10 eV correspond to $1.5 < b < 7$ or 2×10^6cm/s $< \sqrt{m}A/a < 10^7$cm/s. For copper atoms this results in 2.5×10^5cm/s $< A/a < 1.7 \times 10^6$cm/s.

These data may be compared with a ratio $A/a = 2 \times 10^6$cm/s which is usually applied in discussions of photon emission data (116,117,128). It is not clear whether the indicated differences in the transition rates a real. However, it were

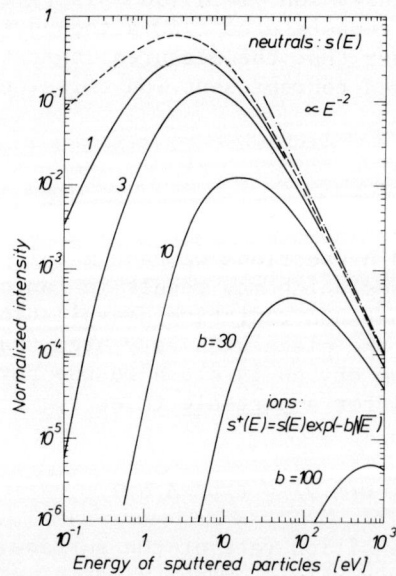

Fig. 22. Normalized energy spectra of sputtered atoms and ions, calculated from Eqs. (7) and (8).

somewhat surprising if destruction of all excited states would be characterized by the same transition rate. Experimental data seem to support the idea that autoionizing states are deexcited less rapidly than other outer shell excited states.

V. DISCUSSION AND CONCLUSIONS

The results summarized in this review article demonstrate that during the past ten years considerable progress has been achieved in the understanding of the mechanismus leading to secondary ion emission from ion-bombarded surfaces. It is obvious, however, that the phenomena observed are usually very complex. In general one has to take into account that more than one mechanism is responsible for the observed secondary ion yields.

Most of the experimental results described in Sect. III have already been discussed in conjunction with current models of secondary ion emission. On the basis of the details given above we summarize what is believed to be the present state of the art. In addition we like to suggest some additional explanations for the phenomena observed.

As far as the excitation mechanisms are concerned kinetic ion emission seems to be understood best. However, details of the ionization process have not been discussed as yet. In Fig. 8 we have demonstrated that the intensity of low energy (<10eV) singly and multiply charged silicon ions increases as a result of the increasing probability for 2p hole production. From Fig. 8 we estimate the minimum energy required for 2p hole production in silicon to be about 2 keV. It is very much unlikely that particles which have experienced a collision with an energy transfer of the order of 1 keV end up in the low energy peak of the secondary ion spectrum. Therefore we have to consider other mechanisms of ionization.

In view of the existence of pronounced energy loss peaks in the secondary electron spectra in the vicinity of the Auger lines, ionization of sputtered particles due to impact of Auger electrons is likely to occur. In a quantitative estimate of this process the following data are importance. (i) The radius if a collision cascade is about 20Å at bombardment energies of a few keV (178). (ii) The distance travelled by a 10 eV silicon atom within the lifetime of a Si 2p hole ($\sim 4 \times 10^{-14}$ s (174)) is only 3 Å. (iii) The electron impact ionization cross sections σ_e for Mg, Al and Si are very large between 15 and 100 eV and peaked around 30 eV, where $\sigma_e \approx 5 \times 10^{-16} cm^2$ (179). On the basis of these figures one can estimate that the probability for ionization of low energy sputtered atoms by Auger electrons emitted from the same cascade may be as high as 10^{-3}. Since measured ionization probabilities are of the same order of magnitude the process discussed should be of some importance in a detailed interpretation of secondary emission. In the high energy part of the secondary ion spectrum on the other hand, one will of course find mainly ions which were produced directly from Auger deexcitation (after 2p hole production).

In case that kinetic secondary ion production does not occur, autoionization seems to be the most effective process (for clean surfaces). Further investigations of this mechanism would be desirable. More information about the excitation process can be expected from experiments in the low energy region (< 1 keV).

Work function models of secondary ion emission do not provide an adequate basis for the interpretation of experimental results. Similarly the band structure model must be viewed with reservation. In particular it is unjustified to assume that the damaged surface can be characterized by the band structure of the bulk crystal. For example, optical absorption spectra of amorphous silicon and germanium show that the main features of the crystal band structure are hardly retained in the amorphous state (180). In addition one has to take into account implantation-induced build-up of impurity concentrations as well as (for compounds) bombardment-induced changes in the original stochiometry.

Secondary ion emission from ionic crystals can be explained within the frame of the bond breaking model. This model also provides an adequate description of enhanced ion emission in the presence of oxygen.

Cluster ion emission is of interest because stability considerations enter into the discussion. Analysis of all available experimental data indicates that the yield of homonuclear dimers M_2^+ is strongly related to the cohesion energy. The bond strength also determines the yield of heteronuclear dimers such as metal oxide ions MO^+. To make the correlation obvious one has to take into account a kinematic factor reflecting the amount of energy which can be transfered in collisions between metal atoms and oxygen atoms.

Various other effects could be discussed. Thereby, however, one would not be led to a better understanding of the secondary ion emission process. In fact, the present state of the art is such that most of the phenomena are known only qualitatively. Improved experimental and theoretical investigations are required for a more quantitative interpretation of secondary ion production.

Acknowledgements

I like to thank Drs. N.H. Tolk and W. Heiland for inviting me to prepare this contribution. Although it has been a time consuming job, the improved understanding of secondary ion production achieved in course of this study will justify the effort.

REFERENCES

1. Fogel, Ya.M., Sov. Phys. Usp. 10, 17 (1967).
2. Carter, G., and Colligon, J.S., "Ion Bombardment of Solids", Chapter 4, Heinemann Educational Books, London, 1968.
3. Castaing, R. et Slodzian, G., J. Microscopie 1, 395 (1962).
4. Liebl, H., J. Appl. Phys. 38, 5277 (1967).
5. Benninghoven, A., and Loebach, E., Rev. Sci. Instrum. 42, 49 (1971).
6. Wittmaack, K., Maul, J., and Schulz, F., Int. J. Mass Spectrom. Ion Phys. 11, 23 (1973).
7. Schubert, R., and Tracy, J.C., Rev. Sci. Instrum. 44, 487 (1973).
8. Wittmaack, K., Rev. Sci. Instrum. 47, 157 (1976).
9. Wittmaack, K., in "Ion Beam Surface Layer Analysis" (O. Meyer, G. Linker, and F. Käppeler, Eds.) Vol. 2, p. 649 Plenum Press, New York, 1976.
10. Socha, A.J., Surface Sci. 25, 147 (1971).
11. Liebl, H., Meßtechnik 12, 358 (1972).
12. Evans, C.A., Jr., Anal. Chem. 44, 67A (1972).
13. Colligon, H.S., Vacuum 24, 373 (1974).
14. Liebl, H., Anal. Chem. 46, 22A (1974).
15. Honig, R.E., in "Advances in Mass Spectrometry", (A.R. West, Ed.), Vol. 6, p.337. Applied Sci.Publ., Barking, 1974.
16. Liebl, H., J. Phys. E: Sci. Instrum. 8, 808 (1975).
17. Liebl, H., J. Vac. Sci. Technol. 12, 385 (1972).
18. McHugh, J.A., in "Methods of Surface Analysis" (S.P.Wolsky and A.W. Czanderna, Eds.), Vol.1. Elsevier, Amsterdam, 1975.
19. Honig, R.E., Thin Sol. Films 31, 89 (1976).
20. Benninghoven, A., Surface Sci. 28, 541 (1971).
21. Benninghoven, A., Surface Sci. 35, 427 (1973).
22. Werner, H.W., Vacuum 24, 493 (1974).
23. Rüdenauer, F.G., Steiger, W., und Portenschlag, R., Mikrochimica Acta, Supp. 5, 421 (1974).
24. Werner, H.W., Surface Sci. 47, 31 (1975).
25. Benninghoven, A., Surface Sci. 53, 596 (1975).
26. Castaing, R., and Hennequin, J.-F., in "Advances in Mass Spectrometry" (A. Quayle, Ed.), Vol.5, p. 419. Applied Sci. Publ., London, 1972.
27. Joyes, P., Rad. Effects 19, 235 (1973).
28. Sigmund, P., Phys. Rev. 184, 383 (1969).
29. Westmoreland, J.E., Sigmund, P., Rad.Effects 6, 187 (197).
30. Van der Weg, W.F., and Rol, P.K., Nucl. Instrum. Meth. 38, 274 (1965).
31. Veksler, V.I., and Tsipinyuk, B.A., Soviet Physics JETP 33, 753 (1971).
32. Fehn, U., Int. J. Mass Spectrom. Ion Phys. 21, 1 (1976).
33. Staudenmaier, G., Hofer, W.O., and Liebl, H., Int. J. Mass Spectrom. Ion Phys. 21, 103 (1976).

34. Hernandez, R., Lanusse, P., Slodzian, G., et Vidal, G., Rech. Aerosp. 1972-6, 313 (1972).
35. Jurela, Z., in "Atomic Collision Phenomena in Solids" (D.W. Plamer, M.W. Thompson, and P.D. Townsend, Eds.), p. 339. North Holland Publ. Comp., Amsterdam, 1970.
36. Honig, R.E., J. Appl. Phys. 29, 549 (1958).
37. Honig, R.E., in "Advances in Mass Spectrometry" (R.M. Elliot, Ed.), Vol. 2, p. 25. Pergamon Press, London, 1963.
38. Blaise, G., and Slodzian, G., C.R. Acad. Sc. (Paris) 266 B, 1525 (1968).
39. Jurela, Z., and Perovic, B., Canad. J. Phys. 46, 773 (1968).
40. Hennequin, J.-F., J. Physique 29, 957 (1968).
41. Werner, H.W., in "Developments in Applied Spectroscopy" (E. Grove, Ed.), Vol. 7A, p. 239. Plenum Press, New York 1969.
42. Staudenmaier, G., Rad. Effects 13, 87 (1972).
43. Dennis, E., and MacDonald, R.J., Rad. Effects 13, 243 (1972).
44. Staudenmaier, G., Thesis, Universität München (1973).
45. Bernheim, M., Blaise. G., et G. Slodzian, Int. J. Mass Spectrom. Ion Phys. 10, 293 (1972/73).
46. Feldman, C., and Satkiewicz, F.G., J. Electrochem. Soc. 120, 1111 (1973).
47. Leleyter, M., and Joyes, P., Rad. Effects 18, 105 (1973).
48. Staudenmaier, G., Rad. Effects 18, 181 (1973).
49. Herzog, R.F.K., Poschenrieder, W.P., and Satkiewicz, F.G., Rad. Effects 18, 199 (1973).
50. Werner, H.W., de Grefte, H.A.M., and Van den Berg, J., Rad. Effects 18, 269 (1973).
51. Jurela. Z., Rad. Effects 19, 175 (1973).
52. Blaise, G., et Slodzian, G., Rev. Physique Appl. 8, 105 (1973).
53. Leleyter, M., et Joyes, P., J. Physique 35, L-85 (1974).
54. Leleyter, M., and Joyes, P., J. Phys. B: Atom. Molec. Phys. 7, 516 (1974).
55. Storp, S., Thesis, Universität Köln (1974).
56. Werner, H.W., Morgan, A.E., and de Grefte, H.A.M., Appl. Phys. 7, 65 (1975).
57. Benninghoven, A., Sichtermann, W., and Storp, S., Thin Sol. Films 28, 59 (1975).
58. Maul, J., and Wittmaack, K., Surface Sci. 47 358 (1975).
59. Wittmaack, K., and Staudenmaier, G., Appl. Phys. Lett. 27, 318 (1975).
60. Wittmaack, K., Surface Sci. 53, 626 (1975).
61. Benninghoven, A., Z. Physik 199, 141 (1967).
62. Können, G.P., Tip, A. and de Vries, A.E., Rad. Effects 21, 269 (1974).
63. Können, G.P., Tip, A. and de Vries, A.E., Rad. Effects 26, 23 (1975).

64. Wittmaack, K., Appl. Phys. Lett., Nov. 1976.
65. Gerhard, W., and Oechsner, H., Z. Physik B 22, 42 (1975).
66. Bradley, R.C., J. Appl. Phys. 30, 1 (1959).
67. Brochard, D., et Slodzian, G., J. Physique 32, 185 (1971).
68. Behrisch, R., in "Ergebnisse der exakten Naturwissenschaften" (S.Flügge und F. Trendelenburg, Hrsg.)35.Band,S.295. Springer-Verlag, Berlin, 1964.
69. Hennequin, J.-F., Rev. Physique Appl. 1, 237 (1966).
70. Wittmaack, K., Int. J. Mass Spectrom. Ion Phys. 17, 39 (1975).
71. Slodzian, G., et Hennequin, J.-F., C.R. Acad, Sc. Paris, 263 B, 1246 (1966).
72. Beske, H.E., Z. Naturforschung, 22a, 459 (1967).
73. Werner, H.W., and de Grefte, H.A.M., Vakuum-Technik 17, 37 (1968).
74. Benninghoven, A., and Müller, A., Phys. Lett. 40 A, 169 (1972).
75. Jurela, Z., Rad. Effects 13, 167 (1972).
76. Jurela, Z., Int. J. Mass Spectrom. Ion Phys. 12, 33 (1973).
77. Blaise, G., and Bernheim, M., Surface Sci. 47, 324 (1975).
78. Hennequin, J.-F., Blaise, G., et Slodzian, G., C.R. Acad. Sc. Paris, 268 B, 1507 (1969).
79. Bayly, A.R., Martin, P.J., and MacDonald, R.J., Nucl. Instr. Meth. 132, 459 (1976).
80. Wittmaack, K., Nucl. Instr. Meth. 132, 381,(1976).
81. Bukhanov, V.M., Yurasova, V.E., Sysoev, A.A., Somsonov, G.V., and Nikolaev, B.I., Sov. Phys. Sol. State 12, 313 (1973).
82. Zwangobani, E., and MacDonald, R.J., Rad. Effects 20, 81 (1973).
83. Bernheim, M., Rad. Effects 18, 231 (1973).
84. Laurent, R., and Slodzian, G., Rad. Effects 19, 181 (1973).
85. Krohn, V.E., Jr., J. Appl. Phys. 33, 3523 (1962).
86. Hortig, G., Mokler, P., and Müller, M., Z. Physik 210, 312 (1968).
87. Hortig, G., and Müller, M., Z. Physik 221, 119 (1969).
88. Andersen,C.A., Int. J. Mass. Spectrom. Ion Phys. 3, 413 (1970).
89. Schulz, F., and Wittmaack, K., Rad. Effects 29, 31 (1976).
90. Abdullayeva, M.K., Ayukhanov, A.K., and Shamsiyev, U.B., Rad. Effects 19, 225 (1973).
91. Adylov, A.A., Veksler, V.I. and Reznik, A.M., Sov. Phys. Sol. State 14, 2696 (1973).
92. Kerkow, H., and Trapp, M., Int. J. Mass Spectrom. Ion Phys. 13, 113 (1974).
93. Hennequin, J.-F., C.R. Acad. Sc. Paris, 264 B, 1127 (1967).
94. Benninghoven, A., Z. Naturforschung 22a, 841 (1967).
95. Benninghoven, A., Z. Physik 220, 159 (1969).
96. Blaise, G., et Slodzian, G., Surface Sci.40, 708 (1973).

97. Maul, J., Thesis, Technische Universität München (1974).
98. Morgan, A.E., and Werner, H.W., Anal. Chem. 48, 699 (1976).
99. Andersen, C.A., Int. J. Mass Spectrom. Ion Phys. 2, 61 (1969).
100. Lewis, R.K., Morabito, J.M., and Tsai, J.C.C., Appl. Phys. Lett. 23, 260 (1973).
101. Tsai, J.C.C., and Morabito, J.M., Surface Sci. 44, 253 (1974).
102. Benninghoven, A., and Storp, S., Appl. Phys. Lett. 22, 170 (1973).
103. Benninghoven, A., Loebach, E., Plog. C., and Treitz, N., Surface Sci. 39, 397 (1973).
104. Benninghoven, A., and Müller, A., Surface Sci. 39, 416 (1973).
105. Müller, A., and Benninghoven, A., Surface Sci. 39, 427 (1973).
106. Benninghoven, A., and Wiedemann, L., Surface Sci. 41, 483 (1974).
107. Müller, A., and Benninghoven, A., Surface Sci. 41, 493 (1974).
108. Stumpe, E., and Benninghoven, A., Phys. Stat. Sol. (a) 21. 479 (1974).
109. Hennequin, J.-F., J. Physique 29, 1053 (1968).
110. Kelly, R., and Kerkdijk, C.B., Surface Sci. 46, 537 (1974).
111. Kerkdijk, C.B., and Kelly, R., Surface Sci. 47, 294 (1975).
112. Wittmaack, K., and Blank, P., in "Proceedings of the Fifth International Conference on Ion Implantation", in press.
113. Nelson, R.S., Rad. Effects 2, 47 (1969).
114. Moline, R.A., Reutlinger, G.W. and North, J.C., in "Atomic Collisions in Solids" (S. Datz, B.R. Appleton, and C.D. Moak, Eds.), Vol. 1, p 159. Plenum Press, New York, 1975.
115. Moline, R.A., and Cullis, A.G., Appl. Phys. Lett. 26, 551 (1975).
116. Van der Weg, W.F., and Biermann, D.J., Physica 44, 2o6 (1969).
117. Tolk, N.H., Simms, D.L., Foley, E.B., and White, C.W. Rad. Effects 18, 221 (1973).
118. Kiyan, T.S., Gritsyna, V.V., and Fogel, Ya.M., Nucl. Instr. Meth. 132, 435 (1976).
119. Martin, P.J., Bayly, A.R., MacDonald, R.J., Tolk, N.H., Clark, G.J., and Kelly, J.C., Surface Sci., in press.
120. Morgan, A.E., and Werner, H.W., Appl. Phys. 11, 93 (1976).
121. Prager, M., Appl. Phys. 8, 361 (1975).
122. Popp, H.-P., Physics Reports (Phys.Lett.C) 16, No. 4, 169 (1975).
123. Andersen, C.A., and Hinthorne, J.R., Anal. Chem. 45, 1421 (1973).

124. Estel, J., Hoinkes, H., Kaarmann, H., Nahr, H., and Wilsch, H., Surface Sci. 54, 393 (1976).
125. Stuart, R.V., and Wehner, G.K., Phys. Rev. Lett. 4, 409 (1960).
126. Terzic, I., and Perovic, B., Surface Sci. 21, 86 (1970).
127. Tsong, I.S.T., Phys. Stat. Sol. (a) 7, 451 (1971).
128. White, C.W., and Tolk, N.H., Phys.Rev. Lett. 26, 486 (1971).
129. Martel, J.G., and Olson, N.T., Nucl. Instr. Meth. 105, 269 (1972).
130. Gritsyna, V.V., Kijan, T.S., Koval, A.G., and Fogel, Ya.M., Rad. Effects 14, 77 (1972).
131. Kerkow, H., Phys. Stat. Sol. (a) 10, 501 (1972).
132. Martel, J.G., and Olson, N.T., Rad. Effects 19, 19 (1973).
133. Jensen, K., and Veje, E., Z. Physik 269, 293 (1974).
134. Koval, A.G., Vyagin, G.I., Bobkov, V.V., Klimovskii, Yu. A., Strelchenko, S.S., and Fogel, Ya.M., Sov. Phys. Tech. Phys. 18, 1105 (1974).
135. Thomas, G.E., and Kluizenaar, E.E., Int. J. Mass Spectrom. Ion Phys. 15, 165 (1975).
136. White, C.W., Simms, D.L., Tolk, N.H., and McCaughan, D.V., Surface Sci. 49, 657 (1975).
137. Kerkdijk, C.B., Thesis, Rijsuniversitet Leiden (1975).
138. Meriaux, J.P., Goutte, R., and Guillaud, C., Appl. Phys. 7, 313 (1975).
139. Kiyan, T.S., Gritsyna, V.V., and Fogel, Ya.M., Nucl. Meth. 132, 415 (1976).
140. White, C.W., Tolk, N.H., Kraus, J., and van der Weg, W.F., Nucl. Instr. Meth. 132, 419 (1976).
141. Kerkdijk, C.B., Schartner, K.-H., Kelly, R., and Saris, F.W., Nucl. Instr. Meth. 132, 427 (1976).
142. Hippler, R., Krüger, W., Scharmann, A., and Schartner, K.-H., Nucl. Instr. Meth. 132, 439 (1976).
143. Hennequin, J.-F., J. Physique 29, 655 (1968).
144. Saris, F.W., and Onderdelinden, D., Physica 49, 441 (1970).
145. Saris, F.W., in "Atomic Collisions in Solids" (S. Datz, B.R. Appleton, and C.D. Moak, Eds.), Vol. 1 p. 343. Plenum Press, New York, 1975.
146. Saris, F.W., Physica 52, 29o (1971).
147. Hennequin, J.-F., Joyes, P., et Castaing, R., C.R. Acad. Sci. Paris, 265 B, 312 (1967).
148. Hennequin, J.-F., et Viaris de Lesegno, P., C.R. Acad. Sc. Paris 272 B, 1259 (1971).
149. Viaris de Lesegno, P., Joyes, P., et Hennequin, J.-F., C.R. Acad. Sc. Paris 257 B, 93 (1972).
150. Hennequin, J.-F., and Viaris de Lesegno, P., Surface Sci. 42, 50 (1974).
151. Viaris de Lesegno, O., Rivais, G., and Hennequin, J.-F.,

Phys. Lett. 49 A, 265 (1974).
152. Viaris de Lesegno, P., et Hennequin, J.-F., J. Physique 35, 759 (1974).
153. Louchet, F., Viel, L., Benazeth, C., Fagot, B., and Colombie, N., Rad. Effects 14, 123 (1972).
154. Colombie, N., Benazeth, C., Mischler, J., and Viel, L., Rad. Effects 18, 251 (1973).
155. Viel, L., Benazeth, C., and Benazeth, N., Surface Sci., 54, 635 (1976).
156. Benazeth, N., Agusti, J., Benazeth, C., Mischler, J., and Viel, L., Nucl. Instr. Meth. 132, 477 (1976).
157. Grant, J.T., Hooker, M.P., Springer, R.W., and Haas, T.W., J. Vac. Sci. Technol. 12, 481 (1975).
158. Kempf, J., and Kaus, G., subm. to Appl. Phys.
159. Schroeer, J.M., Rhodin, T.N., and Bradley, R.C., Surface Sci. 34, 571 (1973).
160. Kopitzki, K., und Stier, H.E., Z. Naturforschg. 17a, 346 (1962).
161. Gries, W.H., Int. J. Mass Spectrom. Ion Phys. 17, 77 (1975).
162. Gries, W.H., and Rüdenauer, F.G., Int. J. Mass Spectrom. Ion Phys. 18, 111 (1975).
163. Šroubek, Z., Surface Sci. 44, 47 (1974).
164. Cini, M., Surface Sci. 54, 71 (1976).
165. Blaise, G., et Slodzian, F., J. Physique 31, 93 (1970).
166. Blaise, G., et Slodzian, G., Rev. Physique Appl. 8, 247 (1973).
167. Blaise, G., Rad. Effects 18, 235 (1973).
168. Blaise, G., et Slodzian, G., J. Physique 35, 243 (1974).
169. Joyes, P., et Hennequin, J.-F., Physique 29, 483 (1968).
17 . Joyes, P., J. Physique 30, 243 (1969).
171. Joyes, P., J. Physique 30, 365 (1969).
172. Fano, U., and Lichten, W., Phys. Rev. Lett. 14, 627 (1965).
173. Lichten, W., Phys. Rev. 164, 131 (1967).
174. McGuire, E.J., Phys.Rev. A3, 587 (1971).
175. Walters, D.L., and Bhalla, C.P., Phys. Rev. A4, 2164 (1971).
176. Sigmund, P., Appl. Phys. Lett. 25, 169 (1974); 27, 52 (1975).
177. Hagstrum, H.D., J. Vac. Sci. Technol. 12, 7 (1975).
178. Sigmund, P., and Sanders, J.-B., in Proc. Int. Conf. on "Applications of Ion Beams"(P. Glotin, Ed.) p. 215. Editions Ophrys, Grenoble, 1967.
179. Lotz, W., Z. Physik 216, 241 (1968).
180. Brodsky, M.H., J. Vac. Sci. Technol. 8, 215 (1971).

APPENDIX

The results shown in the figures of this paper were obtained by different authors by use of various types of secondary ion mass spectrometers. In the following we briefly describe the instruments to ease assessment of the respective data. The instruments are numbered according to their first appearance in the figure captions. Symbols: Θ = angle of beam incidence with respect to the target normal, ϕ = extraction angle with respect to the target normal, V_s = secondary ion acceleration voltage, $m/\Delta m$ = mass resolution, $E/\Delta E$ = energy resolution, p_o = base pressure.

1. Double focussing mass spectrometer with cylindrical energy analyser. $\Theta = 20°$, $\phi = 40°$, $\Theta+\phi = 60°$, V_s variable, usually $V_s = 0$, $m/\Delta m = 72$, $E/\Delta E = 36$, $p_o = 6 \times 10^{-7}$ Torr (42,44).

2. Quadrupole-equipped secondary ion mass spectrometer. The mass filter is preceded by an energy filter in conjunction with an accel-decel system. $\Theta = 0°$, $\phi \approx 20°$, $V_s = 100$ V (typical), $\Delta E = 1$ eV, $m/\Delta m = 300$ (typical). Transmission dependent upon quadrupole setting. $p_o = 4 \times 10^{-9}$ Torr (without baking) (60).

3. Stigmatic imaging, double focussing mass spectrometer. $\Theta = 45°$, $\phi = 45°$, $\Theta+\phi = 90°$, $V_s = 1$ kV, $m/\Delta m = 300$ (typical), 4500 (optimum); $\Delta E = 4$ eV (variable by changing the width of the energy slit), $p_o = 5 \times 10^{-9}$ Torr (46,49).

4. Single focussing mass spectrometer without energy analyser. $\Theta = 70°$, $\phi = 0°$, $V_s = 1.5$ kV, $m/\Delta m < 250$, $p_o = 10^{-10}$ Torr (55).

5. Ion microscope with magnetic prism and electrostatic mirror. $\Theta = 60°$, $\phi = 0°$, $V_s = 3.8$ kV, $m/\Delta m \approx 300$, $\Delta E = 15$ eV, $p_o = 10^{-7}$ Torr (52).

6. Mass spectrometer with circular magnetic field. Energy analysis by retarding field method. Θ, ϕ variable, typical: $\Theta = 0°$, $\phi = 20°$, $V_s = 2$ kV, $p_o = 2 \times 10^{-7}$ Torr (69).

7. Single focussing mass spectrometer without energy analyser. $\Theta = 60°$, $\phi = 0°$, $V_s = 3.8$ kV, $p_o = 10^{-7}$ Torr (67).

8. Single focussing mass spectrometer without energy analyser. $\Theta = 70°$, $\phi = 0°$, $m/\Delta m = 25o$, $p_o = 10^{-10}$ Torr (104). (Basic design the same as for spectrometer type 4.)

optical emission from low-energy ion-surface collisions

C. W. White
Solid State Division, Oak Ridge National Laboratory[*],
Oak Ridge, Tennessee 37830

E. W. Thomas[†]
Dept. of Physics, Georgia Institute of Technology,
Atlanta, Georgia 30332

W. F. Van der Weg
Philips Research Laboratories,
Eindhoven, The Netherlands

N. H. Tolk
Bell Laboratories, Murray Hill, New Jersey 07974

Impact of energetic heavy particles on surfaces gives rise to emission of optical radiation from reflected particles, sputtered particles and also from excited states of the solid. We review the present status of research in this area with emphasis on understanding the basic mechanisms which give rise to formation of excited states. The spectral line shape from ejected atoms may be analyzed to provide information on the distribution of speeds and directions of the excited species; the line intensity provides a measure of the probability for creating the state. Formation of excited species is related both to the collision processes within the solid and also to the interaction of the recoiling ejected species with the target surface. Most ejected species are atomic but important examples of ejected molecules are also discussed. Luminescence induced in the solid itself is related to recombination of electron hole pairs and is related significantly to the presence of defects.

---

[*]Operated by Union Carbide Corporation under contract with the U.S. Energy Research and Development Administration

[†]Supported in part by the Controlled Thermonuclear Research Program of the Energy Research and Development Administration and by the Division of Materials Research of the National Science Foundation

I. INTRODUCTION

One of the most interesting of the low energy inelastic ion-surface collision phenomena is the production of optical radiation in the collision process. Experimental results show that visible, ultraviolet, and infrared optical radiation is a general result of the bombardment of a solid by low energy ions or neutral particles (1-27). The radiation is produced by inelastic processes involving outer shell electrons. In the energy range 30 eV to 100 keV where the beam energy is large compared to both the binding energy of atoms in the solid and to the excitation energy of outer shell electrons, radiation is observed from sputtered particles, (1-16) reflected particles, (17-24) and the solid itself (1,11,15,25-27). This radiation provides direct information on the identity and quantum state of sputtered and reflected particles (including neutral atoms leaving the surface), the velocity distribution of radiating particles, and the transfer of projectile energy to the electrons in the solid. In addition the radiation can provide information on surface chemistry, (28-30) the interaction of excited atoms with solids, (31) and atomic processes which do not result in the emission of electromagnetic radiation (1,2, 19-24, 32-38). Measurements of this collision induced optical raditation is, therefore, a powerful tool for studying fundamental processes which result from the interaction of low energy atomic particles with a solid.

Subsequent sections of this chapter will discuss the following aspects of ion induced optical emission from solids: Section II describes the various types of optical radiation observed in ion-solid collisions. Radiation from sputtered particles is discussed in Section III. Emission of radiation from reflected or backscattered particles is described in Section IV. Section V discusses the intrinsic luninescence which arises from the solid. Section VI discusses the emission from sputtered molecules including the radiative continua which are observed spatially well in front of certain ion bombarded metal targets. Section VII discusses the emission from molecules in the surface including the line spectrum of excited CN produced by ion bombardment of alkali halide crystals and the continuous spectrum from implanted noble gases. Conclusions are presented in Section VIII.

Descriptions of the experimental apparatus used to study optical emission in ion-solid collisions can be found in the literature (2,21). Most of the experiments have been performed in the visible and near UV spectral regions with relatively simple apparatus consisting of an ion beam (a few microamperes intensity), a scattering chamber, a target manipulator, and a monochromater (1-10 Å spectral resolution) with a photomultiplier or a spectrograph with photographic recording.

II. SOURCES OF ION INDUCED OPTICAL RADIATION

In low energy (30 eV-100 keV) ion-solid collisions, at least five distinct kinds of collision induced optical radiations have been identified. The first of these is due to sputtering accompanied by simultaneous electronic excitation which gives rise to radiation from excited states of sputtered surface constituents (1-16). The interaction of the impinging beam with the solid results in the sputtering of atoms, molecules and ions from the surface. A fraction of the particles sputtered from the first few monolayers of the solid escape from the surface in excited states and decay optically well away from the surface where the perturbation of atomic levels by the solid is negligible. This results in the emission of discrete atomic and ionic emission lines as well as molecular emission bands, broadened only by the Doppler effect.

The second type of collision induced optical emission is radiation from backscattered or reflected beam particles (17-24). Discrete atomic or ionic emission lines arising from backscattered particles are observed particularly in those cases where light projectiles (H, He) impact on heavy targets. The excited particles decay optically in vacuum after having experienced violent, momentum changing collisions on the surface or in the bulk of the solid. From Doppler shift measurements, the radiating particles are observed to have energies comparable to the incident ion energy.

There are two distinct kinds of broadband continuum radiation produced by ion-solid collisions. The first of these is the intrinsic continuum luminescence from the solid (1, 25-27). This broadband continuum emission originates in the bulk of the solid and results from the excitation of electrons in the solid or radiative recombination of electron hole pairs (1). Electron excitation or electron-hole pair production results from the inelastic loss of projectile energy to the electrons in the solid (1). The second type of broadband emission is the radiative continuum from sputtered molecular species. Certain ion bombarded transition metal targets give rise to a broadband continuum emission which is observed up to a few millimeters in front of the target (12, 39-43). Evidence indicates that this radiation originates from excited molecular species, possibly metal-oxide molecules, (43) ejected from the solid in the sputtering process. In addition, continuum radiation from excited rare gas dimers has recently been observed in the vacuum ultraviolet spectral region during bombardment of a variety of targets by rare gas ions (44, 61).

The fifth type of collision induced optical emission is radiation from excited molecules on the surface. Recent work (30) indicates that radiation is observed from CN and possibly other molecules bound to the surface. The vibrational bands are observed to be broad as compared to the free radical case and the excitation is believed to be associated with exciton recombination at the surface.

III. RADIATION FROM SPUTTERED SURFACE CONSTITUENTS

A. Spectral Characteristics

Figure 1 shows two examples of radiation arising from excited states of sputtered surface constituents for the case of Ar^+ (4 keV) impacting on surfaces of Cu and Ni.(1). Most of the prominent lines arise from excited states of neutral Cu and Ni sputtered from the first few monolayers of the solid by the impinging beam. Radiation from these excited particles allows identification of atoms and molecules sputtered from the surface and makes this attractive as an analytical tool for surface composition analysis (3, 11, 16, 45).

The width of the lines from Cu and Ni in Fig. 1 are determined by the instrumental resolution used in the measurement ($\sim$ 8 Å) but high resolution measurements (32) at significantly higher bombarding energies show that radiation from atoms sputtered from clean metal targets arises predominately from those atoms which lie in the high velocity tail of the sputtered atom velocity distribution.

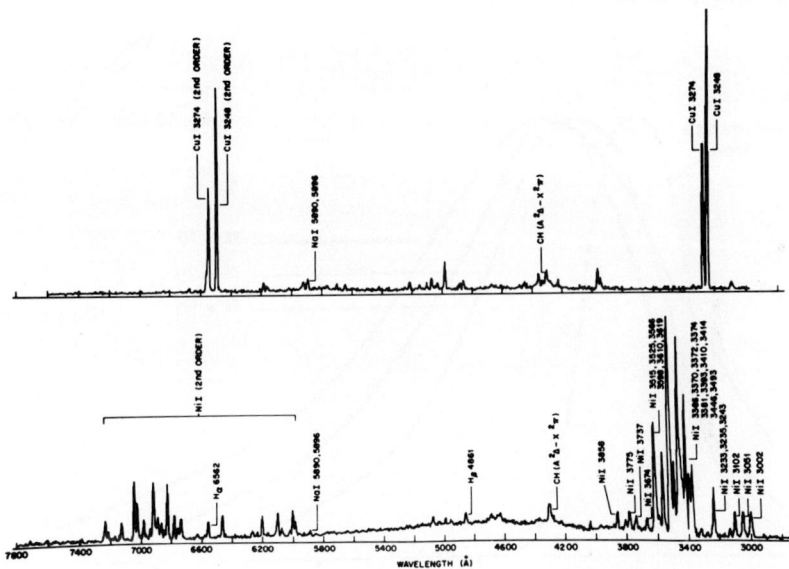

Fig. 1. Spectra of radiation produced by Ar^+ (4 keV) impact on Cu and Ni. Lines arising from excited states of Cu, Ni and contaminants are observed in this wavelength interval. (from Ref. 1)

Figure 2 shows measured emission line profiles of the CuI-3247Å for various angles of the target with respect to the primary ion beam in the case of Ar^+ (80 keV) impacting on Cu (32). The shift in the peak of the emission line (0.6Å) relative to the reference line indicates a mean energy of $\sim$ 1 keV for the radiating atoms. The majority of sputtered atoms have energies < 100 eV and this suggests that the radiating atoms are scattering products arising from violent collisions of beam ions with surface atoms.

The situation is quite different for an oxidized metal target or insulator. In this case, one also observes a line spectrum characteristic of the metal atom, but the intensity is significantly enhanced compared to the pure metal case. In addition, the Doppler width of spectral lines is considerably reduced when using an oxidized target.

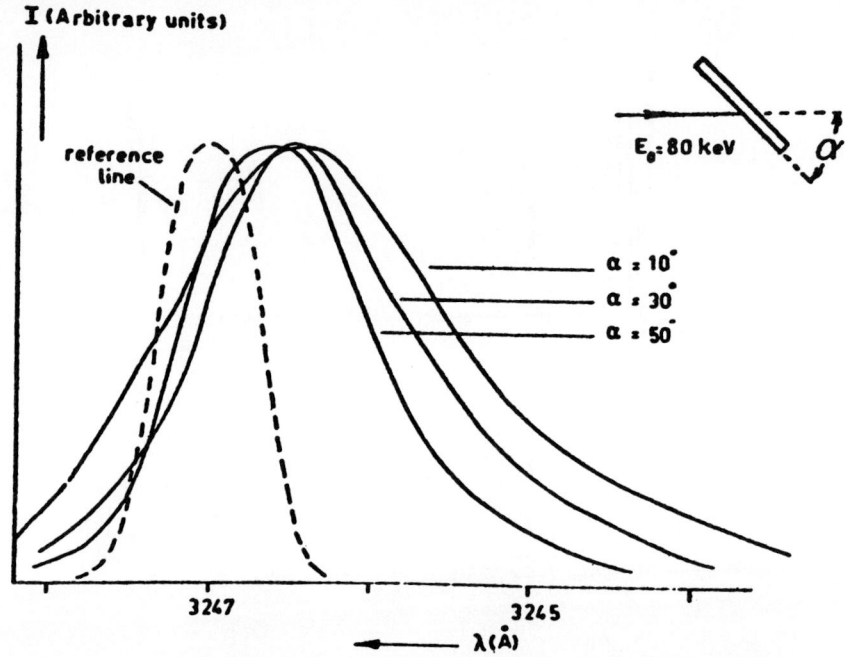

Fig. 2. Line shapes of the CuI-3247Å resonance line for different target orientations, α. The intensities are normalized (from Ref. 32).

This is illustrated in Fig. 3 for the case of Ar^+ (80 keV) on Si and SiO_2 (37). The width of the SiI-2882 Å line is much narrower from the SiO_2 target as compared to the silicon target and this implies that in the oxidized target case the slow particles also contribute efficiently to the emitted radiation. To discuss the optical intensity and velocity distributions of radiating particles in more detail we need first to consider the excitation process and the interaction of the excited atom with the solid.

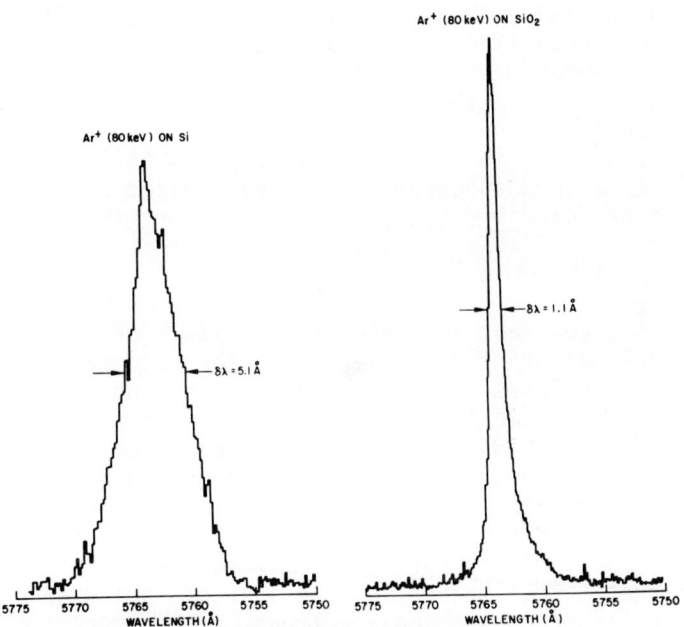

Fig. 3. Emission line profiles of SiI-2882 Å from SiO_2 and silicon. Line-width measurements were made in second order within an instrumental resolution of ~1Å (from Ref.37).

B. The Excitation Process

As compared to gas-phase collisions, it is very difficult to obtain information about the primary excitation process, since atoms that are excited during a collision of a beam particle with a target atom, usually travel some distance through and near the solid before emitting radiation outside the solid. The electron configuration in the outer shells of these moving atoms is necessarily strongly modified by interactions with atoms and electrons of the solid. Some indirect information can be obtained from the emission intensities of the spectral lines as a function of projectile mass and energy. Emission functions for spectral lines of the atomic spectra of Sr, Ni, Cu and Si were measured (2) in the energy range of 30-3000 eV. The results show in many cases a threshold around 100 eV and a subsequent rise in intensity with increasing energy. At primary energies far from threshold (several keV and higher) it appears that all the lower lying levels in the neutral spectra of the target element are populated (10).

Recently the relative population of the levels of several metals under noble and reactive gas bombardment has been studied (46). These authors found that especially transition metals like Fe, Cr, Ni, Cu and Zn emit spectra during ion bombardment which closely resemble an arc discharge, judging from the relative line intensities. A similar conclusion holds for the case of Ar bombardment of Ni (47). In fact, in some cases it appears (46) that line intensities I in a spectrum emitted by excited sputtered atoms can be described by

$$I = \alpha C e^{-E_{exc}/kT} \qquad (1)$$

where α is a constant involving transition probability and statistical weight of the level, C is the concentration of the emitting species, E_{exc} the energy of the upper level of the considered optical transition and T an effective temperature, which depends on the ion-target combination. A tentative conclusion from these findings can be that the excitation mechanism in ion-solid collision is analogous to the one in a discharge, i.e. by collisional excitation with electrons. It is not at all clear, however, in the solid target case, whether interactions of sputtered particles with target conduction electrons or electron promotion via quasimolecular states in the collision complex of target and projectile atom causes the electron excitation. Also, the physical meaning of T in eq. (1) is unclear at present. In the case of alloys, recent results show that the emitting source is not in thermodynamic equilibrium because a single temperature does not fit line intensities from different constituents (82). It has also been suggested that, especially in the case of oxides, the bond breaking or dissociation processes of oxygen-involving molecules during and after the sputtering determine the state of excitation of a sputtered particle, (see Section III-D). Nevertheless eq. (1) may give a phenomenological basis for development of a more quantitative surface analysis tool.

Apart from the line emission intensity, a second striking feature which is possibly related to the excitation process is the velocity distribution of the radiating species when bombarding clean metals. The Doppler profile of spectral lines from sputtered atoms, indicates that the majority of the emitting species have relatively high velocities (Section 3-A). Since the energy distributions of these sputtered atoms are continuous and increasing with decreasing energy, the results for the velocities of emitting atoms are not immediately obvious. One way of explaining these results is by assuming that only relatively fast atoms are excited in the collisions. If one compares to the case of excitation in gases, this assumption seems hardly justified (48). Alternatively, it has often been assumed that all ejected particles are excited, but that slower particles lose their state of excitation because of an interaction with the solid. For oxide targets, however, the line intensities are very much enhanced compared to the clean metal case and the Doppler profiles indicate that the average velocity of emitting particles is considerably lower than with metals (37).

c. Interactions of Excited Atoms with a Solid

Inside a solid a moving excited atom may quickly loose and subsequently capture an electron. Especially when considering highly excited states, it is hard to envisage a survival of such a state over an appreciable distance in the solid. In such cases, it seems more likely that the state of excitation is determined during exit of the particle from the solid and during the time that the ejected particle is still close to the surface (say < 50 Å). At larger distances there is no longer an interaction with the solid and the excited state decays by photon emissions unperturbed by the presence of the solid. There is some evidence that particles in the lower excited states can maintain their state of excitation while moving through the solid. This can be inferred from channeling measurements in which the intensity of several atomic and ionic spectral lines of Cu (excited by 100 keV Ar^+) was measured as a function of the angle of incidence of the beam on the target (31). Angular scans obtained from these measurements are shown in Fig. 4. The fact that a reduction in yield is observed for beam incidence along low index crystal directions indicates that particles coming from deeper than one atomic layer contribute to the light emission. Furthermore, the observation that different spectral lines show different values of minimum yields in the <110> direction is strong evidence for the fact that the different excited states originate from different depths. If excitation would occur upon exit from the surface, the yield curves for the different lines, if normalized at one angle, would coincide for all angles, since in this case the yield curves would simply reflect the number of emerging particles as a function of angle. A closer analysis makes it likely that excited ions originate from very close to the surface, while low excited states in the neutral atom may come from a depth of a few atomic layers. Obviously, there is a need for more measurements and also theoretical calculations to get more information about the question of the lifetimes of excited atoms in solids.

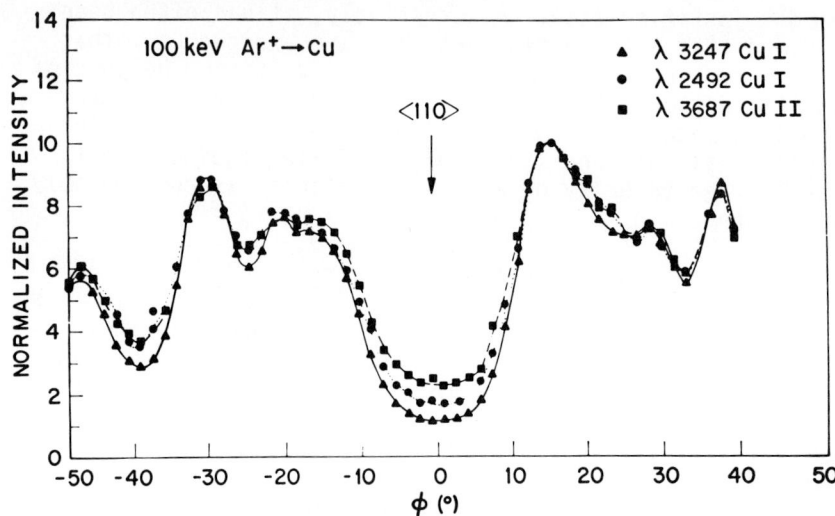

Fig. 4. Intensity of two atomic lines and one ionic line in the copper emission spectrum vs angle φ of rotation of the target around its normal. The curves have been normalized at φ = 15° (from Ref. 31).

We will now consider the interactions between an excited particle that has left the solid state and the surface. From the work of Hagstrum on potential secondary electron emission (49) it has become very clear that electron transitions between slow particles and a surface may occur with very high probability, provided the atom-surface separation is small enough. These transitions can result in Auger-deexcitation or neutralization or resonance neutralization or ionization. A potential energy diagram is schematically indicated in Fig. 5. It has been proposed (50) that these radiationless deexcitation processes influence the beam induced light emission since they compete with deexcitation by photon emission. The transition rate R(s) as a function of metal-atom separation s for resonant deexcitation is often approximated by

$$R(s) = A \exp(-as) \qquad (2)$$

A and a being constants. A reasonable value (49) of a is a ≈ 2 Å^{-1} and A is of the order of $10^{14} - 10^{16}$ sec^{-1}. These numbers show that the transition rate for this process is very large for distances s of the order of a few Angstroms and several orders of magnitude larger than the decay rate for optical emission, which is of the order of 10^8 sec^{-1}.

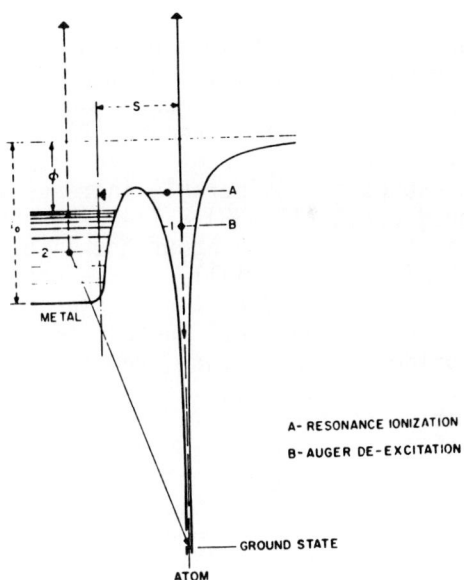

Fig. 5. Schematic representation of the ion-metal energy level system showing two possible nonradiative deexcitation processes, Auger deexcitation and resonance ionization.

It can be shown (50) that the probability P for a particle with velocity component $v_\perp$ perpendicular to the surface to escape without having undergone a resonance transition is given by

$$P = \exp\left[-\frac{A}{av_\perp}\right] \qquad (3)$$

The value of $\frac{A}{a}$, sometimes called the survival parameter, is of the order of 10^7 cm/sec with the numerical values of A and a quoted above. With this value of $\frac{A}{a}$ eq..(2) shows that low energy particles have a small survivial probability. As an example, the survival probability for 10 eV excited Cu atom near a Cu surface would be of the order of 1% (32). The competition of this type of velocity dependent radiationless deexcitation with photon emission, therefore, presents a natural explanation for the observation that mainly the most energetic of the sputtered particles contribute to the radiation. This description has been the basis of a numerical calculation of spectral line profiles, with the aim of determining the parameters A and a (see Section IV).

D. <u>Influence of Surface Oxidation and Other Chemical Effects</u>

It has been observed that the line intensities are very much enhanced when one bombards an oxidized metal, as compared to the clean metal case (8, 32, 37). Also, the admission of oxygen in the scattering chamber during ion bombardment causes a considerable increase of light intensity from sputtered atoms (12, 38). There has been some speculation and also controversy about the nature of this chemical enchancement effect. Initially it was proposed (50) that the enhancement is caused by the change in band structure of the solid upon oxidation. Because of the forbidden band gap in the oxide many resonance transitions of electrons from excited sputtered atoms into the solid would then not be possible any more. This inhibition of the radiationless deexcitation processes caused the number of excited particles to drastically increase upon oxidation of a surface. It is also clear because of the strong velocity dependence of the survival probability (eq. (3)), that especially low velocity particles would have a possibility to radiate, when close to an oxide. This model satisfactorily explained, therefore, not only the enhancement effect, but also the observed change in Doppler profile when oxidizing a metal surface.

One example of a systematic study of the influence of the band structure on the light emission probability can be found in the work on photon emission from clean and oxidized Mg surfaces by Kerkdijk and Kelly (38). The optical spectrum emitted from Mg during Ne^+ bombardment show many MgI and MgII lines, some of which are enhanced, some remain constant and others decrease in intensity upon oxygenation of the surface. A schematic energy level diagram of Mg, MgO and the upper levels of the measured lines is given in Fig. 6. A few representative examples of line intensity behavior can be indicated. Lines from levels 2 and 4 do not change intensity when going from Mg to MgO, lines 14 a, b, c increase when going to MgO, while lines 7 a, b decrease. In terms of the resonance tunneling model, the explanation is quite simple. Levels 2 and 4 for both metal and oxide are opposite empty states, therefore, show no oxygen dependence, level 14 can be emptied, to the metal, but not to the oxide (positive oxygen dependence), and level 7 can be populated from the metal and not from the oxide (negative oxygen dependence). In this manner it was possible to explain the oxygen dependence of 16 lines in the spectrum. There were 5 lines, however, which did not follow the behavior expected from the electron

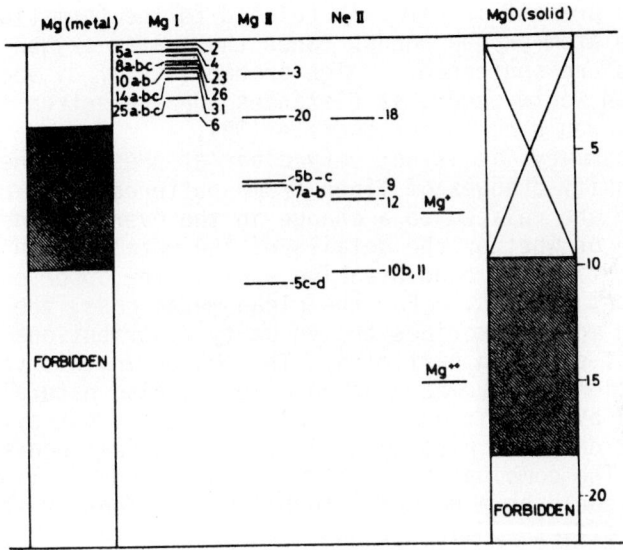

Fig. 6. The energies of excited atoms, ions and excited ions observed when Mg is bombarded by 10 keV Ne^+. The band structures of Mg and MgO are also shown. The scale on the right hand side is in eV (from Ref. 38).

tunneling model. These are all ionic lines, so it seems likely that other mechanisms are operating in excited ion information.

It has been suggested that excited ions are dominantly formed by dissociative excitation of sputtered molecules (51). Recently, some more doubt about the validity of the resonant electron transfer model was raised by a rather sophisticated experiment by G. E. Thomas et al. In this work the light emission from sputtered Cu and Al atoms (using 10 keV Kr^+) was measured during adsorption of Cs, sometimes with co-adsorption of O_2.(52). The Cs coverage changes the work function of the Al or Cu metals and this decrease of 2 eV was simultaneously monitored by laser induced photoelectric emission. In the electron tunneling model, a change in work function should drastically influence the probability for resonance tunneling for those lines which are no longer opposite empty electron states in the solid after a decrease of work function (compare Fig. 5). In the Cs experiment, actually this enhancement effect was expected for the CuI = 3247 Å and AlI 3092 Å lines, but surprisingly enough, no such effect was found. In contrast, O_2 adsorption caused a marked increase of these lines. These results led the authors (52) to conclude that the enhancement effect when oxygen is present is directly related to the formation of substrate atom-oxygen bonds. When these atom-oxygen molecules are sputtered, a high probability of dissociative excitation would occur, at distances where electron tunneling processes would no longer operate (53).

In summary, it is not very clear at present whether the intensity changes of lines from sputtered atoms upon oxidation are related to a change in the overall band structure or whether the details of the metal-oxygen bond formation and bond breaking explain the observed enhancement effects. For the clean metal case, the tunneling model describes the velocity distributions (Doppler shapes) of emitting particles. The change in this velocity distribution when bombarding an oxide is also naturally explained by inhibition of tunneling process, but evidently in the oxide case, also details of the chemical bonds play a role. The combination of SIMS and beam induced light emission could be a powerful technique to identify the chemical effects (54).

IV. RADIATION FROM BACKSCATTERED PARTICLES

A. General Characteristics

A small fraction of an incident projectile beam is backscattered as excited atoms (or ions) and gives rise to emission of Doppler broadened spectral lines; this was shown for the first time by McCracken and Erents (18). Detailed analysis of the line shape provides information on the distribution in speeds and direction of the backscattered species. Measurement of the total line intensity gives a measure of the probability for backscattering in the relevant excited state defined conveniently as the number of excited atoms scattered out per atom (or ion) incident. As with sputtered excited particles an important factor in understanding the backscattered flux of excited atoms is the interaction of the recoiling species with the target surface; again radiationless de-excitation processes such as Auger or resonance transitions can materially alter the proportion of the recoils which emerge in an excited state.

Most of the published research is concerned with the impact of H^+ and He^+. These species involve simple and well understood electronic energy levels; moreover at the energies commonly utilized in such experiments (1 to 100 keV) the Doppler line width is sufficiently broad to readily permit detailed recording and analysis. There are recent studies of backscattered H^+ and H^* formed when the molecular species of H_2^+ and H_3^+ are incident, which indicate that the backscattered flux of excited atoms and ions may be related to the molecular state of the incident species. This is surprising since the binding energy of the molecule in these experiments is quite insignificant compared with the impact energy.

B. Line Shape Analyses

When a projectile beam, such as H^+ or He^+ is directed at a target, it is observed that the atomic projectile lines are Doppler broadened. This broadening reflects the speed and directional distribution of the backscattered ions. Fig. 7 shows such a Doppler broadened line from work by Baird et at. (21).

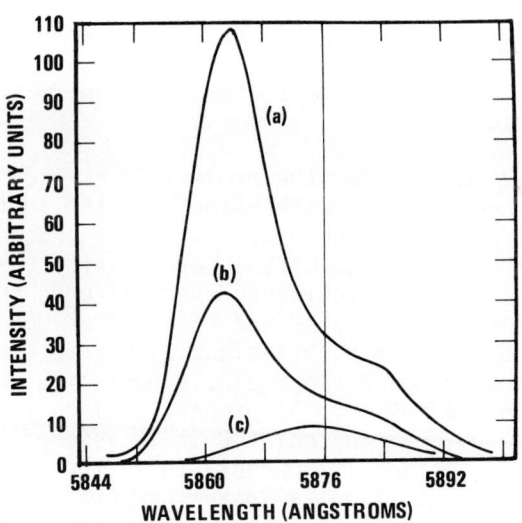

Fig. 7. Line shape of the 5876-Å ($3^3D \rightarrow 2^3P$) HeI emission induced by 30-keV He^+ impact on niobium at an incidence angle ϕ of (a) 60°, (b) 45°, and (c) 0°. Only a smoothed curve is shown. (From Ref. 21).

A significant feature is that the intensity at zero shift is quite small indicating that few recoils with low velocity are occuring.

One can model the line shape using a backscattering theory of McCracken and Freeman (55). Let us refer to the two dimensional Fig. 8.

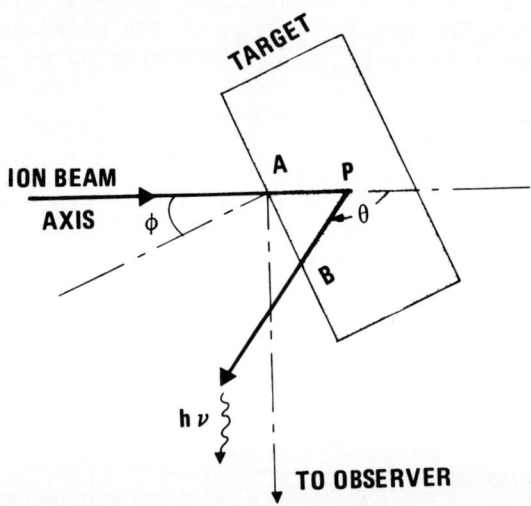

Fig. 8. *Geometry of the scattering problem shown in two dimensions only.*

McCracken and Freeman consider the projectile as incident at point A with some angle ϕ, with respect to the surface normal, and penetrating to the point P where it undergoes a close encounter with an atom of the target lattice and scatters through an angle θ to return eventually to the surface at B. Along the paths AP and PB it is assumed that the projectile undergoes only collisions with free electrons causing loss of energy but no appreciable deviation; the rate of energy loss is taken to be proportional to the square root of energy. At the point P it is assumed that the collision cross section is appropriate to the interaction of the two nuclei, i.e., a Rutherford cross section is used. Based on this picture one may formulate an expression for the probability that a projectile will emerge at B with a velocity component perpendicular to the surface lying between $v_\perp$ and $v_\perp + dv_\perp$ and moving into some element of solid angle $d\omega$. This probability may be denoted as:

$$P(v_\perp) = N(v_\perp) \, dv_\perp \, d\omega \qquad (4)$$

There is no information on the proportion of the scattered projectiles which might be neutralized into a specific excited state. In analyses performed to date it is assumed that this proportion, F, is independent of the emergent particle's energy and direction.

The absence of low energy recoils is believed to be due to either resonance ionization or Auger neutralization as the excited particle recoils from the surface. For either mechanism the probability of a particle with velocity component $v_\perp$ escaping without radiationless decay is given by eq. (3). Those excited particles escaping without radiationless decay will decay by photon emission and a certain fraction F' will be detected.

Thus, combining these various terms we have the probability $P(v_\perp)$ of detecting a photon from an emergent particle of velocity component $v_\perp$ scattered into a solid angle $d\omega$, which is

$$P(v_\perp) = F'F\exp(-A/av_\perp)N(v_\perp)dv_\perp\, d\omega \qquad (5)$$

The wavelength of the photon can be simply calculated by the Doppler-shift formula.

With this information one can use eq. (5) to calculate the spectral line shape in relative terms taking F and F' to be constant but unknown. It is necessary to take into account light reflected by the target towards the observer (21). One has no prior knowledge of the magnitude expected for the survival parameter A/a. Computations are performed for various A/a and the value selected which gives the best measured shape. In practice the fit is governed primarily by the position of the peak in intensity.

We shall not trouble to list all the values of A/a now quoted in the literature. Generally for He triplet states on various metals (Aℓ, Cu, Nb, Mo, Ag, W) A/a comes out to be about 1 to 3 x 10^8 cm/sec (21, 22). For backscattered H atoms from the same metals a value of around 8 x 10^7 cm/sec is found by the Georgia Tech group for the H(n = 3 state). The FOM group also determined a value of A/a for the H(n = 4) state by studying only the wavelength at which the maximum intensity occured; (56,57) for hydrogen on Cu they get a value of 1.5 x 10^8 cm/sec. Small variations of A/a with target material are found but no systematic trends are clearly indicated. It is worth noting that Cobas and Lamb (58) have predicted the two constants for both resonance ionization and Auger neutralization of metastable helium. As interpreted by Hagstrum (49) the ratio would be 1.3 x 10^8 for the Auger effect and 4.8 x 10^{10} for the resonance effect; obviously the Auger figure is consistent with the data. An exemplary fit is shown as Fig. 9.

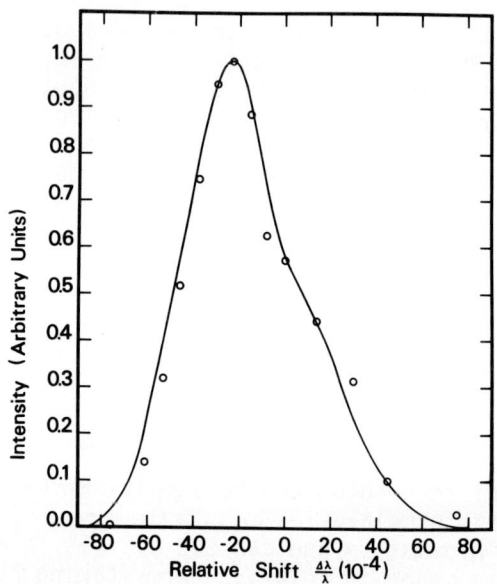

Fig. 9. Measured and predicted line shape of the 6563 -Å (n=3 to n=2) H emission induced by 25-keV H⁺ impact on Mo at an incidence angle φ of 60°. Intensity is shown as a function of relative wavelength shift, defined as the shift (Δλ) from 6563 -Å line divided by the wavelength of that line, 6563 Å (λ). Circles, experimental data points; solid line prediction by the above model with survival coefficient chosen for best fit to data points (A/a = 7.2 x 10^7 cm/sec). (From Ref. 22).

This subject would now benefit from some detailed predictions of the factors A and a. The existing predictions of Cobas and Lamb (58) are quite crude; for example they use hydrogenic wavefunctions to describe He atoms and conduction band electrons. Clearly this can be improved by application of our knowledge of wavefunctions. Analysis of the problem has proceeded on the assumption that the energy levels of the atom and metal are unperturbed by the interaction. This is quite wrong. At the low collision energies used here we need a treatment similar to an adiabatic approximation in atomic collisions where the interaction perturbs the energy levels. Such treatments are now in progress.

The analysis of this problem ignores the question of how and where the excited recoils were formed; it is simply assumed that excited state formation probability is an unknown factor independent of recoil velocity. If formation occurs at the surface then a very rapid mechanism must be involved since the radiationless de-excitation must also subsequently occur within a few Å of the surface. A further possibility is that the excited state formation occurs inside the solid; although the projectiles in the solid are generally stripped there seems to be no reason why a few percent should not be in an excited neutral state at any one time. It has been concluded (22) that excited atoms represent less than 1% of all recoils.

C. Line Intensity Measurement

The total intensity in a spectral line from a back-scattered particle can also be measured and is related to the flux of backscattered excited atoms. It is convenient to introduce a symbol γ_j to represent an "excitation coefficient" for the formation of atoms in the excited state j; we may define this as the number of atoms backscattered in the excited state j for each projectile incident. Similarly we can define an "emission coefficient" γ_{jk} as the number of photons in the transition j → k emitted for each projectile incident. The coefficients for excitation and emission are of course closely related. In a case where only the excited state j is formed then

$$\gamma_{jk} = \frac{A_{jk}}{\sum_{k<j} A_{jk}} \gamma_j \qquad (6)$$

where A_{jk} is the transition probability for the decay from j to k. In general the state j will also be populated by cascade through radiative decay from higher states j; the magnitude of the cascade can in principle be estimated by measuring all emission coefficients for transitions from states i above j into j. In the event that cascade is negligible then eq. (6) holds and γ_{jk} is proportional to γ_j. Most published work presents measurements of γ_{jk} and predictions of γ_j; it is frequently found that cascade is indeed negligible so that these two quantities are proportional to each other. To estimate the excitation coefficient one simply integrates eq (5) over all recoil velocities

$$\gamma_j = \int_{V_\ell}^{V_u} F\cdot F \exp - (\frac{A}{av_\perp}) N(v_\perp) \, dv_\perp d\omega \quad (7)$$

Since one does not know F it is possible only to compute a relative value of γ_j as a function of the incoming variables of projectile energy and incidence angle. The upper limit of integration is related to projectile scattering at the surface where no penetration occurs; as a lower limit one may assume that any projectile whose energy is reduced to 20 eV will be trapped in the lattice and will not escape.

It is useful now to compare experimental and theoretical values of γ_j, as a function of projectile energy. Fig. 10 shows (23) such a comparison for He(3^3D); the data is absolute and theory is normalized to it.

First note that small changes to A/a cause enormous changes to the functional dependence of γ_j on energy. Secondly note that a prediction with A/a = 3 x 10^8 cm/sec is in essential agreement with experiment from 3 to 30 keV energy. This value of A/a is the same as that derived from analysis of line shape for this transition.

A further factor we can explore is the dependence of γ_j on the target. If one assumes the formation of excited atoms to be independent of target material then the variation can occur only in the backscattering factor $N(v_\perp)$ of eq. (7). This of course is given by the backscattering prediction of McCracken and Freeman (55), the functional dependence of γ_j on projectile and target atomic numbers Z_1 and Z_2 is given by:

$$\gamma_j \propto Z_1^{5/6} Z_2 (Z_1^{2/3} + Z_2^{2/3})^{3/2} \quad (8)$$

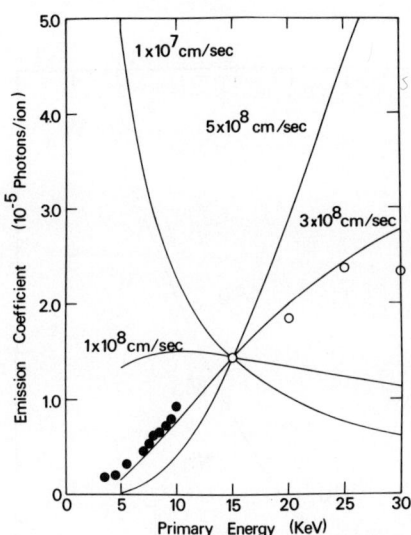

Fig. 10. Calculated energy dependency of γ_j for He^+ incident on Cu at 45°; calculations are for the values of A/a indicated. All data are arbitrarily normalized to a value of 1.43×10^{-5} photons/ion at 15 keV impact energy. Also shown is the measured value of γ_{jk} for the HeI $3^3D \rightarrow 2^3P$ transition. (From Ref. 23).

The dependencies on Z arise from the expressions for Rutherford scattering and for electronic stopping. In Fig. 11 is shown the measured dependance (22) of γ_j on target Z for backscattered He (3^3D); also indicated is a line given by eq 8 and normalized to the Mo point. There is rather good agreement.

One of the more sweeping assumptions of this analysis is that the probability of excited state formation (factors F of eq. 5) is independent of recoil velocity and of target material. The excellent agreement between theory and experiment in energy dependance (Fig. 10) and target atomic number dependance (Fig. 11) shows that this assumption is reasonably valid.

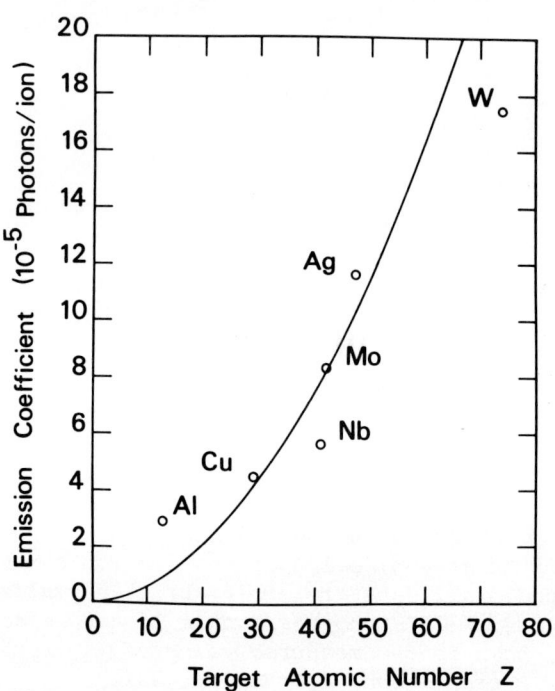

Fig. 11. Emission Coefficient of the He $3^3D \rightarrow 2^3P$ Transition Shown as a Function of Target Atomic Number. (Experimental conditions: 25 KeV He^+ incident at an angle of $60°$. Circles, experimental data points; solid line, predicted dependance. The predicted curve has been normalized to the data at the Mo data point.) (From Ref. 22).

D. Impact of Molecular Species

There is recent activity in the study of emission from backscattered H atoms when targets are bombarded by the molecular species H_2^+ and H_3^+. In general qualitative terms the emission is very similar to that observed when the monatomic species H^+ is incident; the spectral lines are Doppler broadened by the distribution in speeds and directions of the backscattered H atoms. Since molecular binding energies are very small compared to incident collision energies, the reflection coefficient for an atom initially bound in a molecule must be essentially unchanged from that of a free atom. Thus there will be three times as many backscattered nucleons for the H_3^+ case per incident H_3^+

than for H^+ impact. It was expected that the probability of forming an excited nucleon would be independent of the incident molecular state so that equivelocity H_3^+, H_2^+ and H^+ ions should produce intensities in the ratio 3: 2:1.

Rausch and Thomas have recently published (57) relative measurements of the emission coefficient γ_{ik} for Hβ emission (H(=4) → H(=3)) induced by impact of H^+, H_2^+ and H_3^+ on various targets. The target environment was a vacuum of 2×10^{-9} torr and targets were cleaned by preliminary bombardment with Ar^+; angle of projectile incidence on the surface was 60° with respect to the surface normal. Fig. 12 shows the measured emission coefficient per nucleon; thus the measured emission coefficient for H_n^+ impact at energy E is plotted divided by n at an energy of E/n. The data for H^+, H_2^+ and H_3^+ impact are all the same when plotted in this manner and therefore show the expected behaviour. The solid lines in Fig. 12 are computed

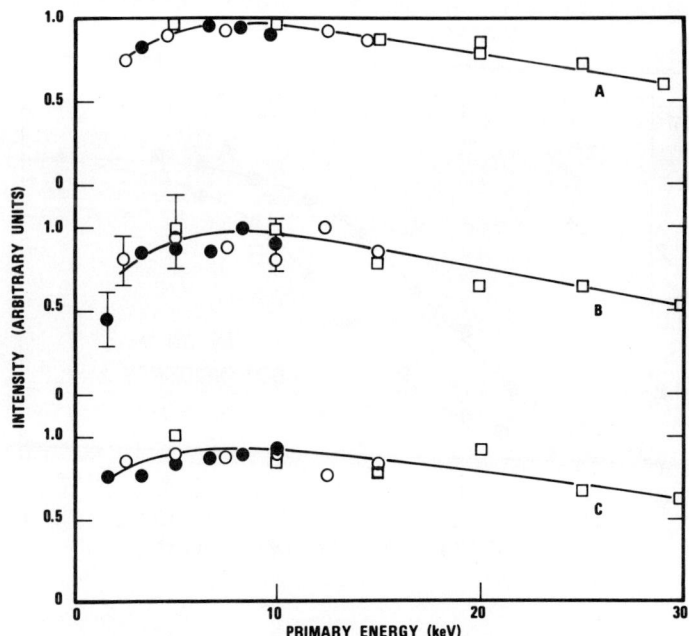

Fig. 12. Energy dependence of Balmer - β line emission from backscattered H* atoms. A, Molybdenum target; B, Stainless Steel type 304 target; C, Copper target. Squares show data for H^+ impact; open and closed circles are, respectively, data for H_2^+ and H_3^+ impact plotted as described in the text. (From Ref. 59).

values of γ; obtained from eq. 6 and normalized to the experimental data. The prediction utilizes a survival coefficient A/a of 1.8×10^8 cm/sec which is in acceptable agreement with the value of A/a determined by Kerkdijk et. al. (57) from analysis of line shape (1.5×10^8 cm/sec). Baird et. al. (22) had also shown earlier that the Hα line shape for H_2^+ impact on metals was the same as for equivelocity H^+ impact; this again shows that the same survival coefficient A/a governs scattering of both monatomic and diatomic incident ions. Very recently Rausch and Thomas (60) have investigated the influence of ambient oxygen on the Hβ emission induced by H^+, H_2^+ and H_3^+ impact at 10 to 30 keV energies on metals. For an Mo target the Hβ intensity increases by 50% as ambient O_2 pressure is increased from 2×10^{-9} torr to 2×10^{-7} torr; this is illustrated in Fig. 13. The change in intensity is shown to be consistent

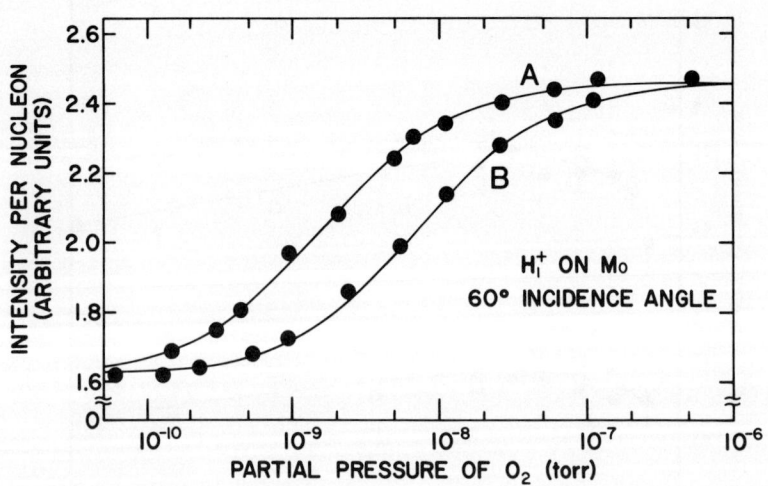

Fig. 13. Relative value of Hβ emission as a function of ambient O_2 pressure for 20 keV H^+ impact on Mo at $60°$ to the target surface normal. Line (a) is for a beam current density of 30 µA/cm^2 and line (b) for 140 µA/cm^2. The solid lines are a semi-empirical prediction. (From Ref. 60).

with the coefficient A/a changing from 1.8×10^8 cm/sec at 2×10^{-9} torr to 1.2×10^8 at 2×10^{-7} torr. For small incidence angles on Mo a similar behaviour is observed but the form of the pressure dependence is not the same for the species H^+, H_2^+ and H_3^+ even when studied at the same impact velocities and nucleon fluxes. Thus there is evidence that the molecular state of the incoming projectile has some influence on the probability of forming a backscattered excited atom in the case of an Mo target. For a Cu target the emission is little effected by the presence of oxygen and no difference between molecular species is observed. These observations are consistent with similar measurements of Leung et. al. (24),in the study of the H_α (Balmer alpha) line radiation arising from excited neutral atoms backscattered at normal incidence during bombardment of molybdenum surfaces with beams of H^-, H^+, H_2^+ and H_3^+ performed at lower energies (500 ev to 6 keV).

V. INTRINSIC CONTINUUM LUMINESCENCE FROM SOLIDS

A. General Spectral Features

In addition to lines and bands from scattered and backscattered atoms, ions, and molecules, in many cases one also observes very intense broadband emissions which originate in the bulk of the solid. These broadband emissions from the bulk of the solid are produced in a wide variety of insulating targets (1, 11, 16, 27) and have also been reported from metal targets (26). Similar broadband emissions are produced by other types of radiation such as electron and x-ray bombardment and comparison of the ion bombardment results to cathode luminescence and x-ray fluorescence results provide insight into the collision mechanisms responsible for producing the emissions.

An example of continuum emission from the bulk of the solid is shown in Fig. 14 for the case of He° (5 keV) impacting on CaF_2 (1). The broadband emission from CaF_2 extends over ~ 2000Å with a pronounced peak at ~ 2800Å. The sharp lines in Fig. 14 arise from excited states of sputtered Ca and Ca^+. A similar continuum luminescence is produced by electron bombardment at low (1) and high (55) energies, and evidence is available which indicates that this emission arises from radiative recombination of electrons with the self-trapped V_k center of the crystal (56, 57). In the case of the He° impact it is likely that the production of electron-hole pairs results from the inelastic transfer of projectile energy to the bound electrons in the solid (1). The luminescence then results from electron-hole recombination. A similar luminescent continuum has been reported

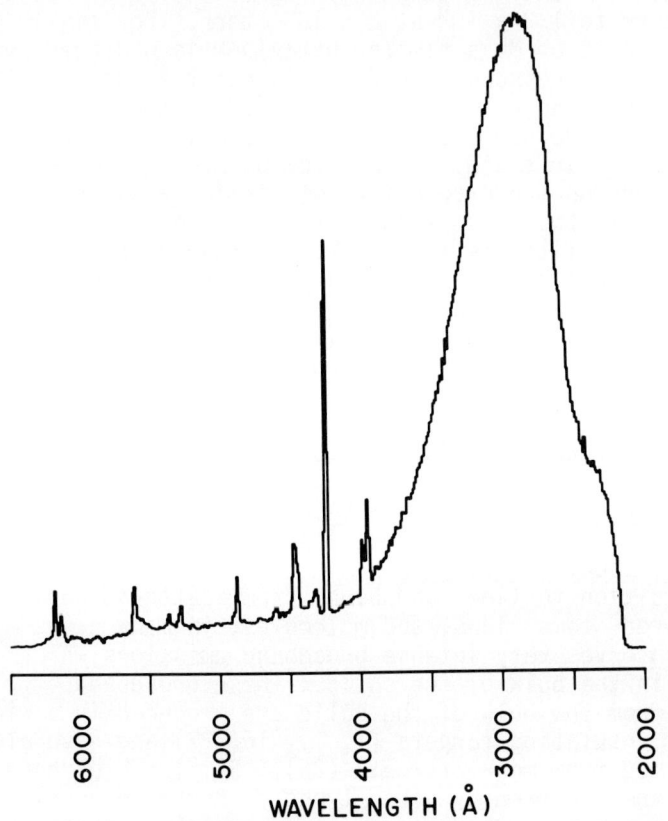

Fig. 14. Spectra of radiation produced by impact of He^0 (5 keV) on CaF_2 (From Ref. 1).

in the case of He° impacting on MgF_2 (15).

The intensities of luminescent continuum from the solid are quite sensitive to the velocity of the impinging projectile. At impact energies where elastic collisions with target nuclei is the dominant energy loss mechanism, the intensities are quite low. Continua of this type, therefore, may be useful to provide information on energy loss mechanisms as atomic particles move in solids.

B. Luminescence from Alkali-Halide Crystals

Bazhin et al., (27) have investigated luminescence of certain alkali halides induced by the impact of H^+ and He^+ ions with energies of 5 to 25 keV. Targets of principal interest were NaCl, KCl and KBr; various target temperatures in the range -160 to 200°C were employed. The high density of electron excitation close to the target surface causes a high efficiency for defect generation in the cation sublattice and the formation of hole centers that are stable at room temperature. The observed spectra were broad bands of luminescence with some line emission in the Sodium-D when crystals containing Na were employed; this sodium emission is from sputtered atoms in excited states. Preliminary experiments were performed with neutral hydrogen beams which have precisely the same spectra as for H^+ and He^+ impact. Various precautions were taken to thermally anneal defects and remove the effects of contaminants; these are described in the original publication (27).

In Fig. 15 are shown sample spectra induced by ion beam impact on targets of KCl, and KBr, at low temperatures; the reader is referred to the captions for the precise conditions of bombardment in each case. The spectra are corrected for the variation of relative sensitivity with wavelength but have not been placed on an absolute scale. The spectra are typical for these samples and remain

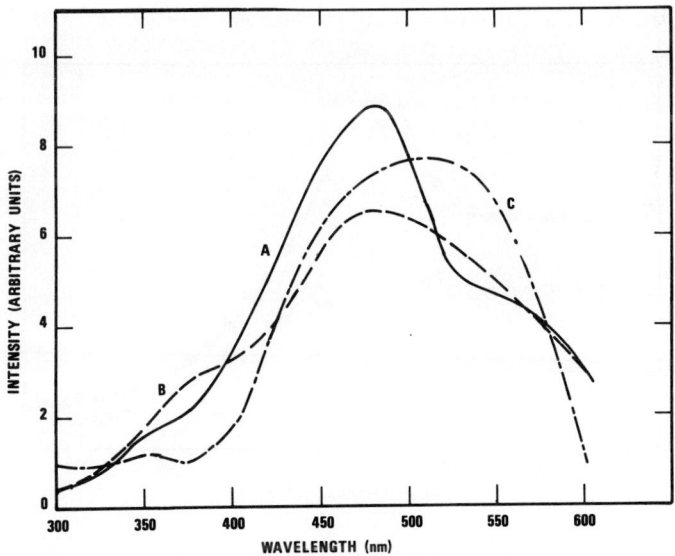

Fig. 15. Luminescence spectra of KCl_2 and KBr at low temperature under bombardment by a $10\mu A/cm^2$ beam of 25 keV H^+ ions. A, KCl at -29°C; B, KCl at -50°C; C, KBr at -150°C. (From Ref. 27).

essentially unchanged in basic shape if one alters beam energy or interchanges protons and He+ ions; such changes however, do alter the intensity of the emission. The spectra are wide luminescent bands varying in shape between the various crystals.

The most significant features of the KCl and KBr spectra are peaks at 480 nm and 500 nm respectively; it will be shown later that these are due to recombination of electrons from the conduction band with V_3 centers. At low temperatures (Fig. 15) for these same targets a weak peak is observed at lower wavelengths, 370 nm for KCl and 350 for KBr; it will be shown later that these peaks are due to recombination of conduction band electrons with V_4 centers. The low wavelength peaks are strongly overlapped by the more intense high wavelength peak and show up on the spectra as only weak shoulders; they are, however quite reproducible.

In Fig. 16 are measurements of luminescent intensity as a function of temperature for KCl and KBr crystals. The intensity is measured at 480 nm for KCl and 500 nm for KBr with a spectral resolution of 4.8 nm; these wavelengths correspond to the peaks of intensity in these spectra.

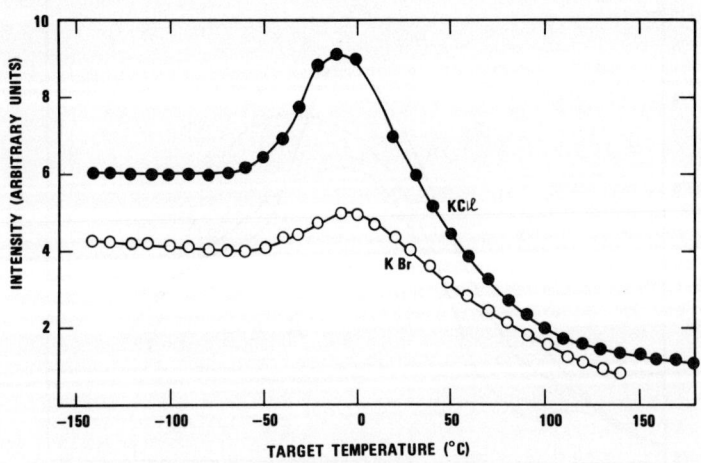

Fig. 16. Dependence of the luminescence intensity on target temperature. Targets are KCl and KBr; bombarding beam is 25 keV H^+ at a density of $10\mu A/cm^2$. Intensity is measured at the wavelength of maximum intensity which for KCl is 480 nm and for KBr is 500 nm. (From Ref. 27).

A continuous record of intensity as a function of time, and therefore dose, was performed. Fig. 17 shows the results for KBr and KCl crystals at room temperature. In the studies of dose dependence (Fig. 17) there are regions where intensity increases as a function of dose; these suggest that the centers responsible for luminescence are created by ion bombardment. However, the intensity of luminescent centers are already present in the unbombarded crystal despite the preliminary annealing procedure. The fact that intensity is a function of temperature shows that the formation, and the stability, of the defects responsible for luminescence is temperature dependent.

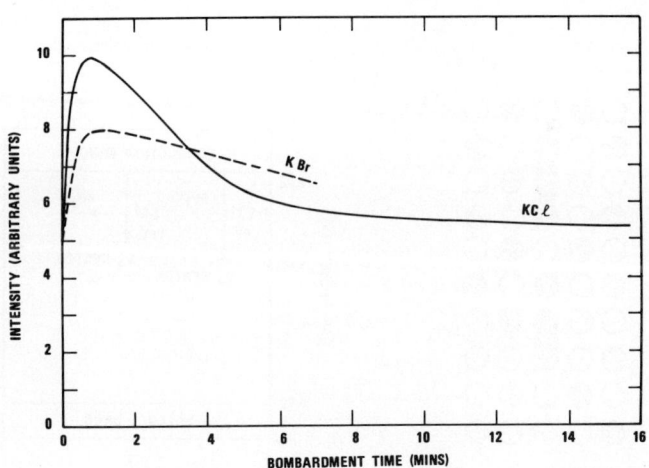

Fig. 17. Dependence of luminescence intensity on dose for targets at room temperature (22°C). Targets are KCl and KBr; bombarding beam is 25 keV H^+ at a density of $10\mu A/cm^2$. Intensity is measured at the wavelength of maximum intensity which for KCl is 480 nm and for KBr is 500 nm. (from Ref. 27).

The intensity of the ion-induced luminescence was unchanged by an attempt at F center bleaching. The crystal was also exposed to unfiltered white light which should bleach more complicated electron centers; again no effect was observed. It was concluded that the centers of luminescence can be V_3-centers which are known (63) not to be bleached by white light.

It was proposed that the V_3 center is responsible for the principal luminescent intensity at 480 nm in KCl and 500 nm in KBr; it is known (63, 64) that the V_3 center is stable up to 200°C. The small peak observed at 370 nm in KCl and 350 nm in KBr (See Fig. 15) for low temperatures was identified as due to the V_4 center. The low wavelength peak does not appear to be connected with impurities and as temperature is raised the feature is no longer apparent. It is known that at temperaures below -30°C the V_4 center occurs in addition to the V_3 center. However, at all temperatures considered in this work the luminescence identified as due to V_3 centers predominates. In Fig. 18 are shown

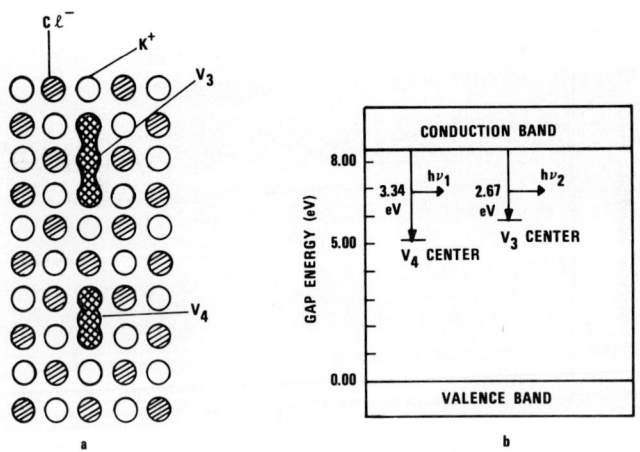

Fig. 18. Diagram of the structure ascribed to V_3 and V_4 centers with an energy level diagram for KCl showing the position of the energy levels associated with such centers. (From Ref. 17).

the structures postulated for the V_3 and V_4 centers (64). The X_3^- molecule occupies a divacancy to form a V_4 center, and the X_3^- molecule occupies one cation and two anion vacancies to form the V_3 center. The centers produce localized energy levels in the band gap as shown in Fig. 18. The emission is due to excited electrons from the conduction band decaying into the localized V_3 and V_4 energy levels. The V_3 and V_4 energy levels have been previously identified by analysis of photoabsorption spectra. If the recombination of conduction band electrons with V_3 and V_4 centers is responsible for the emission then the sum of the emitted photon energy and the previously measured photoabsorption energy should equal the band gap. Fig. 18b shows an energy level scheme appropriate to KCl with the suggested radiative recombination transitions. In table I is listed the band gap energies for KCl, KBr and NaCl with the energy of photon emission which is identified as due to recombination of electrons with the V_3 and V_4 centers; also listed are the photon energies of maximum absorption ascribed by others as due to the V_3 and V_4 centers. In KCl the band gap is quoted as 8.5 eV and the energies for photo absorption by V_3 and V_4 centers are 5.85eV and 5.16eV respectively; thus the predicted energies of photon emission in recombination of electrons with V_3 and V_4 centers are 2.65 and 3.34 eV respectively. Hence emission is expected at 468 nm for the V_3 center and 371 nm for the V_4 center; quite close to the observed peaks at 480 nm and 370 nm. This reinforces the conclusion that the observed luminescence is due to recombination of electrons with V_3 and V_4 centers.

It is difficult to make precise statements about the width of the observed luminescence bands since there is considerable overlap between the prominent V_3 band, and the weak V_4 band. However, a rough estimate of the V_3 band width at half maximum for KCl (Fig. 15) is 1.1 eV; this is consistent with the width of the V_3 absorption band which is about 1.2 eV in the measurements of Seretlo (63).

TABLE I

Table of known band gap energies and known photon energies for absorption maxima in the V_3 and V_4 bands presented with the photon energies, (and wavelengths) of the peak intensity observed in the present luminescence studies. Band I is the high wavelength band ascribed to the V_3 center and Band II is the low wavelength band we ascribe to the V_4 center. (From Ref. 27)

Sample	Energy Gap (eV)	Absorption Band Maxima (eV)		Observed Luminescent Band Maxima			
				Band I		Band II	
		V_3	V_4	Energy (eV)	Wavelength (nm)	Energy (eV)	Wavelength (nm)
NaCl	8.6[a]	3.9[b]	5.56[c]	2.76	450	3.31	375
KBr	7.8[a]	5.35[c]	4.5[d]	2.48	500	3.54	350
KCl	8.5[a]	5.85[c]	5.16[e]	2.58	480	3.35	370

References
a. "American Institute of Physics Handbook," D. E. Gray (Ed.) (McGraw-Hill Book Co. Inc., N.Y. 1963) p. 9-22
b. R. Casler, P. Pringsheim and P. Yuster, J. Chem. Phys. 18, 1564 (1950)
c. M. Dorendorf, Z. Physik, 129, S317 (1951)
d. J. D. Kingsley, J. Phys. Chem. Solids, 23, 949 (1962)
e. J. Faraday, and W. D. Compton, Phys. Rev., 138 A893 (1965)

With the identification of the emission as being due to electron recombination with V_3 and V_4 centers one should be in a position to understand the dependance of intensity on temperature and irradiation dose. The intensity of the ion-induced lunimescence will undoubtedly depend on the rate of electron excitation to the conduction band and on non-radiative decays of such electrons; also intensity will depend on the density of relevant recombination centers through their rates of creation and rates of annealing. Thus, the relationship of luminescent intensity to dose and temperature is more complex than the relationship of directly measured V_3 and V_4 center density to these same experimental parameters. It would be presumptuous, and possibly misleading, to attempt a modeling of the phenomenon based only on the measured parameter of intensity; supporting information on the directly measured defect-center concentrations would be necessary in order to develop a reliable quantitative understanding.

C. Luminescence from Metals

Light ion (H^+ and He^+) bombardment of clean metal surfaces produces continuous spectra located at the target surface. Fig. 19 shows certain of these continua from Aℓ, Mo and Cu targets from the work at Georgia Tech (65). It is observed that Cu and Mo give similar spectra and essentially the same is found for Nb and W. The most striking spectrum is for Aℓ, it is quite intense (0.02 photons per incident 25 keV H^+ ion) and peaks around 5000 Å.

The aluminum luminescence has been subject to a variety of experimental tests. Its intensity is linear with beam current. The dependance of intensity on beam incidence angle is appropriate to an emission source on the surface or in a region of 100-200 Å depth. The emission is not due to ejected materials. The emission exhibits no polarization; this indicates the emission is not due to a plasmon effect. The intensity of emission (0.02 photons per 25 keV H^+ ion) is seven orders of magnitude higher than expected for bremsstrahlung or transition radiation.

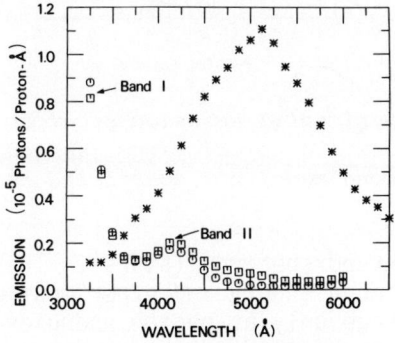

Fig. 19. Luminescence of Aℓ, Mo, and Cu induced by 25-keV H^+ impact at an incidence angle ϕ of 45^0. Stars, Aℓ data; squares, Mo data (x25); circles, Cu data (x20). (From Ref. 65).

It was concluded that the emission is due to electron hole recombination (65). The incident ion excites some electrons to excited electronic bands and these in turn decay radiatively back to the valence band. One might argue that for a metal the various valence and excited electronic bands will produce a continuum of energy levels and that excited electrons will decay by photon emission. This is in general true. However, at certain points in the Brillouin zone (notably the W point) there are excited levels from which only radiative decay is possible (65). The spectral distribution for radiative decay has been computed for a uniform distribution of excited electrons and holes. In practice the only significant contributions to emission occur close to the symmetry planes where the energy bands deviate appreciably from the parabolic form. In Fig. 20 is shown the computed spectrum (dashed

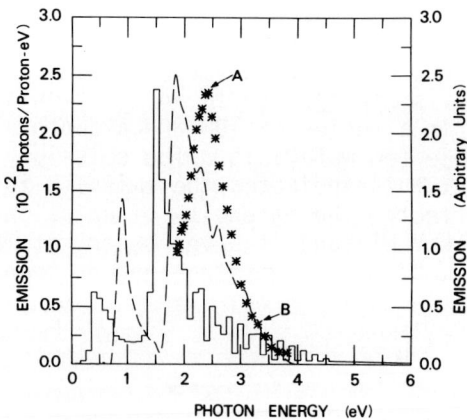

Fig. 20. Experimental emission spectrum (stars) compared with theoretical predictions (dashed line). Histogram is a second model not considered here. (From Ref. 65).

line) and the observed spectrum (stars). There is substantial agreement between theory and experiment with intensity predicted to be large only at photon energies where significant intensity is indeed observed. (The histogram of Fig. 20 is a second model and does not show such agreement). There are obvious disagreements in detail, but these could be due to the arbitrary assumption of a uniform electron and hole distribution. If it is assumed that the cross section for electron excitation varies over the valence band then one could force a better agreement between theory and experiment.

There have been a number of experiments on "cathodoluminescence" of metals under electron bombardment. In particular the work of Papanicolaou et al. (66); concerning the spectrum induced by 0.3 to 1 KeV electron impact on copper shows a continuum peaking at 320 nm; this is essentially the same as that shown if Fig. 21 for H^+ impact. Papanicolaou et al., (66) conclude that the 320 nm peak is due to direct radiative recombination of excited electrons from the X_4 point with holes at X_5; this is in effect an electron transition from a p state to a d state. We should also note that in cathodoluminescence of copper (66) there is a second peak at 520 nm whenever a surface oxide layer is present.

It would seem that the study of the intrinsic continuum of metals should yield information on the mechanisms of energy loss as ions traverse solids. Moreover the experience with cathodoluminescence suggests that oxide films will also contribute to such spectra.

D. Transition Radiation, Bremsstrahlung and Plasmons

There are a number of phenomena which are expected to give appreciable radiation when very high velocity ions interact with solids in a "collective" manner; they represent interactions with all the electrons or ions in the target rather than with single electrons or atoms. Some of these have been found for high energy electron impact on solids; there is however very little work with ion impact. Most of these emissions will be quite weak at low impact energies and may very probably be obscured by emission from backscattering and sputtering.

(1) Transition Radiation

When a charged particle enters a metal surface, the annihilation of the dipole formed by the charge and its image causes the emission of a very weak pulse of electromagnetic radiation. This is known as transition tradiation. Numerous calculations on this phenomenon have been performed and a simple summary is to be found the the work of Goldsmith and Jelly (67). The emission will be a continuum and clearly will exhibit complete polarization in the plane of the dipole. Extensive experimental and theoretical studies for electron impact on metals are to be dound in the work of DiNardo and Goland (68) as well as Arakawa and co-workers (69). (See also references cited in these papers).

The only experimental work with heavy particles would seem to be that of Goldsmith and Jelly (67) using 1 to 5 MeV H^+ incident on Aℓ, Ag and Au. They monitor only the total intensity of the emission and find it to increase linearly with projectile energy; the observed photon yield is roughly consistent with theoretical computations (67). A very rough estimate of photon yield is that one proton at an energy of a few tens of KV will produce 10^{-9} photons in the visible and near u-v regions.

(2) Bremsstrahlung

Bremsstrahlung is generated when energetic particles move in the field of ion cores within a material and undergo scattering. The resultant radiation is a smooth contiuum but due to the fact that it emanates from within the target the observed spectrum is modified by the refractive and absorptive properties of the medium. The Bremsstrahlung will have both parallel and perpendicular polarization components with respect to the incoming beam's direction. Bremsstrahlung is readily observed for high energy electron impact on metals; it will accompany transition radiation and exhibit a similar intensity (68, 69). Separation of the two radiation components is performed through their characteristic polarization (68); unpolarized emission is Bremsstrahlung and the polarized component is transition radiation.

There are not published attempts to study Bremsstrahlung from non-relativistic heavy particles. The effect should occur but it will be very weak.

(3) Plasmons

The volume plasmon is a quantum of electron (or hole) density oscillation. The energy of oscillation is related to the "free electron" densities in the target.(70). A bounded electron gas can support natural oscillations in electron density at its surface; these are called surface plasmons. Plasmons are generally observed by studying energy loss in electrons backscattered from surfaces; they are well known for most common materials.

It is predicted that (71) surface plasmons in thin films should radiate; coupling to the external vacuum environment occurs via periodic surface irregularities. Some attempts have been made to excite surface plasmons by electron impact on Ag films; (72, 73) the predicted photon emission energy is 3.75 eV giving an emission wavelength of about 330 nm. Unfortunately this is not a clear cut case since silver has a high transmission band close to this wavelength and an intensity peak is also expected from bremsstrahlung. A situation of less ambiguity is $A\ell$ where decay of surface plasmons should occur at about 80nm; this emission has been observed by electron bombardment (74).

In general surface plasmons will radiate in the vacuum ultraviolet region. No attempts have been made to observe them by ion impact; the coupling of pojectile energy to electrons is inefficient for heavy projectiles due to the mass difference, therefore the intensity will be very low.

VI. RADIATION FROM SPUTTERED MOLECULES

Sputtering of a target gives rise to emissions from ejected molecules as well as from the ejected atoms (and ions) discussed earlier in Section I. Two types of molecular emission have been recognized with the distinctions being made principally from an empirical standpoint. Impact of heavy ions (or atoms) on certain transition metals (notably Mo, Ta, W and Nb) gives rise to broad band continuum emissions from a region extending many millimeters in front of the target (15, 39-43). This emission has no readily identifiable features and is not recognized as a spectrum normally observed in conventional spectroscopic sources; the spatial extent of the emitter implies a state with a lifetime considerably in excess of that expected from a normal electronically excited state (of the order 10^{-8} sec). In these cases there is first the problem of identifying the species before any attempt can be made to understand the mechanism leading to their creation. The other group of observations is of spectra emanating from a region close to the surface and exhibiting many features, such as lines and bands, that are found in conventional spectroscopic sources; (28, 29) the species giving rise to these spectra are often unambiguously identified by reference to published spectroscopic data and the limited spatial extent implies a conventional lifetime of the order 10^{-8} seconds.

For both types of emission the radiating species is identified as being an ejected molecule by the fact that the source of emission is located in front of the target. In a later discussion (Section VII) we shall consider certain molecular emissions that emanate from the surface itself and therefore result from excited molecules located on or below the surface.

A. Radiative Continua from Sputtered Molecules

This type of radiative continuum is produced when certain transition metal targets are impacted by low energy ions or neutral atoms (15, 39-43). Examples are shown in Figs. 21 and 22 for the case of Ar^+ (8 keV) impacting on Mo, Ta, W and Nb (15, 40). These three targets give rise to an intense broadband continuum emission extending over several thousand angstroms. In addition, in each case there are numerous atomic lines arising from excited states of sputtered neutral atoms. A spectral resolution of ∼ 8Å was used for the measurements shown in Figs. 21 and 22, but higher resolution scans over selected wavelength intervals show no more detailed structure down to a resolution of ∼ 0.75 Å (40). This strongly suggests that this radiation is not the result of the superposition of many atomic lines.

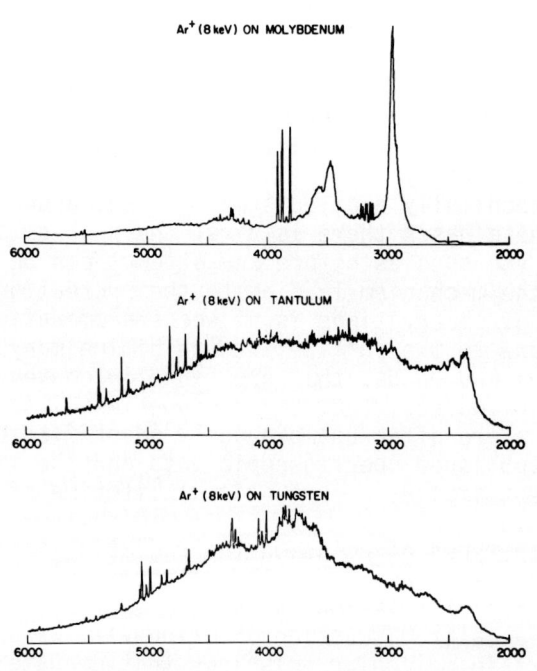

Fig. 21. Spectra of radiation produced by Ar^+ (8 keV) impact on Mo, Ta and W (From Ref. 40).

In the case of molybdenum with its very distinct continuum features, detailed measurements have been made of the spatial distribution of both continuum and line radiation as a function of distance in front of the target and these measurements are shown in Fig. 23 (40). As indicated in Fig. 23 two segments in the molybdenum radiative continuum

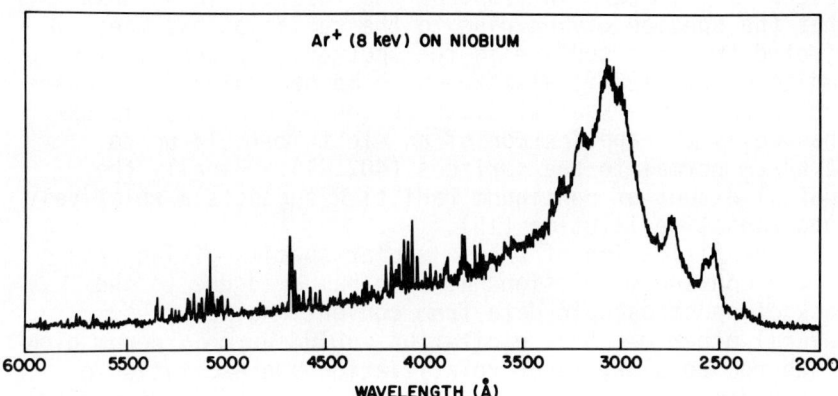

Fig. 22. Spectra of radiation produced by Ar^+ (8 keV) impact on Nb (From Ref. 15).

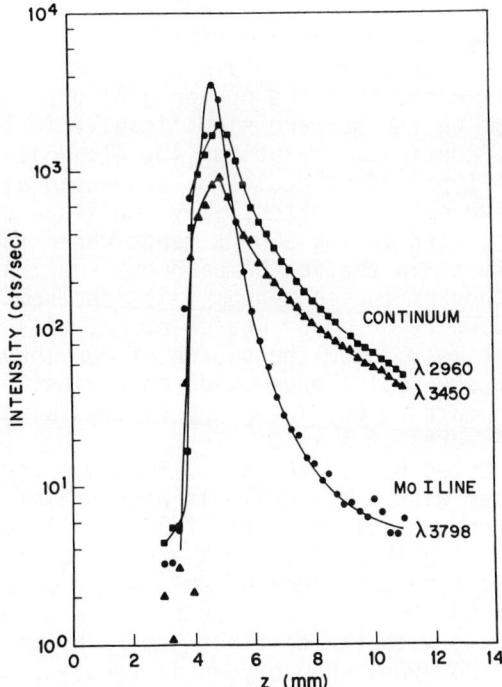

Fig. 23. Spatial distribution of radiation from Ar^+ (50 keV) impacting on Mo. The target surface is at approximately 5 mm. (From Ref. 40).

are observed with appreciable intensity to distances of several millimeters in front of the target. This shows that the species giving rise to the radiation has been ejected from the surface in the sputtering process. The radiating species is also known to be neutral and not charged, because the spatial distribution of continuum is unchanged by the application of an electric field up to ±4 kV/cm normal to the surfaces (40, 41). Finally the spatial extent of continuum radiation suggests a relatively long radiative lifetime (15).

Identification of the molecular species giving rise to the continuum emission has been hampered due to the lack of spectroscopic data from conventional sources. Several other means of excitation including molybdenum glow discharge results, laser volatiziation and excitation of a molybdenum target, and gas phase atomic collisions using Mo^+ and MoO^+ projectiles produce only well resolved line radiation with little or no evidence of continuum radiation in the wavelength region where significant continuum features appear in the ion bombardment case (40). The molecular species giving rise to continuum radiation may be composed of metal atoms only or it may contain one or more oxygen atoms since it is known that the presence of oxygen in the target chamber or on the surface significantly influences the intensity of continuum radiation (40, 41, 43). The spectra of these types of molecules by conventional means of excitation have not been extensiyely studied and little or nothing is reported in the 3000 Å range where the most distinctive features in the ion bombardment spectra appear.

A recent study of the continuum emission from molybdenum has been reported by Rausch et al. (43) which suggests that for molybdenum the source of continuum emission is sputtered molybdenum oxide molecules. Previous studies had been carried out in vacuum systems with residual gas pressures of 5×10^{-8} Torr or higher (40, 41), but the work reported by Rausch et al. (43) was performed in a vacuum system with a residual gas pressure of 3×10^{-9} Torr and an oxygen partial pressure of 5×10^{-11} Torr. The most significant results of that work are shown in Fig. 24. Spectrum 24a is for Mo bombarded by Ar^+ in the presence of 2×10^{-8} Torr O_2; the spectrum has the form noted previously (40, 41) with the distinct "sharp continuum" band at 296 nm. However, when oxygen is removed, spectrum 24b is observed; the continua are absent and the remaining features are identified as sputtered Mo atoms. Now if an MoO_3 target is bombarded (Fig. 24c) the same spectrum is observed as for Mo metal in the presence of O_2 (Fig. 24a). Figure 25 is the intensity of the 296 nm peak as a function

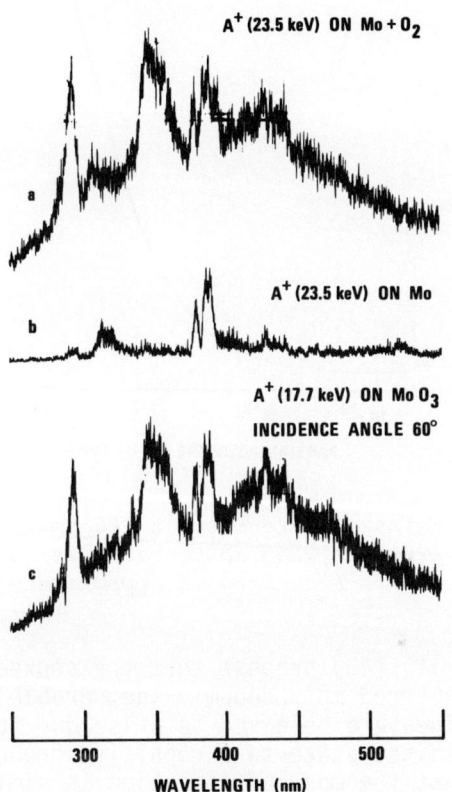

Fig. 24. Spectra recorded for Ar^+ impact on Mo and MoO_3 targets. (a) 23.5 keV Ar^+ on Mo in the presence of 2×10^{-8} torr of O_2. (b) 23.5 keV Ar^+ on Mo at chamber base pressure of 2×10^{-9} torr (oxygen partial pressure of 2×10^{-11} torr). (c) 17.7 keV Ar^+ on MoO_3 at chamber base pressure of 2×10^{-9} torr. (From Ref. 43).

of O_2 pressure when Mo is bombarded; the peak achieves a significant intensity at O_2 pressures above 10^{-9} Torr. The continuous spectra are also observed when Mo is bombarded with Ar^+ in the presence of CO; also when Mo is bombarded with O^+ with no ambient oxygen present. The general conclusion is that oxygen must be present as a sorbed species or as an incident beam, in order for the continuum to occur.

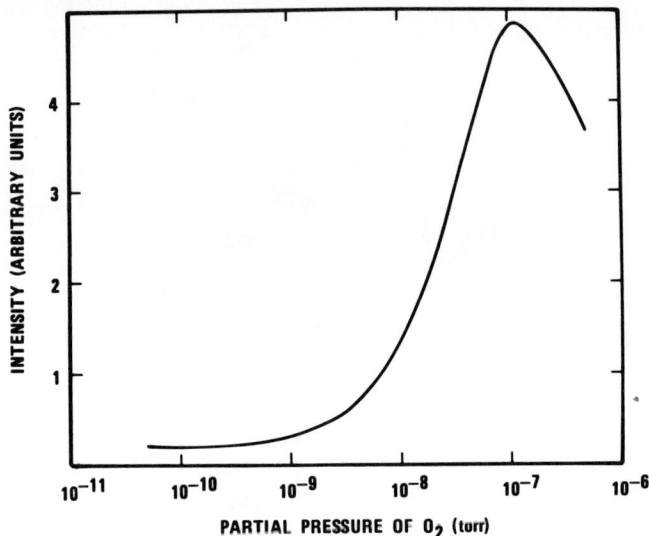

Fig. 25. Emission intensity at 296 nm as a function of oxygen partial pressure when an Mo taret is bombarded by 23 keV Ar^+ at a current density of approximately $10\mu A/cm^2$. (From Ref. 43).

Rausch et al. (43) propose that the source of the emission is sputtered molybdenum oxide, probably excited MoO. The continua are beleived to arise due to transitions from a bound excited state to a repulsive ground state. They suggest that the mode of formation is similar to that invoked by Können et al., (75) for metal dimer formation; Mo and sorbed O are separately sputtered by the impact of a single projectile and emerge with a sufficiently small relative velocity that they are bound. On the basis of this simple model the measured pressure dependence of the continuum intensity can be predicted (43).

The evidence for the emission being due to ejection of excited metal oxide can be be summarized as follows (43):
(1) Presence of oxygen is necessary.
(2) Pressure dependence of emission can be predicted with a model involving sputtering of Mo and sorbed oxygen. Analysis of the pressure dependence gives an oxygen sputtering coefficient of 0.4 atoms/Ar^+ ion incident.
(3) The time decay of continuum emission after oxygen is removed is consistent with the sputtering model if an oxygen sputtering coefficient of 0.4 atom/ion is used.

(4) The dependence of intensity on projectile beam current is consistent with the model.
(5) The continuum is observed also for MoO_3 targets.
(6) The large spatial extent of the emission is explicable if the emitter is a Mo atom combined with a metastable oxygen atom; such a state will be long lived.
(7) Published analyses of metal oxide spectrum conclude that the MoO spectrum is a continuum (76, 77). However, these analyses do not extend to the region below 3500 Å where the most distinctive features in the ion bombardment spectra appear.

Only limited attention has been given to other metals for which continua are observed but their behavior is also consistent with the proposition that these continuum also result from radiative decay of sputtered metal-oxide molecules.

In another experiment, (40) a single crystal of molybdenum was used for channeling studies and the intensity of both continuum and line radiation was measured in the random and channeling directions. Continuum intensity was found to decrease by less than 10% in the channeling direction, but the intensity of line radiation decreased by a factor of 2.2 in the channeling as compared to random directions. The fact that the line radiation shows a pronounced channeling effect implies that particles originating from a depth greater than one monolayer contribute to line radiation. The absence of any significant channeling effect for continuum radiation shows that particles giving rise to continuum radiation must originate much nearer to the surface than particles giving rise to atomic line radiation. Recent computer simulations of single crystal sputtering have shown (78) that dimers are formed principally from atoms which were next nearest neighbors on the crystal face and they are not formed by atoms moving in channeling directions. The above observations on the continua observed with single crystals of Mo are therefore completely similar to the predictions of a theory for dimer formation; this again lends support to the identification of MoO as the source of the continuum.

B. Line Spectra from Sputtered Molecules

G. E. Thomas (Philips Research Labs) has produced a number of studies of spectra from metal (Aℓ and Cu) and Si surfaces in the presence of reactive gases (28, 29). Fig. 26 shows the spectra observed when Aℓ is in the presence of O_2 and H_2O (29). Shown on the figure are positions of molecular bands expected from AℓH and from AℓO. There is some apparent correlation between the observed spectra and the expected position of the molecular bands. Studies were also made with Si surfaces carrying chemisorbed oxygen, nitric oxide, water, ammonia, methanol, and methyl mercaptan; the incident projectile was 10 keV Kr^+ ions (28). In the latter cases the observed spectra included features identifiable as due to SiH, OH, HN, CH and possibly also SiO and SiS. Although many features are identifiable there are distinct bands in all these studies that do not appear in conventional spectroscopic sources.

Introduction of a reactive gas onto the metal or Si surfaces produces some spectral features which may often be correlated with spectra of isolated molecules, observed in more conventional spectroscopic sources. However, with these ion induced emissions one does not get the sharp distinct spectral features that are expected from an isolated molecule.

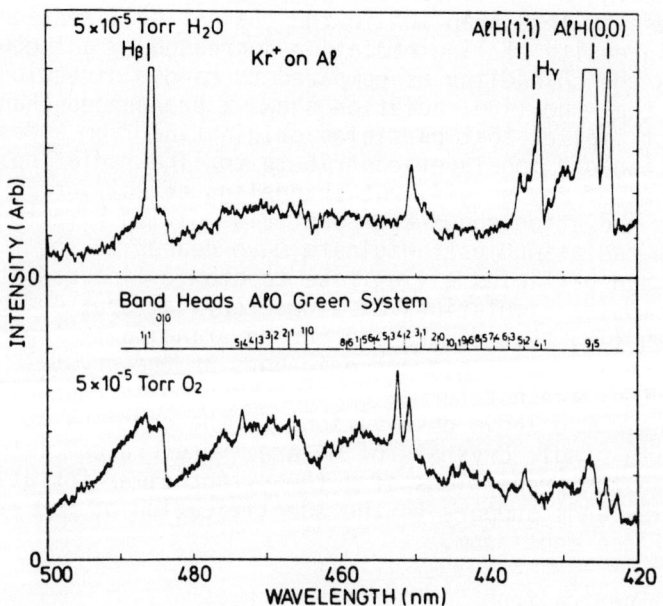

Fig. 26. Detailed scans of the long wavelength region of the Al spectrum with O_2 (bottom trace) and H_2O present. The band heads of the AlO green emission system are indicated on the O_2 spectrum. The length of the line showing the band-head wavelength is proportional to the intensity. (From Ref. 28).

There are two possible explanations for the absence of distinct features. First, there will of course be a Doppler broadening due to appreciable velocity of sputtered species. Secondly, it is quite possible that the excited molecules are in high rotational and vibrational states; this would lead to many lines unresolvable in the equipment used by Thomas. Studies using higher resolutions (preferably on systems of lesser spectral complexity) are needed to solve this problem.

VII. RADIATION FROM EXCITED MOLECULES ON THE SURFACE

There are two experiments where molecular emissions are observed from species on or below the surface. We shall discuss first the identification of the CN spectrum from impurity molecules or alkali halides. This spectrum exhibits regular line features associated with a bound molecular state. The second phenomenon is a broad luminescence observed in the vacuum u-v when rare gas ions are used to bombard metals. In this case there are no line features and the molecule must have a repulsive ground state; the source of emission is believed to be due to rare gas dimers formed in the target by implantation of the bombarding species.

A. The CN Spectrum on Alkali Halides

In studying luminescence of alkali halides under H^+ and He^+ bombardment Bazhin et al., (30) observe a series of regularly spaced broad lines shown in Fig. 27. The bands in the region 200 to 350 nm are spaced equally in energy at about 0.25 eV. The background pressure in the target chamber was 10^{-9} torr. These bands are generally found only on a "new" crystal that has been recently exposed to air and/or handled. On heating the sample, or sputter cleaning with Ar^+, the bands go away and one gets a spectrum like line b. These bands have all the characteristics of a diatomic molecular vibrational spectrum; the only simple molecule with a vibrational spacing of 0.25 eV is CN. To test whether CN is the origin of the band Bazhin et at., deliberately introduced various molecular gases; cyanogen caused the bands to increase markedly in intensity but other gases (H_2, N_2, O_2, CO_2) had no effect. Doping the crystal with CN (e.g. adding NaCN to $NaC\ell$) also gave these bands. Identical bands have been seen in the u-v induced fluorescence of NaBr doped with NaCN (79).

These bands are identified as the $D^2\pi \rightarrow X^2\Sigma^+$ transition of CN. In discharges there have been observations of bands

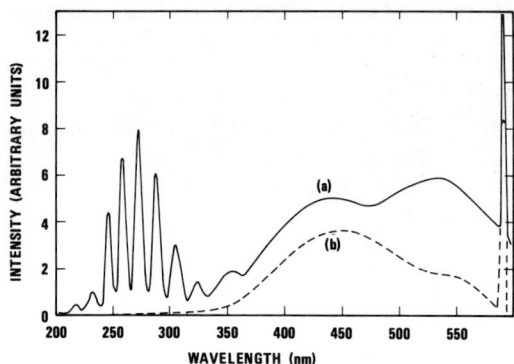

Fig. 27. Luminescent spectra induced by impact of 20 keV He^+ on NaCl at 26°C. Line (a) is for a crystal exposed to H_2O at 10^{-3} Torr and is typical also for all samples which have been handled in air. Line (b) is for the same sample after subsequent annealing to 450°C. (From Ref. 30).

at 288 and 274 nm ascribed (80) to the $0 \to 10$ and $0 \to 9$ transitions of CN and which correspond to two of the lines in this spectrum. The observed lines are very wide leading to the conclusion that their source is not a free CN radical but rather a CN molecule bound to the surface.

Bazhin et al., believe that the CN is excited by energy from recombination of exitons (electron-hole pairs); this will give an excitation energy approximately equal to the band gap of the crystal (30). From the known characteristics of CN it is estimated that the required excitation energy to give the observed lines is about 7 eV. For KBr the band gap is only 6.71 eV; for this case the CN bands are not generated. For all other cases considered (LiCl, NaCl, KCl) the band gap is 7.8 eV or higher and the CN emission is observed.

There remains the question as to how the CN is formed. One certainly expects C to be a contaminant and also perhaps N or nitrogen compounds. It is known that irradiation of gas phase N_2-hydrocarbon mixtures with u-v light or electrons causes formation of HCN; a similar reaction may be occuring on the crystal surface during ion bombardment. To further understand the mechanism leading to CN formation it clearly is desirable to perform studies with simultaneous monitoring of surface composition by Auger or photoelectron spectroscopy.

B. Rare Gas Dimers

Radiative continuum from well-known excited rare gas molecules has been observed in two very recent experiments (44, 81) in the vacuum ultraviolet spectral region when noble gas ions bombarded a variety of targets. Fig. 28 shows results which were obtained by bombarding Al and Si

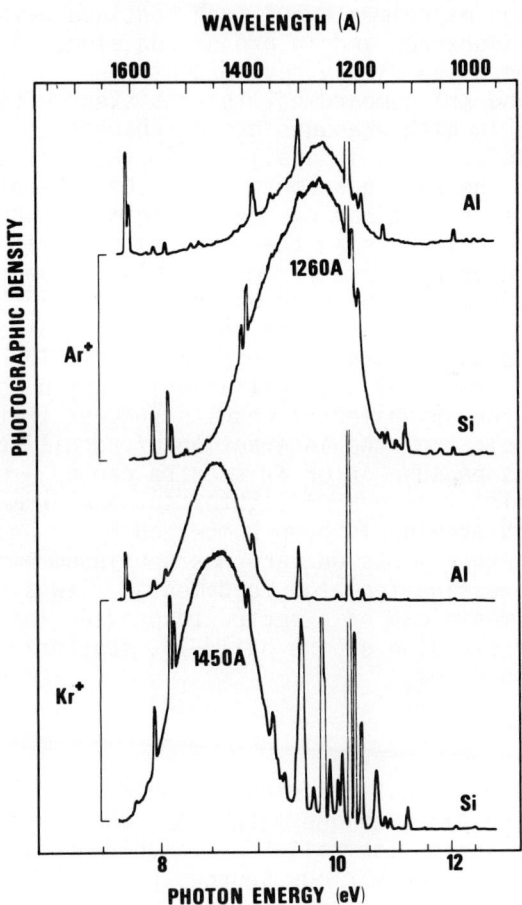

Fig. 28. Spectra of radiation observed in the impact of Ar^+ (100 keV) and Kr^+ (100 keV) on Al and Si (From Ref. 44).

targets with Ar^+ and Kr^+ ions at an energy of 100 keV (44).
The shape and wavelength of these continua depend on the
identity of the impinging ion and not the target species.
The continua position and general shape are strikingly similar to continua observed from Ar and Kr excited by various
other means. They arise due to transitions from transient,
bound, excited (excimer) states of noble gas molecules to
the separated-atom repulsive ground state. These rare
gas continua have not previously been reported in ion-
solid collision experiments but such continua have been
seen in gas discharges and in proton and electron bombardment
of noble gases. The singly peaked continuum spectrum observed for Ar^+ and Kr^+ bombardment is consistent with the spectrum produced in high pressure gas discharges.

Little information is available on the excitation
mechanism for the rare gas continua in the ion-solid
collision case, and it has not been determined whether
these continua originate in the bulk of the solid or from
excited molecules sputtered from the solid. However,
sputtering is known to play an essential role in that the
rare gas continua are not observed until the target has
been sputtered to a depth which is in reasonable correspondence with the range of the initially implanted ions (44).
In addition, the experimental results suggest that these
emissions result from the interaction of recoiling earlier
implanted ions because Ar or Kr spectra can be produced
during the initial stages of lighter ion bombardment if
the target has previously been bombarded by Ar and Kr (44).
These recent experiments on rare gas continuum emission
are very interesting from the fundamental viewpoint and
future experiments can be expected to provide insight
into the understanding of the build-up, sputtering and
excitation processes.

VIII. CONCLUSION

Studies of optical emission produced in low energy ion-solid collisions is a unique method to study fundamental
inelastic low-energy ion-solid collision processes. Radiation arising from the various sources shows that the participants in the collision process including sputtered particles, reflected projectiles, species excited on the surface,
and the solid itself, can be left in excited states following
the collision. Since a significant portion of these excited states decay by the emission of radiation, much
can be learned about the detailed processes of the collision
as well as about the interaction of excited particles.
with the solid and the nature of the coupling between the

surface and excited states of near by particles. Future optical emission experiments can be expected to provide more detailed information on the various excitation mechanisims, the atomic processes which influence the ejection of positive ions, negative ions, and excited neutral species, and fundamental sputtering mechanisms. These experiments should contribute greatly to our basic understanding of inelastic low-energy ion-solid collision phenomena.

I. REFERENCES

1. Tolk, N. H., Simms, D. L., Foley, E. B., and White, C. W., Radiation Effects 18, 221 (1973).
2. White, C. W., and Tolk, N. H., Phys. Rev. Letters 26, 486 (1971).
3. White, C. W., Simms, D. L., and Tolk, N. H., Science 177, 481 (1972).
4. Fluit, J. M., Friedman, L., Van Eck, J., Snoek and Kistemaker, J., Proc. of the Fifth International Conf. on Ionization Phenomena in Gases, ed. by Maecker, H., North Holland, Amsterdam (1962).
5. Snoek, C., Van der Weg, W. F., and Rol, P. K., Physica 30, 341 (1964).
6. Terzic, I., and Perovic, B., Surface Sci. 21, 86 (1970).
7. Meriaux, J. P., Guttierrez, J. M., Schneider, Ch., Goutte, R., and Guilland Cl., Nouv. Rev. d'Optique Appliquee 2, 81 (1971)
8. Tsong, I. S. T., Phys. Stat. Solidi A 7, 451 (1971)
9. Kerkow, H., Phys. Stat. Solidi A 10, 501 (1972).
10. Braun, M., Emmoth, B., and Martinson, I. Physica Scripta 10, 133 (1974).
11. White, C. W., Simms, D. L., and Tolk, N. H., Characterization of Solid Surfaces, ed. by Kane, P. F. and Larrabee, G. R., Plenum Press, New York, Chap. 23 (1974)
12. Van der Weg, W. F., and Lugujjo, E., Atomic Collisions in Solids, p. 511, ed. by Datz, S., et al. Plenum Press, New York (1975)
13. Kyan, T. S., Gritsyna, V. V., Fogel´, Nucl. Inst. and Methods 132, 435 (1976).
14. Bayly, A. R., Martin, P. J., MacDonald, R. J., Nucl. Inst. and Methods 132, 459 (1976).
15. White, C. W. and Tolk, N. H., J. Nucl. Materials (in press).
16. Tolk, N. H., Tsong, I. S. T. and White, C. W. Analytical Chemistry 19, 16A (1977).

17. Gritsyna, V. V., Kijan, T. S., Koval´, A. G., Fogel´, Ya. M., Phys. Letters 27 A 292 (1968).
18. McCracken, G. M., and Erents, S. K., Phys. Letters 31 A, 429 (1970)
19. Kerkdijk, C., and Thomas, E. W., Physica 63, 577 (1973).
20. Kerkdijk, C. B., and Thomas, E. W., Radiation Effects 18, 241 (1973).
21. Barid, W. E., Zivitz, M., Larsen, J., and Thomas, E. W., Phys. Rev. A 10, 2063 (1974)
22. Barid, W. E., Zivitz, M. and Thomas, E. W., Phys. Rev. A 12, 876 (1975).
23. Baird, W. E., Zivitz, M. and Thomas, E. W., Nucl. Inst. and Methods 132, 445 (1976)
24. Leung, S. Y., Tolk, N. H., Heiland, W., Tully, J. C., Kraus, J. S., and Hill, P., To be published.
25. Chaudhri, R. M., Chaudhri, M. M. Kabir, S. M., and Rafigue, M., Proc. of the Seventh International Conf. on Phenomena in Ionized Gases, Vol. 1, ed. by Perovic, B. and Tosic D. (Gradevinska Knjiga Publ. House, 1966)
26. Zivitz, M., and Thomas, E. W., Nucl. Inst. and Methods 132, 411 (1976)
27. Bazhin, A. I., Rausch, E. O., Thomas, E. W., Phys. Rev. B 14, 2583 (1976)
28. Thomas, G. E., de Kluizenaar, E. E. and Beerlage, M., Chem. Phys. 7, 303 (1975).
29. Thomas, G. E., and de Kluizenaar, E. E., Int. J. Mass. Spectrom. and Ion Phys. 15, 165 (1974)
30. Bazhin, A. I., Rausch, E. O., Thomas, E. W., J. Chem. Phys. 65, 3897 (1976)
31. Van der Weg, W. F., Tolk, N. H. White, C. E., and Kraus, J. M., Nucl. Inst. and Methods 132, 405 (1976)
32. Van der Weg, W. F., Bierman, D. J., Physica 44, 206 (1969)
33. Gritsyna, V. V., Kijan, T. S., Goutte, R., Koval´, A. G., and Fogel´, Ya.M., Bull. Acad. Sciences USSR 35, 530 (1971)
34. Gritsyna, V. V., Kijan, T. S., Koval´, A. G. and Fogel´, Ya. M., Radiation Effects 14, 77 (1972).
35. Koval´, G. A., Vyagin, G. I., Bobkov, V. V., Klimovskii, Yu. A., Strel´chenko, S. S. and Fogel´, Ya. M., Zh. Tekh. Fiz. 43, 1753 (1974)
36. Kelly, R. and Kerkdijk, C. B., Surface Sci. 46, 537 (1974)
37. White, C. W., Simms, D. L., Tolk, N. H., and McCaughan, D. V., Surface Sci. 49, 657 (1975)

38. Kerkdijk, C. B., Thesis, Leiden, Chapter 4 (1975).
39. Tolk, N. H., White, C. W., and Sigmund P., Bull. Am. Phys. Soc. 18, 686 (1973)
40. White, C. W., Tolk, N. H., Krauss, J. and Van der Weg, W. F., Nucl. Inst. and Methods 132, 419 (1976)
41. Kerkdijk, C. B., Schartner, K. H., Kelly, R., and Saris, F. W., Nucl. Inst. and Methods 132, 427
42. Kiyan, T. S., Gritsyna, V. V., and Fogel´, Ya. M., Nucl. Inst. and Methods 132, 415 (1976)
43. Rausch, E. O., Bazhin, A. I., and Thomas, E. W., J. Chem. Phys. 65, 4447 (1976)
44. Hill, K. W., Comas, J., Nagel, D. J., and Knudson, A. R., submitted for publication.
45. Tsong, I. S. T. and McLaren, A. C., Nature 248, 43 (1974).
46. Martin, P. J., and MacDonald, R. J., submitted for publication. Communicated to the conference by Kelly, J.
47. Van der Weg, W. F., Proceedings of Summer School on Materials Characterization Using Ion Beams, Corsica, 1976, to be published by Plenum Press.
48. Thomas, E. W., Excitation in Heavy Particle Collisions (Wiley Interscience, N.Y. 1972)
49. Hagstrum, H., Phys. Rev. 96, 336 (1954)
50. Van Der Weg, W. F., and Rol., P. K., Nucl. Inst. and Methods 38, 274 (1966).
51. Jesen, K., and Veje, E., Zeitschr. f. Physik (to be published).
52. Thomas, G. E. and de Kluizenaar, E. E., Nucl. Inst. and Methods 132, 449 (1976)
53. Thomas, G. E., to published.
54. Bayly, Ar. R., Martin, P. J., and MacDonald, R. J., Nucl. Inst. and Methods 132, 459 (1976)
55. McCaracken, G. M., and Freeman, N. J., J. Phys B 2, 661 (1969)
56. Olander, D. R., Kerkdijk, C. B., and Smits, C., Surface Sci. 49, 28 (1975)
57. Kerkdijk, C. B., Smits, C. M., Olander, D. R., and Saris, F. W., Surface Sci. 49, 45 (1975)
58. Cobas, A. and Lamb, W. E., Phys. Rev. 65, 327 (1944)
59. Rausch, E. O., and Thomas, E. W., Phys. Rev. A, 14 1912 (1976)
60. Rausch, E. O., and Thomas, E. W., (to be published).
61. Petroff, I. and Viswanathan, C. R., Phys. Rev., B 4 799 (1971).
62. Verbeek, H., Eckstein, W., and Datz, S., J. App. Phys. 47, 1785 (1976).

63. Seretlo, J. R., *Phys. Stat. Sol.* (a) 10, 639 (1972).
64. Christy, R. W., Phelps, D. H., *Phys. Rev.*, 124, 1053 (1961)
65. Zivitz, M. and Thomas, E. W., *Phys. Rev.* B 13, 2747 (1976)
66. Papanicolaou, B. G., Chen, J. M., Papageongopoulos, C. A., *J. Phys. Chem. Solids* 37, 403 (1976)
67. Goldsmith, P., and Jelley, J. V., *Philos. Mag.* 4, 836 (1959).
68. DiNardo, R. P., and Goland, A. N., *Phys. Rev.* B 4, 1700 (1971)
69. Arakawa, E. T., Emerson, L. C., Hammer, D. C., and Birkhoff, R. D., *Phys. Rev.* 131, 719 (1963)
70. Ritchie, R. N., *Surface Sci.* 34, 1 (1973).
71. Ferrell, R. A., *Phys. Rev.* 111, 1214 (1958)
73. Boersch, H., Dobberstein, P., Fritzsche, D., and Sauerbrey, G., *Zeits fur Physik* 187, 97 (1965).
74. Arakawa, E. T., Merickhoff, R. J., and Birkhoff, R.D., *Phys. Rev. Letts.* 12, 319 (1964)
75. Konnen, G. P., Tip, A., and de Vries, A. E., *Radiation Effects* 21, 269 (1974), and *Radiation Effects* 26, 23 (1975).
76. Gatterer, A., Junkes, J., and Salpeter, W., "Molecular Spectra of Metallic Oxides," Specola Vaticana, Citta del Vaticano (1957).
77. Hartley, W. N., and Ramage, H., *Trans. Dublin Soc.* (2) 7, 339 (1901).
78. Harrison, D. E., and Delaplain, C.B., *J. Appl. Phys.*, 47, 2252 (1976).
79. Von der Heyden, E., and Fisher, F., *Phys. Stat. Solidi* B 69, 63 (1975)
80. Douglas, A. E., and Rontty, *Astrophys. J. Suppl.* 1, 295 (1955).
81. Braun, M. and Emmoth, B., private communication.
82. I. S. T. Tsong, *Spectrochim. Acta B*, to be published.

electron pickup by fast ions in solids

M. C. Cross
*Bell Laboratories**

The theory of electron pickup by fast protons in solids is reviewed. The important physics involved has not been well discussed in the literature. It is shown here that the charge states for high energy protons may be simply understood in terms of capture into and subsequent loss from bound states on the proton within the solid. Collective screening and the short lifetime of the bound state are not important. The surface does not play an important role. For high enough proton velocities, capture and loss cross-sections for single atoms may be directly transferred to the solid. The solid behaves as a gas scaled to higher densities. A comparison with other theories is given, and the intermediate energy range is briefly discussed.

I. INTRODUCTION

The study of charge changing processes of fast, heavy ions passing through a solid or gas has long been of interest. It developed from the importance of the effective charge of an ion as a parameter in the energy loss, and consequent stopping, of fission fragments in matter. Although all the details of these processes may still not be understood the basic effects and the formalism for treating the charge state seem clear. It is also apparent that the complexity of the problem - in particular the many excited states available to the outer electrons of the ion, and the possibility of inner shell vacancies - makes a first principles theoretical calculation impractical. The review of Betz (1) gives a good guide to the current state of the art. As a solid-state physicist approaching the subject from the point of view of gaining information about the solid, I shall therefore concentrate on the charge state or protons interacting with solids.

---

*Postdoctoral research fellow.

It was immediately striking, on investigating the recent literature, that the concepts involved in electron pickup by protons are less well developed than for heavy ions: the basic questions of where the electrons come from and how the neutral hydrogen is formed have not been convincingly answered. I shall briefly review the theories that have been published; most of these, it will be seen, do not contain the important physics involved. This article is therefore largely an attempt to answer the basic questions of electron pickup by fast protons. Of necessity, any discussion of fascinating but more complicated phenomena must be omitted. Fortunately, other speakers at this workshop will cover some of these topics. In particular, a thorough understanding of the spin polarized electron pickup studied recently by Rau (see these proceedings), but in particular by Kaminsky (2) in channeled transmission experiments, may throw important light on the electron pickup process. These experiments may offer a unique opportunity to study where the captured electrons come from.

The types of experiment we are trying to understand are illustrated in Fig. 1. At intermediate energies (10-200 keV)

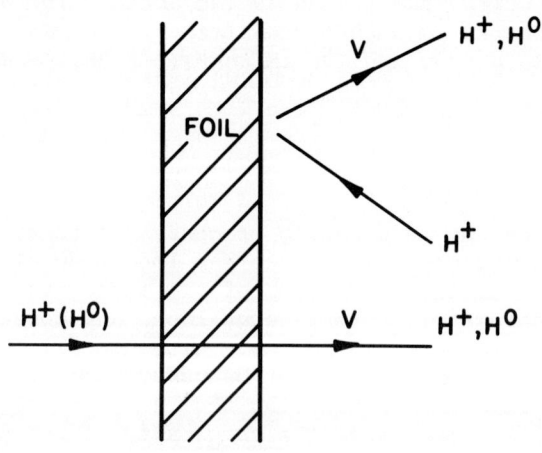

Fig. 1. *Experiments on charge fractions.*

protons are fired at a solid, and the charge fractions (positives, neutrals and a few negatives) of either transmitted or reflected beam are measured as a function of exit velocity. These experiments have been reviewed by Buck (these proceedings). At higher energies transmission experiments with incident protons have been performed for energies 300 keV to 16 MeV (3,4), and recently by Poizat and Remillieux

with incident neutrals (unpublished). The dividing energy at
about 200 keV between the energy ranges is not sharp, but
will turn out to be convenient.

To introduce the theoretical concepts involved, it is useful
here to review the formalism used in the heavy ion problem as
it would be applied to a hydrogen beam in a solid. For fast
protons we may neglect the possibility of negative ions. If
in addition we neglect the complication of excited states and
excitation processes, then for a beam of N_p protons and N_o
neutrals the differential equation for the rate of change of
N_o along the beam direction is:

$$\frac{dN_o}{dx} = (N_p \sigma_c - N_o \sigma_\ell) n \qquad (1)$$

Here σ_c (σ_ℓ) is the capture (loss) cross-section per target
atom (of density n per volume) for a bound state around the
proton. Equation 1 simply describes the change in the number
of neutrals as the difference between the number of capture
and loss events. It would apply both to the approach to
equilibrium, and to the dynamic equilibrium given by setting
dN_o/dx to zero. The equilibrium neutral fraction $N_o/(N_p+N_o)$
is then

$$\Phi_o = \frac{1}{1 + \sigma_\ell/\sigma_c} \qquad (2)$$

Recent theories (5,6,7,8) have however started from an opposite statement: No bound state exists on a proton in a solid.
Capture processes at the surface then become essential. These
theories are based on two assumptions:

(1) For low velocities $v \lesssim v_o$ (v_o is the Bohr velocity
2.2×10^8 cm/sec, corresponding to a proton energy 25 keV)
the collective screening of the ion potential by the valence
electrons does not permit a bound state.

(2) For higher velocities, v greater than a few v_o:
"Collision broadening makes the bound state unstable" (9).

Although we are mainly interested in the region covered by
the second statement, it will be useful to look briefly at
screening in solids. This is done in section II. As might be
expected, for v greater than a few v_o collective screening is
not important in the electron capture problem.

In section III we return to the second statement above. The
central conclusion of this article is that for fast protons
this too is not important. The charge state of protons may
be treated by the formalism of capture into and loss from

bound states and equations (1) and (2) do indeed describe the charge states of fast protons in solids.

Section IV reviews other theories, and section V briefly discusses the intermediate energy range, where no detailed theoretical predictions have been made, and the methods to be used to make such predictions remain uncertain.

II SCREENING IN SOLIDS

An ion placed in a solid distorts the surrounding electron distribution. This in turn modifies the electrostatic potential felt by an electron, and hence may change the possibility of a bound state existing on the ion. In the usual dielectric screening theory the effect of the solid is described by a wavevector and frequency dependent dielectric function $\varepsilon(q,\omega)$ (10). The resultant potential ϕ_{tot} around an ion in a solid is then

$$\phi_{tot}(q,\omega) = \frac{1}{\varepsilon(q,\omega)} \phi_{ext}(q,\omega) \qquad (3)$$

where $\phi_{ext}(q,\omega)$ is the Fourier transform of the bare Coulomb potential. Similarly, the induced charge density ρ_{ind} is related to the ion point charge density, after Fourier transforming, by

$$\rho_{ind}(q,\omega) = \left[\frac{1}{\varepsilon(q,\omega)} - 1\right]\rho_{ext}(q,\omega) \qquad (4)$$

ϕ_{tot} is the electrostatic potential due to the total charge density $\rho_{ind} + \rho_{ext}$.

These results depend on an important assumption: that the perturbation of the solid by the ion is small, and only the first order perturbation need be calculated ("linear response theory"). A necessary condition for this to be valid is clearly that the induced charge density ρ_{ind} should be everywhere small compared with the unperturbed electron density. With this assumption in mind, we may consider cases of interest.

A. Slow Ions, $v \lesssim v_o$

For slow ions the static dielectric function $\varepsilon(q,\omega)$ may be used.

The simplest approximation for a nearly free electron metal (we may take the alkali metals or aluminum as good examples),

$\varepsilon(q) = 1 + \lambda^2/q^2$ gives the well known Thomas-Fermi screening (10): the potential around an ion charge Z is then

$$\phi(r) = \frac{Ze}{r} e^{-\lambda r} \qquad (5)$$

For typical metal conduction electron densities the Thomas-Fermi screening length λ^{-1} is between 0.9 and 1.5 Bohr radii. The potential Eq. 5 for an ion charge $Z = 1$ would then be sufficiently strong to bind an electron in free space (see Ref. 9 and references therein).

The potential around an ion in an insulator, on a similar level of approximation, takes the form:

$$\phi(r) = \frac{Ze}{\varepsilon r} \qquad (6)$$

with ε the static dielectric constant. The binding of an electron in this model is reduced by a factor ε^2 from the hydrogen atom result. ε is large in semiconductors (Si:12, Ge:16, C:5) but smaller in large bandgap insulators.

What do the results Eqs. 5,6 tell us about binding on a proton in a solid? In the models leading to these equations the potentials $\phi(r)$ must be added to the mean potential of the crystal, and so would predict a weakly bound state pulled below the conduction bands, and therefore stable. It is clear that the exchange interaction gives an additional attractive interaction to the region of enhanced electron density around the proton, and increases the binding energy.

However, the existence of a bound state, even on a stationary proton, is a difficult problem involving the self-consistent interaction of many electrons over distances comparable to the atomic structure of the solid. The above arguments are not adequate to determine with confidence the presence or absence of a stable bound state. Friedel (11) from simple arguments has suggested a different criterion for metals, which should also be applicable to those metals (e.g., transition metals) which are not "nearly free electron" like. He simply compares the energy E_c of the bottom of the conduction band measured from the vacuum level, with the free space hydrogen atom binding energy. For $|E_c| < 13.6$ eV he suggests a bound state should exist. With more modern knowledge of E_c this predicts a bound state in common metals. Again the validity of such a simple model to a complicated problem remains in doubt. Experiments and more detailed calculations on hydrogen dissolved in transition metals (e.g., Nb, Pd) look likely to settle the question for these metals, but as yet seem contradictory (12).

It is worth emphasizing that the simple models do predict a bound state[a]: it cannot yet be assumed that bound states do not exist on slow protons in solids. Furthermore, it is not obvious that the absence of a <u>stable</u> bound state would necessarily imply that electron pickup by a slow ion is from the "tail" of the electron distribution at the surface, with no influence from the state of the electrons around the ion in the bulk solid. Linear response theory is certainly not valid for a unit charge at typical metal electron densities. If we think of linear dielectric screening and screening by a bound state as opposite extremes, the true picture may be somewhere in between. The electron state in the bulk solid may then influence the emerging neutral fraction.

B. Fast Ions $v \gtrsim v_o$

The time dependence of the perturbation must now be taken into account, and the full dielectric function $\varepsilon(q,\omega)$ must be used. Much recent work has used the "plasmon pole" approximation (13)

$$\varepsilon(q,\omega) = (1 - \omega_p^2/\omega^2) \quad \textit{for all wavevectors} \qquad (6)$$

This assumes that the plasmon, at a single frequency ω_p, is the dominant excitation for all wavevectors. Such an approximation must break down for short wavelengths of atomic dimensions (wavevector q_c say), where collective effects no longer occur. The validity of Eq. 6 at <u>small</u> wavevectors, in particular the dominance of a single frequency ω_p, can be judged from plots of the "loss function", $-\text{Im } \varepsilon^{-1}(o,\omega)$ from electron loss or optical measurements (14,15,16). If Eq. 6 is good there is a single peak at $\omega = \omega_p$. Some examples are shown in Fig. 2. It may be seen (14) that the validity of Eq. 6 at small wavevectors does not depend on the conduction properties of the material, but rather on the absence of interband transitions near the plasma frequency ω_p. In some "good" metals, e.g., Cu, interband transitions from d states mix with the plasma oscillation to give a broad range of frequency responses. Many semiconductors, on the other hand, show well defined plasma oscillations.

---

[a] This disagrees with the conclusion of Ref. 9, which is based on incorrect results or doubtful arguments: - His Ref. 15 includes the exchange interaction only in calculating the screening charge and neglects the important attractive exchange interaction with this charge. The argument of his Ref. 16 is not explained. The conclusion of his Ref. 17 is based on a too large E_c for copper.

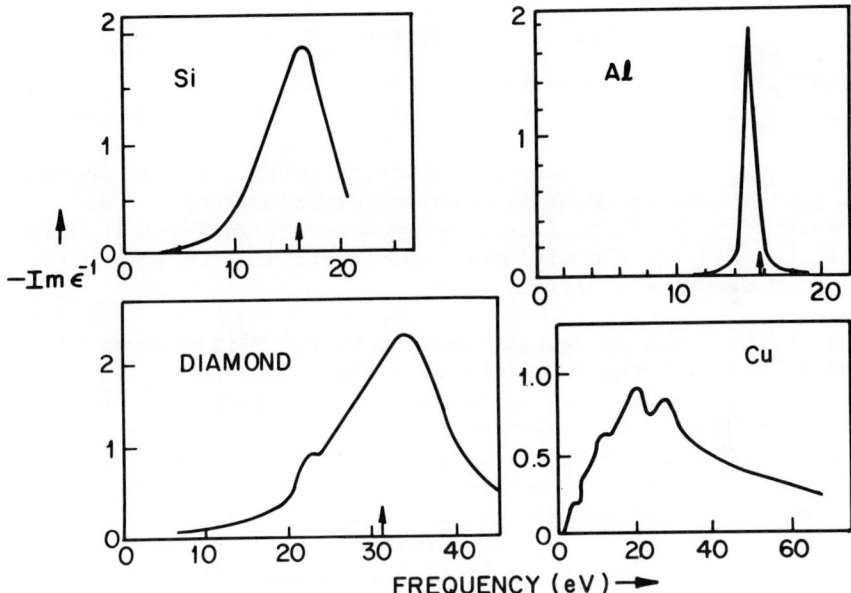

Fig. 2. *Dielectric loss function of selected elements. The arrow on the frequency abcissa is the plasma frequency calculated for the valence electrons treated as free electrons.*

It is easy to calculate the induced charge density with the approximation Eq. 6. The result is a <u>line</u> charge behind the ion of charge per unit length oscillating with a wavelength $2\pi v/\omega_p$ (see Fig. 3).

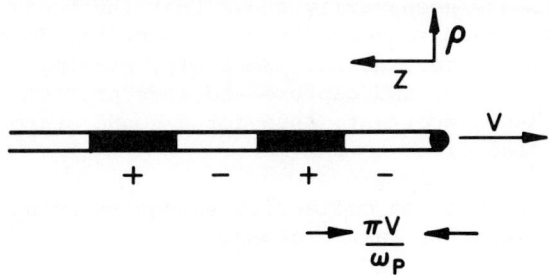

Fig. 3. *Screening charge behind a fast ion in a solid.*

The charge density is:

$$\rho_{ind} = -Ze\frac{\omega_p}{v}\sin\frac{\omega_p z}{v}\delta_2(\rho) \qquad (7)$$

with $\delta_2(\rho)$ the two dimensional delta function. The potential ϕ_{tot} is then the sum of the potentials due to the "screening" charge Eq. 7 and the ion charge. The number of well defined oscillations in the wake and resultant potential will depend in part on the range of frequencies of importance at zero wavevector: if there is a wide range of frequencies involved (cf. Cu, Fig. 2) the different components will interfere to rapidly damp the oscillation.

The infinite charge density implied by Eq. 7 is unphysical, and arises from the oversimplified form Eq. 6 neglecting any dispersion. As a first approximation this unphysical result may be corrected by smearing the line charge near the ion over a transverse distance q_c^{-1}. The important result to us, that the collective screening charge is "left behind" a distance v/ω_p, which becomes many Bohr radii for $v \gg v_o$, should be independent of these details. Screening is then unimportant in considering bound states about the ion.

Bound states in the potential of the oscillating wake require a much more careful treatment: the transverse charge distribution here becomes very important. The large binding energy originally suggested (17) arose from the logarithmic divergence of the binding potential due to the unphysical infinite charge density. It has been shown (18) using a semiclassical approximation to the dielectric function, including dispersion, that the binding energy in the wake behind a proton is for practical purposes, zero. The calculated binding energy behind more highly charged ions is significantly reduced. In any case, a prediction of a bound state based on linear screening arguments must be viewed with caution. Firstly, such a bound state necessarily shows that the fundamental assumption of linear response theory has broken down, and the conclusions cannot be trusted. Secondly, binding in such a state will be weaker, and capture and loss processes will make such a state less important, than for a bound state in the Coulomb potential close to the ion.

We may conclude that the collective screening is not important for electron pickup by fast protons.

III. BOUND STATES, CAPTURE AND LOSS

As screening is unimportant for velocities greater than a few

v_o the suggested absence of a bound state on fast protons must depend on the second statement on page 266. We now look at this in greater detail.

It is a truism that due to loss processes, an electron state around a fast proton in a solid has a finite lifetime, and in this sense is unstable. This is equally true for the electrons involved in charge exchange on heavy ions. There is one way, however, that protons may be thought a special case: using typical single atom loss cross-sections and a density of scattering centres appropriate to a solid it turns out that the lifetime of the electron state may be so short as to be comparable to the reciprocal of the binding energy of the state (at least for proton energies up to a few MeV). It is presumably in this sense the absence of a bound state is meant: the uncertainty in the energy due to the short lifetime is comparable to the binding energy. It will be shown that this is not in fact important, and that the charge states of protons, as of heavy ions, may be calculated in terms of capture into and loss from bound states.

A. Experimental Motivation

First consider two experimental plots that promote this idea.

Figure 4 shows a charge state ratio for ions (nuclear charge Z) from protons to neon emerging from aluminum foil, as a function of the velocity scaled to the K shell velocity Zv_o. Φ_Z is the fraction of bare nuclei; Φ_{Z-1} the fraction with a single captured electron. The points are the heavier ions: B, C, N, O, F, Ne; the line is hydrogen {see the original plot (19) for more details}. The success of such scaling to both hydrogen and the heavier ions suggests protons in solids are not to be treated in an essentially different way from the heavier ions.

Figure 5 is a plot of the neutral hydrogen fraction emerging from solid carbon as a function of exit energy per nucleon E. The heavy lines are from experiments on protons at high energies (3) and on deuterons at low energies (20). The dashed line is from measurements of protons on "dirty gold" (21): at the lower energies these are found to be identical to carbon data (20). The points are calculated using Eq. 2:

$$\Phi_0 = \left(1 + \frac{\sigma_\ell}{\sigma_c}\right)^{-1}$$

where σ_ℓ, σ_c are cross-sections measured for single carbon atoms (22). These were found by measuring cross-sections for

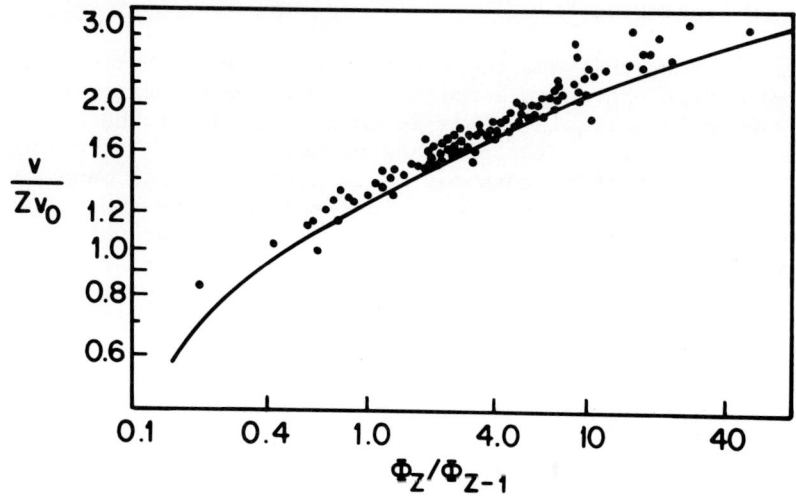

Fig. 4. Ratio of fraction of bare nuclei Φ_z to fraction of nuclei with one bound electron Φ_{z-1} emerging from aluminum for nuclear charges z from 1 to 10, as a function of velocity scaled to the K shell velocity. The points are for elements B, C, N, O, F, Ne; the line is hydrogen. (After Northcliffe: Ref. 19.)

the gaseous oxides and four hydrocarbons, and then subtracting the appropriate number of oxygen or hydrogen cross-sections. The agreement between the two sets of data is self-evident. (The error bars shown are estimated from those in reference (22) for the individual cross-sections. A large part of the scatter there may be seen to come from the oxide measurements: the oxygen cross-section dominates and the total cross-section provides a poor measurement of the carbon cross-section. The hydrocarbon data give much less scatter.)

B. Theoretical Basis

How may we understand the obvious success of Eq. 2 in describing charge states of fast protons in solids? I put forward the following simple theory:

1. The charge state is produced by capture into and loss from bound states (discrete hydrogen atom eigenfunctions) on the moving proton with cross-sections for the processes σ_c and σ_ℓ. The approach to equilibrium is described by Eq. 1 and the

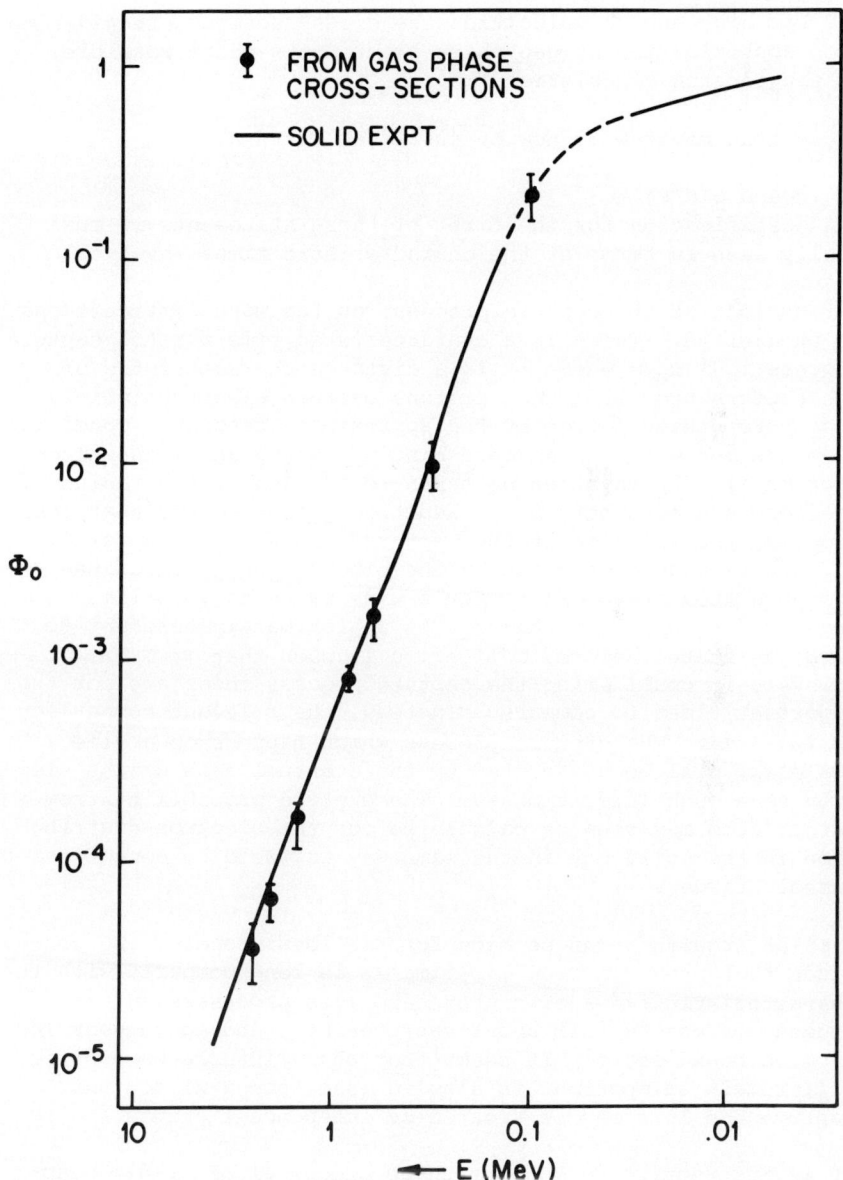

Fig. 5. Equilibrium neutral fraction Φ_0 of hydrogen emerging from solid carbon as a function of proton energy E (see text for details).

equilibrium fractions by Eq. 2.

2. For high enough velocities the cross-sections are given to good approximation by gas phase experiments where possible, or single atom calculations.

These statements will now be justified in turn.

1. Bound States

The justification for the first of these statements is most easily seen in terms of the characteristic times involved.

Let us look at the capture process, as the more difficult one to understand. There is a characteristic time for the capture process $\tau_c \sim r_c/v$ where r_c is a distance characteristic of the capture process. Fast protons capture electrons mainly from core states (given by the approximate resonance condition $E_c = 1/2\ mv^2$ with E_c the core binding energy and m the electron mass). r_c may then be taken to be roughly the radius of the core state. There is in addition a time to the next loss process, the lifetime of the captured state, $\tau_\ell \sim (n\sigma_\ell v)^{-1}$. A third timescale is given by the binding energy E_0 of the hydrogen atom state $\hbar/E_0$. For a wide range of velocities above v_0 it turns out that $\sigma_\ell v$ is approximately constant to give, as stated before, a lifetime τ_ℓ such that $\tau_\ell \sim \hbar/E_0$. However, in considering the capture process these are not the important times to compare. Instead, the relevant comparison is $\tau_c \ll \tau_\ell$. The capture process, which happens on a time scale τ_c, will be unaffected by the eventual loss of the electron at a much later time τ_ℓ. The capture probability from a target atom may then be calculated for the electron distribution in the solid but in the same way as would be done for a gaseous target.

Similar arguments can be made for the loss process. We conclude that providing the lifetime τ_ℓ is long compared with the characteristic times of capture and loss processes, it is indeed correct to talk about capture into, and subsequent loss from, a bound state. It seems that phase interference effects will remain unimportant in a solid (see Appendix) so that capture and loss may be treated as independent events.

It is instructive to compare three processes (Fig. 6): capture and loss in a gas; capture and loss in a solid; and capture to give a neutral emerging from the solid. It is convenient to define a distance $\lambda = v\tau_\ell$ over which the proton travels with a bound electron.

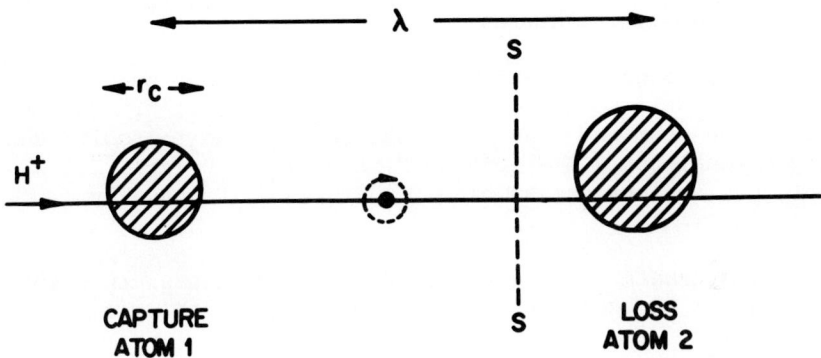

Fig. 6. *Capture and Loss.*

In a gas the density of scattering centres is low, and the time between capture and subsequent loss is large ($\hbar/\tau_\ell \ll E_o$). The electron completes many orbits around the proton, and the energy is well defined.

In a solid if $\hbar/\tau_\ell$ is comparable to E_o not even one orbit is completed, and the energy is not well defined. Nevertheless if $\lambda \gg r_c$ it is still correct to calculate capture into and loss from the orbit: as Fig. 6 suggests the capture and loss are then separate, well defined processes. We may say that capture and loss, involving mainly interaction with core electrons, take place within target atoms and are rather independent of the exterior, including the separation of the events $\lambda \gg r_c$.

Now suppose the capture event took place within the distance λ from the surface. In Fig. 6 this corresponds to a surface at SS and the absence of scattering centres to the right. There is no further interaction with the solid, and the electron completes many orbits to be detected in a neutral with the orbital state into which it was originally captured. The surface is important only as the absence of further interactions: no change of the electron wavefunction is needed on passing through the surface.

These arguments in terms of the characteristic times include the essential physics. They can be rewritten more precisely in terms of the quantum mechanical description of capture and loss. This is described in the Appendix. The result found there can be stated that for fast protons capture and loss depend on the wavefunction of the electron state around the proton, but not on the binding energy and a fortiori not on the uncertainty in this energy.

The energy range over which this description may be reasonable is conveniently seen from the distance $\lambda = (n\sigma_\ell)^{-1}$. Values for Carbon from the measured cross-section and the solid density n equal to 17.6×10^{22} cm^{-3}, are given in Table I. λ continues to increase at higher energies. (Eventually τ_ℓ

TABLE 1
The "loss length" λ as a function of proton energy in Carbon. The figures in brackets are the number of interatomic spacings.

Energy MeV	$\lambda = (n\sigma_\ell')^{-1}$
2	19 Å (12)
1	10 Å (7)
0.1	3 Å (2)

becomes proportional to v^{-2} and the lifetime τ_ℓ also becomes long.) At high energies, where λ is many interatomic spacings and capture and loss, involving core states, are characterized by a distance equal to some core radius R_c, then the lifetime λ/v is large compared to the process times R_c/v and the description is good.

For proton energies greater than 100 keV, $\sigma_\ell \gg \sigma_c$. λ is then also the length needed for the charge state to approach equilibrium[b]. This "equilibrium length" becomes long for MeV protons.

2. *Use of Gas Cross-Sections*
The gas capture cross-section may be used if the electron is

---

[b] Equation 1 is solved to give $\Delta\Phi = \Delta\Phi^0 \exp[-(\sigma_c+\sigma_\ell)nx]$ where $\Delta\Phi$ is the deviation from equilibrium of either neutral or positive fraction at x and $\Delta\Phi^0$ the corresponding value at x = 0.

captured from an inner shell. These are unaffected by binding in the solid. The minimum velocity at which this may be assumed can be seen from shell by shell calculations on various atoms (23).

Figure 7 shows the result for Lithium. For $v > 2v_0$ inner (K) shell capture dominates. For inert gases a larger v is

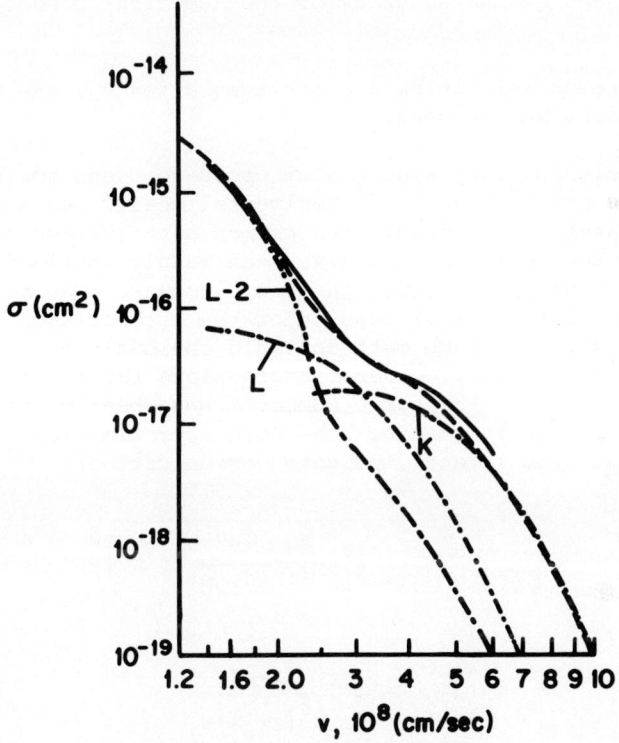

Fig. 7. Capture cross-section of protons in Lithium as a function of proton velocity v (after Nikolaev: Ref. 23); continuous curve - experimental data, dashed - calculated; the curves K, L and L-2 are respectively calculated capture cross-sections from the lithium K and L shells to the hydrogen ground state, and from lithium L shell to the first excited hydrogen shell.

required: they may however be considered exceptional as having full, tightly bound outer shells. We may take $2v_0$ as an optimistic lower limit.

The loss cross-section is not expected to be sensitive to the valence electron states for proton velocities v greater than a few v_o. For example, one model of the loss cross-section is that it is the total scattering cross-section for an electron of velocity v (with small corrections from the velocities in the original orbit) - see the Appendix. This total electron scattering cross-section is independent of the valence electron distribution for $v \gg v_o$ (25). The conclusion should be independent of the accuracy of the numerical predictions of the model.

Thus for proton velocities greater than a few v_o, gas atom cross-sections may be used.

Unfortunately not many single atom cross-sections for commonly used solids are available. Experimental measurements on relevant gases are limited: the carbon data (22) seems the most complete. Calculations have been mainly confined to inert gases and other common gaseous elements. The reader is referred to (26) for a discussion of the methods used and their success. The same methods could obviously be repeated for gold, carbon, etc. General expressions for the cross-sections of atoms with atomic number Z have been quoted (27). These are apparently based on the Born approximation and a Thomas-Fermi atom (Brandt, private communication):

$$\sigma_c = \pi a_o^2 \frac{2^{18}}{5} \frac{z^5}{v_1^6 \left(v_1^2 + \frac{2^6}{3\sqrt{40}} z^{14/9} \right)^3}$$

$$\sigma_\ell = \frac{\pi a_o^2 z^{2/3}}{z^{2/3}+v_1} \frac{4z^{1/3}(z+1)}{4z^{1/3}(z+1)+v_1}$$

(7)

where $v_1 = v/v_o$ and a_o is the Bohr radius. A comparison with gas data (between 100 keV and 10 MeV protons) shows the capture cross-section in general parallels the measured values, but is a factor 2 or 3 small; the loss cross-section is in general too small, and decreases with increasing energy too slowly (see Fig. 10 for argon). It is known that the Born approximation does not predict very well loss cross-sections in this energy range (28). Nevertheless the expressions Eq. 7 give a ratio σ_ℓ/σ_c that apparently account reasonably well for the trend with Z observed in solid foils (4) for proton energies above 1 MeV.

C. Experimental Tests

1. *Equilibrium Fractions*

The remarkable agreement between carbon foil measurements and predictions from cross-sections measured in the gas phase has already been shown, Fig. 5. Measurements on other solids for proton energies 0.3 to 16 MeV (4) show reasonable agreement with predictions from the ratio of the Brandt-Sizmann cross-sections (Eq. 7) but the disagreement increases at the high energies (to at worst a factor of 2.5 in Φ_0 for gold). No more accurately calculated cross-sections for these elements are, to my knowledge, available.

2. *Nonequilibrium Charge States*

There is experimental evidence for the long equilibrium lengths for MeV protons. Figure 8 shows the approach to equilibrium for 7 MeV protons fired at thin silver and gold

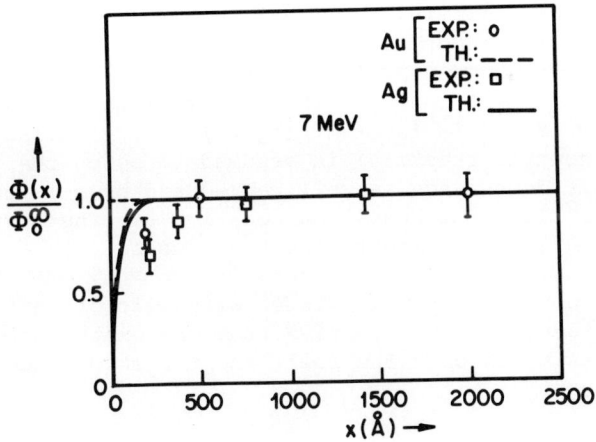

Fig. 8. *Emerging neutral fraction $\Phi(x)$ of 7 MeV protons after passing through foils thickness x of silver and gold. (3). The theoretical curves shown are from Eq. 1 using the approximate BS cross-sections Eq. 7.*

targets (3). For Carbon foils, extrapolating the measured loss cross-section would given an equilibrium length at 7 MeV of about 25 interatomic spacings. This would correspond to 70 Å for gold, of the correct order of magnitude, but somewhat smaller than implied by the experiment. The approach to equilibrium is not well characterized in Fig. 8, and the uniformity of the foils is not discussed. The measurement must be considered preliminary.

Recent experiments (Poizat and Remillieux, unpublished) use neutral hydrogen atoms incident with energy 1 to 2 MeV on thin Carbon foils. From the approach to equilibrium with increasing thickness σ_ℓ is found, and from the equilibrium neutral fraction σ_c is then calculated. A preliminary comparison shows these cross-sections to be close to those measured in the gas phase.

At lower energies (500 keV/nucleon) the equilibrium length for Helium ions in Carbon is measured to be less than 5 or 10 Å (29), consistent with the predictions of Table I.

At still lower energies (less than 100 keV protons) the extreme sensitivity to surface condition suggested by Table I is well known (21).

3. Dependence on Exit Angle

For 1.4 MeV protons emerging from gold there is a reduction in the neutral fraction by about 25% when the exit angle is increased to 60° (3). The angle dependence for aluminum is much smaller. The theory of capture and loss in the bulk would predict no angle dependence for exit from a <u>clean</u> surface at constant <u>exit</u> energy. The gold results can be accounted for if a thin layer of light impurity (having a lower equilibrium neutral fraction) is assumed to be on the surface. The exit angle dependence is well reproduced by a layer of 10^{16} atoms/cm^2 (3 Å thickness) Carbon, Fig. 9. The state of the surface in the experiment is not discussed.

It should be noted an <u>increase</u> in Φ_o with tilt of foil at constant <u>incident</u> energy may arise from the extra energy lossed in passing through the foil, and the strong dependence of Φ_o on energy.

IV. OTHER THEORIES

In this section we will briefly review, in terms of the ideas of section III, theories for fast protons by Trubnikov and Yavlinskii (TY) (6), Kitagawa and Ohtsuki (KO) (8), and Yavlinskii, Trubnikov and Elesin (YTE) (5). It will also be useful to look at their low velocity results.(c)

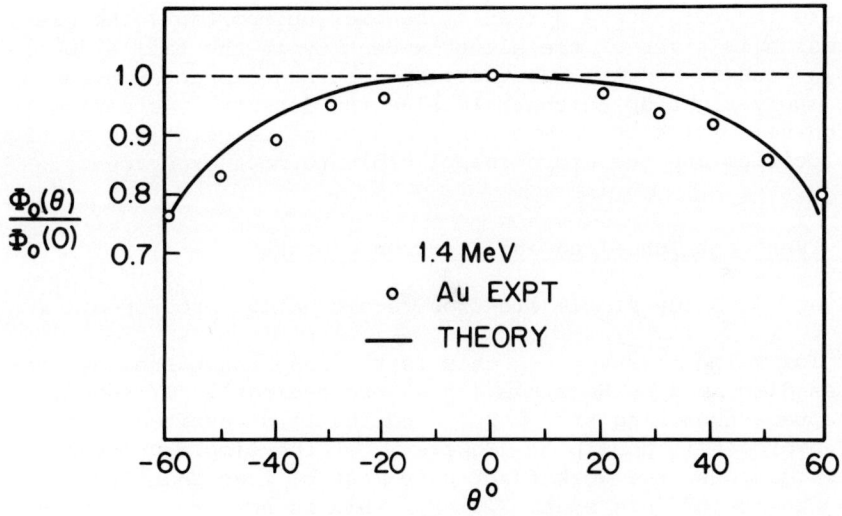

Fig. 9. Equilibrium neutral fraction $\Phi_0(\theta)$ as a function of exit angle to the normal θ. Experimental points are from Ref. 3; for theory see text.

(c) A comparison between the discussion in section III and the theory of Brandt and Sizmann must await a publication of that theory. From brief descriptions (4,7,9,30) the emphasis of the theory appears to be on the absence of bound states on the proton in the solid and instead, a buildup of a "correlation in speed and direction" between proton and electron. If the correlation persists to the surface, capture may then occur (9). This correlation has been suggested (9) to be reminiscent of the "charge exchange to a continuum state". In contrast, in the theory of section III, charge exchange to a continuum state is a separate process from capture to a bound state, and gives rise, exactly as in gas collisions (31), to the cusp of unbound electrons at the proton exit velocity (32, see also the discussion by Meckbach, these proceedings).

Nevertheless, Eqs. (2) and (7) are used to describe the process. The discussion of section III in terms of capture and loss and bound states is more informative and simpler than the "correlation" argument, and more correct in that no additional interaction at the surface is implied to be necessary.

These three papers assume that screening prevents a bound
state at all velocities, and that electron pickup must occur
in the tail of the electron distribution outside the surface
where the collective screening becomes unimportant. No discussion is given of the electron density in the tail at which
this occurs, and the calculations are inconsistently assumed
to apply right up to the bulk electron density. Screening at
high velocities is unimportant, and these calculations at high
velocities are therefore inapplicable to most experiments.

A. Resonant Tunneling and Electron Capture

TY and KO study single electron recombination processes.

At low velocities $v \ll v_0$ this is the resonant tunneling process discussed by Hagstrum (these proceedings). TY suggest
resonant tunneling from tungsten to the first excited state
of hydrogen is possible. However, for the simple process
they discuss, the work function ϕ must be less than the binding energy of this state 3.4 eV. This is not true for heavy
metals (e.g., tungsten ϕ = 4.5 eV) but may be for light metals
(e.g., calcium ϕ = 2.7 eV). As Hagstrum pointed out, the
important processes for a slow ion leaving the surface will
take place at small distances, and the perturbation of the
solid by the ion will not be small. The calculation of the
transition rate is then a difficult problem and the simple
first order perturbation theory of TY is unlikely to be valid.
The low velocity form for the emergent neutral fraction
$1 - e^{-v^1/v}$ with v^1 some characteristic velocity is however
independent of the details of the calculation.(d)

For $v \gg v_0$ their calculation is essentially a calculation of
the capture probability from the tail of the conduction electrons outside the surface. They therefore neglect the process
important at high velocities, namely capture from the target
atom cores over the larger distance λ. It should also be
noted that their high velocity asymptotic forms for the capture probability (v^{-8} by TY and v^{-4} by KO compared with v^{-12}
(or v^{-11}) for capture from an atom in first (second) Born

(d) It may be obtained only assuming
1. The neutralization of the proton takes place in the surface tail.
2. The transition is irreversible and the rate depends on the
distance from the surface but not on the ion velocity.
3. The ion velocity is constant at v during the process.

approximation) are spurious. It is clear from simple Born approximation (23) that the asymptotic form depends crucially on the high momentum components of the wavefunctions of the initial and final electron states. These in turn are governed by the discontinuities of the wavefunctions. TY and KO use an unrealistic wavefunction for the metallic electron, with discontinuous second derivatives, thus building in unphysical high momentum components. A more realistic wavefunction is essential to calculate electron capture from the tail, which may be important in the glancing angle reflection experiments (33).

B. Triple Recombination

At low velocities YTE treat a recombination process involving more than one electron somewhat analogous to the Auger process (Hagstrum, these proceedings). YTE, however, suppose the excess binding energy of the captured electron is diffused away by many collisions: the process is treated as the inward diffusive flow of electrons under the driving force of the ion potential. Such a continuum treatment cannot be correct for the low densities in the electron tail where the ion potential varies rapidly over the electron separation distance, and the dynamics of degenerate Fermions is not correctly treated in terms of a collision cross-section calculated for a distinguishable particle. The success of the theory of Ion Neutralization Spectroscopy in terms of Auger processes (34,35) confirms the physically more reasonable picture of a single collision scattering one electron into the bound state and one to higher energies. The form found by YTE:
$\Phi_0 = 1 - \exp(-v^1/v)$ is also given by an Auger calculation. However, the rate for the Auger process depends on the energy level diagram of the solid (the availability of electron states such that energy is conserved) and not on conduction properties as suggested by YTE. Any difference in rates between metals, semiconductors and insulators follows from a study of bandgaps, densities of states, etc., and is not an immediate consequence of the presence or absence of conduction electrons.

Triple recombination should be less important than quasi-resonant capture at high velocities: the asymptotic form of YTE (v^{-2}) is hard to understand.

V. INTERMEDIATE ENERGIES

The intermediate energy range is difficult to treat, even in gas atom collisions. The Born approximation is not valid. Neither are the simple resonant process calculations (Auger or resonant tunneling) expected to be good, since the width of the resonance (or order $\hbar v/R_i$ as in section III) is many electron volts for $v \sim v_0$. Similarly two level schemes (see Tully, these proceedings) are not adequate.

In solids there is the additional difficulty that the collective screening by the valence electrons is becoming of increasing importance as the velocity is reduced. The validity of the simple dielectric screening ideas might be expected to be most in doubt at this "turn on" point - the simple criterion of Brandt based on calculations by Ebel (9) may not be sufficiently accurate.

Most experiments have therefore been compared with interpolations between $v \ll v_0$ and $v \gg v_0$ expressions, themselves of doubtful validity (section IV). This procedure cannot hope to account for the small but complicated dependence on material and surface cleanliness reviewed by Buck (these proceedings).

There is also an expression of Brandt and Sizmann (7)

$$\Phi_0 = \frac{1}{1 + 40E}$$

with E the proton energy in MeV. No derivation is given. They describe the result as applying for $v < 3v_0$ where they suggest screening means electron pickup must be from the tail of the electron distribution. They further say the neutral fraction should be nearly the same for all surfaces.

An interesting alternative is that the ideas used at high energies retain some usefulness at these intermediate energies. It may be noted that the loss cross-section of hydrogen in various gases (28) goes through a broad maximum at energies 50 to 100 keV, and then decreases at lower energy (e.g., Fig. 10). The value for λ, the loss distance, at 100 keV (Table I) is about the minimum to be expected. The lifetime τ_ℓ increases below 100 keV as σ_ℓ and v both decrease. It is not unreasonable to expect Eq. 2 to still be useful. The change in the cross-sections due to the distorted valence electron wavefunctions in the solid must be calculated: the magnitude of this effect is unknown. The use of Eq. 2 depends on the unimportance of screening effects, and Auger neutralization processes.

It has been pointed out before (36) that argon cross-sections give a reasonably good guide to solid neutral fractions. The

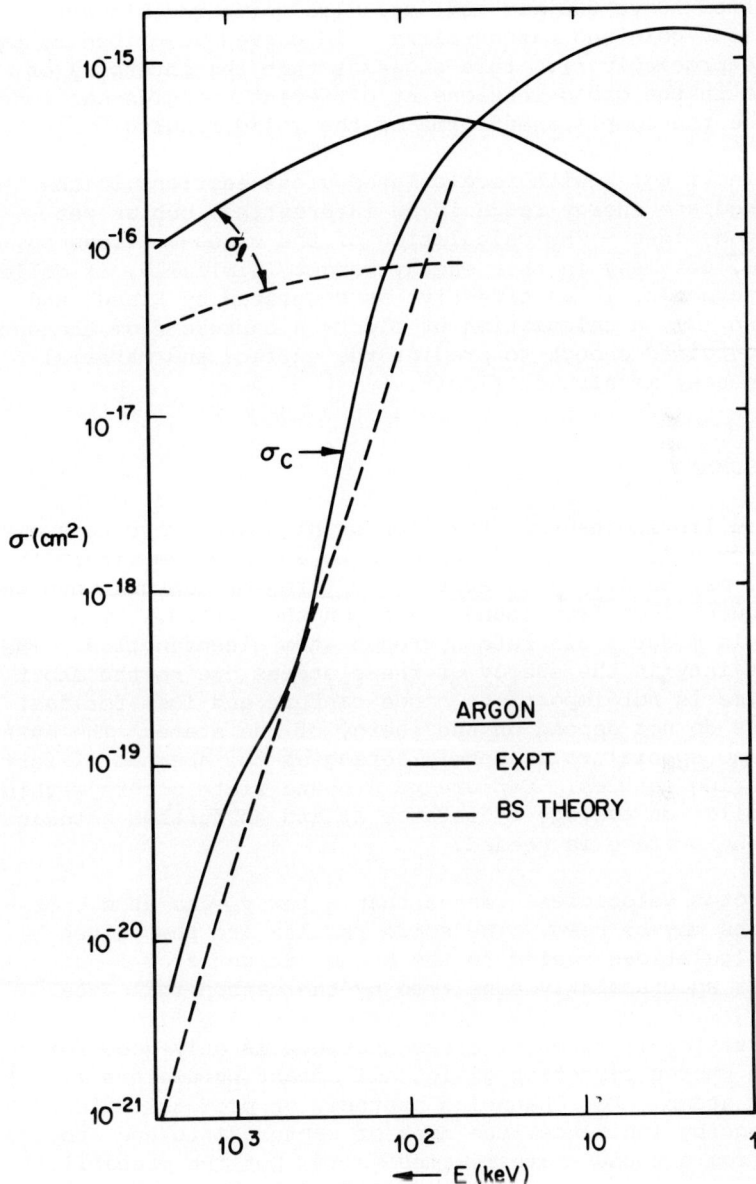

Fig. 10. Capture and loss cross-sections for protons in argon, as a function of proton energy E. Experimental points are from Ref. 37. For comparison the approximate BS cross-sections Eq. 7 are plotted.

relative shape of hydrogen-neutral fractions in argon and neon from 10 to 200 keV (37) are also quite reminiscent of the clean gold and carbon-dirty gold curves described by Buck (these proceedings): this suggests that the interplay of maxima in the cross-sections at different energies can indeed produce the complicated forms of the solid results.

The use of Eq. 2 with recalculated cross-sections in the intermediate energy range is an interesting, but as yet untested, idea. The calculation of the cross-sections is, of course, not easy in this energy range. Similarly, if collective screening is as effective as suggested by Brandt and Sizmann (7), a calculation of electron capture from the surface detailed enough to predict the surface and material dependence, is also difficult.

VI. SUMMARY

The equilibrium neutral fraction of high energy protons emerging from solids, and the approach to equilibrium within the solid, may be simply understood in terms of capture into and subsequent loss from bound states on the proton. By a bound state is meant a discrete hydrogen atom eigenfunction. The uncertainty in the energy of these states due to the short lifetime is not important, since capture and loss for fast protons do not depend on the energy of the state. The surface plays no special role, merely acting as the absence of further scattering centres. Capture to a bound state occurs within the solid (on average a distance λ) and no further interaction with the surface is needed.

For proton velocities greater than a few v_0 gas atom cross-sections may be used. The solid results are then given by gas calculations scaled to the higher target atom density. This is spectacularly confirmed by the carbon foil data.

A discussion in terms of cross-sections is only good for a random proton direction giving all impact parameters with the target atoms. For channeled protons, or protons reflected at glancing incidence, the idea of capture into and subsequent loss from a bound state remains valid, but the probabilities must be calculated for the impact parameters appropriate to the proton path. The lifetime of the bound state may well be longer because of the reduced interaction with the solid for a well channeled proton.

The intermediate energy regime is perhaps more important - giving higher neutral fractions - and more interesting, perhaps giving information specific to the solid or solid surface. It is unfortunately a difficult energy to treat theoretically, and no theory has explained the experimental results. These may possibly be accounted for in terms of recalculated capture and loss cross-sections. Alternatively if collective screening is important electron pickup at the surface must be calculated. Both calculations may be expected to be difficult, and the intermediate energy regime may remain an uncertain transition region, where high energy processes are turning off and low energy processes turning on.

VII. APPENDIX

In this appendix it will be shown that the uncertainty of the energy of the captured electron state is unimportant in the capture and loss process.

The result for the loss process is very straightforward. For proton velocities v greater than a few v_0 the loss cross-section may be calculated as the total scattering cross-section of a target atom for scattering an electron of velocity $\underset{\sim}{v}$: essentially all scattering processes result in loss of the electron. Instead of a single velocity $\underset{\sim}{v}$, a velocity distribution around $\underset{\sim}{v}$ given by the momentum distribution of the initial electron wavefunction may be used. The uncertainty in the energy is clearly not important. This is the classical impulse theory (24). The conclusion should be independent of the exact validity of the model.

The capture process is quasi-resonant, and must be treated more carefully. On a Born approximation, the cross-section for capture from a state i on atom A to a state f on atom B is given by

$$\sigma_{if} = (2\pi\hbar^2 v)^{-2} \int d^2 p_\perp |\langle f|V|i\rangle|^2 \qquad (A.1)$$

where the matrix element is

$$\langle f|V|i\rangle = \int \exp\left\{\frac{-i}{\hbar}\left[\left(\frac{E_i - E_f}{v} + \frac{mv}{2}\right) z_b + \underset{\sim}{p}_\perp \cdot \underset{\sim}{\rho}_b\right]\right\} \psi_f^*(\underset{\sim}{r}_b) V(\underset{\sim}{r}_a, \underset{\sim}{r}_b)$$

$$\psi_i(\underset{\sim}{r}_a) \exp\left\{\frac{i}{\hbar}\left[\left(\frac{E_i - E_f}{v} - \frac{mv}{2}\right) z_a + \underset{\sim}{p}_\perp \cdot \underset{\sim}{\rho}_a\right]\right\} d\underset{\sim}{r}_a d\underset{\sim}{r}_b \qquad (A.2)$$

$\underset{\sim}{r}_a$ is the electron position measured from nucleus A, with components z_a parallel and ρ_a perpendicular to the proton velocity v. Similarly r_b is the position measured from B. E_i and E_f are the binding energies of initial and final eigenfunctions $\Psi_i(r_a)$ and $\Psi_f(r_b)$. $V(r_a, r_b)$ is the interaction potential: the form chosen determines the approximation used. These equations may be derived from either a wave or impact parameter approach (26): the integral over perpendicular momentum transfer $p_\perp$ in Eq. A.1 corresponds to an integral over impact parameters in that approach.

The total cross-section for capture to the state f is the sum $\sum_i \sigma_{if}$ over occupied states i, which will in general include core states with $E_i \gg E_f$.

It is a little difficult to analyze the contributions according to Eq. A.2 of each i because both Ψ_i and E_i change with i. It may be seen that, independent of the form of V, the integral over z_a gives a broad resonance condition: the maximum contribution to the integral would be expected from the region of stationary phase $E_i - E_f \sim 1/2\, mv^2$; however, the width of the resonance is given by the space over which the momentum components of Ψ_i decrease. This gives E_i between the limits

$$E_i - E_f \sim \frac{1}{2} m v^2 \pm \hbar v / R_i \qquad (A.3)$$

where R_i is some characteristic radius of the orbit i. For hydrogen like orbits $E_i \sim 1/2\, mv^2 \sim \hbar v/R_i$, and the resonance is very broad.

An uncertainty $\Delta E_f \ll \hbar v/R_i$ broadens the resonance by an amount small compared with the width $\hbar v/R_i$, and will have little effect on the cross-section. The integral over z_b is not resonant, and for $\Delta E_f \ll 1/2\, mv^2$, E_i (both of order $\hbar v/R_i$) the cross-section is again insensitive to ΔE_f.

It is also clear from Eq. A.2 that the energy E_f itself is irrelevant for $v \gg v_o$, and may be left out from that equation. The cross-section depends on the <u>wavefunction</u> Ψ_f but not the <u>energy</u> E_f of the final state. This is particularly obvious if the Brinkman-Kramers (38) approximation for the interaction is used: $V = -e^2/r_b$. This does not give good answers for the overall value of the cross-section, but is often used to calculate relative values between different cross-sections. The capture cross-section from a full hydrogen-like shell with nuclear charge Z and quantum number n* to the states in a shell n on a proton is given in closed form.

$$\sigma(Z,n^*;1,n) = \frac{2^{18}}{5}\pi a_o^2 \frac{1}{n^3}\frac{Z^5}{n^{*3}}\left(\frac{v}{v_o}\right)^8$$

$$\times \left[\frac{\left[\frac{1}{2}mv_o^2\right]^2}{\left(E_i - E_f - \frac{1}{2}mv^2\right)^2 + (\hbar v/R_i)^2}\right]^5 \quad (A.4)$$

with $R_i = n^*a_o/Z$, just the Bohr radius of the initial shell divided by the quantum number n^*. The quasi-resonant nature and the unimportance of E_f (and therefore ΔE_f) much less than hv/R_i or $1/2\,mv^2$ is clearly displayed. The dependence on the final state is the well known n^{-3} prefactor, which comes simply from the dependence of the high momentum components in the wavefunction on the shell quantum number n (23).

It is instructive to consider once again the capture into and loss from bound states. We calculate capture into the complete set of states f which are all the hydrogen atom eigenfunctions (bound and continuum) centered on the moving proton. The contribution of the ground state is largest. Capture into the continuum states may be detected as the cusp at velocity v in the unbound electron distribution emerging from the solid (32). In a gas each state evolves in time:

$$|\phi(t)\rangle = e^{-iE_f t/\hbar}|\phi(o)\rangle \quad (A.5)$$

and the phases $E_f t/h$ pass through all values before the loss process. If however loss took place immediately

$$|\phi(\tau_\ell)\rangle = |\phi(o)\rangle \quad (A.6)$$

we could have calculated capture into and loss from a different complete set of states f'. However it is easier to use the eigenstates f. Also in a solid it remains more correct, because then the time evolution that does occur in the time τ_ℓ is included as in Eq. A.5. Furthermore, if capture takes place near the surface so that there is no loss process, then the state detected at a much later time is automatically given by Eq. A.5, and an electron captured in the state f will be detected in that state.

Finally we may discuss one apparent difficulty with the theory. From the pictorial description in Fig. 6 it would appear that since the captured electron does not complete an orbit the correlation between the position of the electron, relative to the proton, produced in the capture process on the one hand, and the proton path on the other, may influence the subsequent loss process. Capture and loss could not then be treated as independent.

The quantum mechanical treatment shows this is probably not important. Capture to fast protons is mainly to the ground 1s state, and to excited states is again mainly to s states (39). There is little <u>directional</u> information from the capture process to influence the loss process. Phase interference effects between the s states will in any case be small because of the different amplitudes, and, averaging over impact parameters, will presumably be washed out. It is then sufficient to calculate capture and loss probabilities, as in Eq. A.1, and to neglect any influence of the capture process on the loss process.

VIII. REFERENCES

1. Betz, H.D., <u>Rev. Mod. Phys.</u> 44, 465 (1972).
2. Kaminski, M., <u>Phys. Rev. Lett.</u> 23, 819 (1969).
3. Chateau-Thierry, A., and Gladieux, A., in "Atomic Collisions in Solids" (S. Datz, B.R. Appleton, and C.D. Moak, Eds.), p. 307, Plenum Press, New York, 1973.
4. Chateau-Thierry, A., Gladieux, A., and Delaunay, B., <u>Nucl. Inst. Meth.</u> 132, 553 (1976).
5. Yavlinskii, Yu.N., Trubnikov, B.A., and Elesin, V.F., <u>Bull. USSR Acad. Sci. Phys.</u> 30, 1996 (1966).
6. Trubnikov, B.A., and Yavlinskii, Yu. N., <u>JETP</u> 25, 1089 (1967).
7. Brandt, W., and Sizmann, R., <u>Phys. Lett.</u> 37A, 115 (1971).
8. Kitagawa, M., and Ohtsuki, Y.H., <u>Phys. Rev.</u> B13, 4682 (1976).
9. Brandt, W., as reference 3, p. 261.
10. Kittel, C., "Introduction to Solid State Physics" (4th Ed.) p. 708, John Wiley and Sons, New York, 1971. For an account at a more advanced level see reference 14, pp. 121-156.
11. Friedel, J., <u>Phil. Mag.</u> 43, 153 (1952).
12. Recent references to act as an entry to this large literature are: Zbasnik, J., and Matnig, M., <u>Z. Physik</u> B23, 15 (1976); Gilberg, E., International Conference on the Physics of X-ray Spectra, 1976, at Gaithersburg, Maryland; Erckmann, V., and Wipf, H., <u>Phys. Rev. Lett.</u> 37, 341 (1976).
13. Neufeld, J., and Ritchie, R.H., <u>Phys. Rev.</u> 98, 1632 (1955).

14. Pines, D., "Elementary Excitations in Solids", ch. 4, Benjamin, New York, 1964.
15. Raether, H., Springer Tracts in Modern Physics 38, 84 (1965).
16. Daniels, J., v. Festenberg, C., Raether, H., and Zeppenfeld, K., Springer Tracts in Modern Physics 54, 136 (1970).
17. Neelavathi, V.N., and Ritchie, R.H., as reference 3, p. 289.
18. Day, M.H., Phys. Rev. B12, 514 (1975).
19. Northcliffe, L.C., Ann. Rev. Nucl. Sci. 13, 67 (1963).
20. Berkner, K.H., Bomstein, I., Pyle, R.V., and Sterns, J.W., Phys. Rev. A6, 278 (1972).
21. Buck, T.M., Feldman, L.C., and Wheatley, G.H., as reference 3, p. 331.
22. Toburen, L.H., Nakai, M.Y., and Langley, R.A., Phys. Rev. 171, 114 (1968).
23. Nikolaev, V.S., JETP 24, 847 (1967).
24. Massey, H.S.W., Burhop, E.H.S., and Gilbody, H.B., "Electronic and Ionic Impact Phenomena", Vol. 4, p. 2503, Oxford, 1974.
25. See reference 24, Vol. 2, p. 666.
26. See reference 24, Vol. 4, ch. 23.
27. Brandt, W., and Sizmann, R., quoted by reference 4.
28. See reference 24, Vol. 4, p. 2831.
29. Lurio, A., and Ziegler, J.F., in "Beam Foil Spectroscopy", (I. Sellin and D. J. Pegg, Eds.), Vol. 2, p. 665, Plenum, New York, 1976.
30. Gaillard, M.J., Poizat, J.-C., Remillieux, J., Chateau-Thierry, A., Gladieux, A., and Brandt, W., Nucl. Instr. Meth. 132, 547 (1976).
31. Crooks, G.B., and Rudd, M.E., Phys. Rev. Lett. 25, 1599 (1970).
32. Dettman, K., Harrison, K.G., and Lucas, M.W., J. Phys. B. 7, 269 (1974).
33. Rau, C., and Sizmann, R., as reference 3, p. 295.
34. Hagstrum, H.D., Phys. Rev. 96, 336 (1954).
35. Appelbaum, J.A., and Hamann, D.R., Phys. Rev. B12, 5590 (1975).
36. Buck, T.M., Wheatley, G.H., and Feldman, L.C., Surf. Sci. 35, 345 (1973).
37. Allison, S.K., Rev. Mod. Phys. 30, 1137 (1958).
38. Brinkman, M.C., and Kramers, H.A., Proc. Acad. Sci. Amst. 33, 973 (1930).
39. See reference 24, Vol. 4, p. 2467.

processes of charge exchange into the continuum of ionic projectiles interacting with gases and solids*

W. Meckbach and R. A. Baragiola

Centro Atómico Bariloche
Comisión Nacional de Energía Atómica

ABSTRACT

A critical survey is given of research on electron ejection into continuum states of fast ionic projectiles bombarding gaseous and solid targets. Emphasis is laid upon the shape of the energy and angular distributions which is evaluated in terms of the vectorial electron-ion velocity difference, and the yield of the process which is discussed as a function of the velocity of the ion. The results are analyzed in comparison with theories and the discrepancies pointed out. The role of "wake riding" electron states in solids is also discussed.

I. INTRODUCTION

Much attention has been given recently to the detailed study of electron ejection from atoms induced by collisions with ionic projectiles. Since 1963, a considerable amount of experimental as well as theoretical research has been directed towards the study of cross sections, differential in energy E_e and ejection angle θ of the electrons, $\sigma(E_e,\theta)$. This research has been reviewed by Ogurtsov (1), Kim (2), Manson et al. (3), Rudd (4) and Rudd and Macek (5).

Graphically, $\sigma(E_e,\theta)$ may be represented for a fixed E_e as a function of θ, or for a fixed θ as a function of E_e. Fig. 1 shows angular distributions obtained with 300 keV protons incident on Ne and H_2 targets (6). They are compared with scaled Born approximation calculations using hydrogenic wave

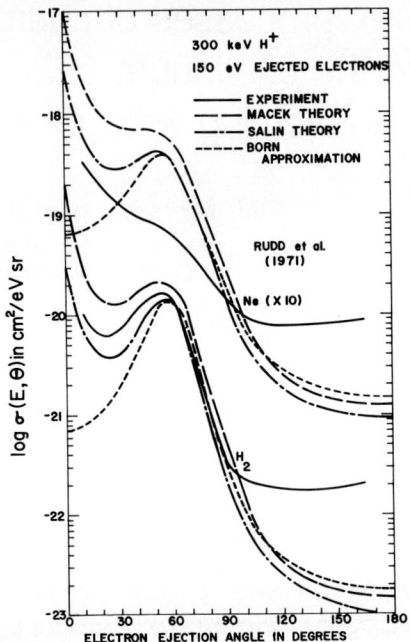

Fig. 1. Double differential cross sections $\sigma(E_e,\theta)$ *of 150 eV electrons as a function of ejection angle* θ, *for 300 keV* H^+ *on* H_2 *and Ne (from: Rudd et al., ref. 6; Born approximation, ref. 35; Macek theory, ref. 8; Salin theory, ref. 9).*

functions for the target. These calculations show a peak, due to binary proton-electron collisions, broadened by the momentum distribution of the initially bound target electrons. For H_2 there is excellent agreement between theory and experiment from $\theta=60°$, the angle of the binary peak maximum, up to about 90°.

We will now focus our attention to the large discrepancies seen at smaller angles. At $\theta=10°$ the experimental cross section is about an order of magnitude larger than the cross section given by the Born approximation calculation. This is also observed in Fig. 2 which shows energy spectra of electrons emitted in the direction ($\theta=0°$) of a 300 keV proton beam bombarding He (7). The cusp shaped peak, observed at $E_e=E_{eq}=m_eE_i/m_p=163$ eV, the electron equivalent ion energy, is not reproduced by the Born approximation cross section, also shown in Fig. 2.

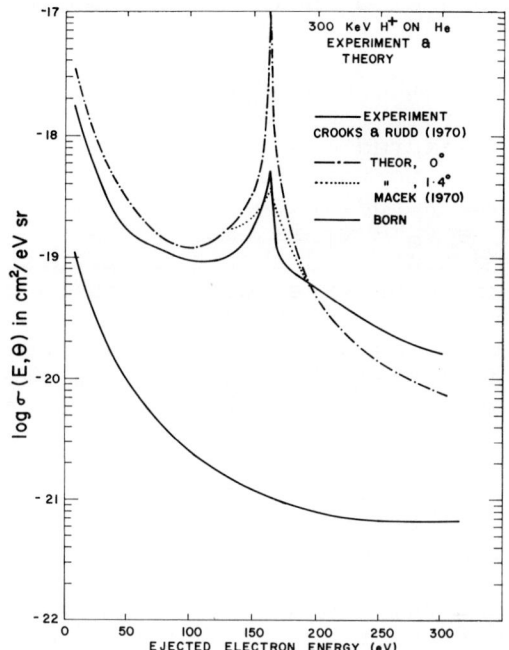

Fig. 2. Energy distribution of electrons ejected at Θ=0°±1.4°, in collisions of 300 keV H⁺ on He (from: Crooks and Rudd, ref. 7) Macek theory (ref.8) at Θ=0° and Θ=1.4°.

In this theory, the final state describes the emitted electron as moving only in the field of the ionized parent atom. This is a good approximation for slow electrons, but for electrons initially ejected from the parent atom with velocities not too different in magnitude and direction from the ion velocity, a post collisional long-range Coulomb interaction between these electrons and the moving ion is predominant. As a consequence, the electron will be "dragged along" in a continuum state of the moving ion. Accordingly the process bears the name "charge exchange into the continuum" (CEC), or may be called, more specifically, "electron capture into the continuum" (ECC).

The above mentioned interaction has been accounted for by different theoretical approaches (Macek (8), Salin (9), Band (10), Dettmann (11)), treating the three-body problem represented by the system: moving ion-electron-residual ion. In Figs. 1 and 2 the theoretical cross sections calculated by Macek and by Salin are also shown. Both theories result in cusp-shaped peaks as a function of angle and energy, centered at $\theta=0°$ or $E_e=E_{eq}$ respectively.

We see that the cross section for electron ejection is sharply enhanced when the velocity vector $\vec{v}_e$ of the ejected electron is, in magnitude and direction, close to the velocity $\vec{v}_i$ of the moving ion or in other words, when the vectorial velocity difference $\vec{v}' = \vec{v}_e - \vec{v}_i$ is small.

The above mentioned theories all discuss the electron ejection under single collision conditions, given experimentally by the interaction of an ion beam with rarefied gas. Surprisingly, the same cusp shaped peak characteristic for ECC has been observed behind thin solid foils traversed by energetic ions (19).

The theory of Dettmann (11) and Dettmann et al. (12,13) actually deals with this beam-solid foil case Considering that energetic secondary electrors produced by collisions inside the solid are strongly scattered in angle and attenuated in energy on their way to the surface, then those which emerge in the high-energy part of the spectrum are assumed to originate from single collisions with atoms of the last one or two layers of the target. This "last-layer hypothesis" of Dettmann can be valid only if any collision of energetic secondary electrons which occurs further inside the solid target is assumed to be violent enough to produce deflections in angle and/or energy losses large enough to throw these electrons completely out of the peak. This "go or not go" assumption for electrons originating from further inside must be looked at with caution because the angular straggling and energy degradation of electrons in a solid can not be attributed only to such violent collisions.

Another obvious mechanism for the production of electrons with velocities close to the velocity of the moving ions ($\vec{v}_e \approx \vec{v}_i$) is electron loss from bound states of the projectile, equivalent to collisional ionization in the moving system. As there is abundance of electrons emitted with low velocities with respect to the parent ion, this mechanism must be considered as very effective.

Experiments performed by Burch et al.(14) and Stolterfoht et al.(15) with 17 to 41 Mev O^{n+} on Ar and O_2 show a strong peak in the electron energy spectra measured at $\theta=90°$ and $25°$ respectively, the intensity of which decreases as n increases. It would be both interesting and desirable to pursue this peak and study its shape down to lower angles, $\theta \to 0°$.

For solid targets there is still another effect which lends itself to the production of "$\vec{v}_e \approx \vec{v}_i$ electrons" emerging from the downstream surface of a foil. According to calculations of Neelavathi et al. (16, 17), fluctuations in the valence electron density induced by the moving ion produce a periodic sequence of potential troughs which are stationary in the moving system. Electrons may be bound in quantum states in these troughs and obliged to "wake-ride" with velocity v_i. The depth of these troughs is proportional to the ion charge and inversely proportional to v_i. When the ion emerges through the solid surface, the wake disappears leaving the electrons with velocities centered about $\vec{v}_i$ and with a velocity distribution in the moving system given initially by the Fourier transform of the emerging wave packet.

While experiments determine double differential electron distributions in energy E_e and angle θ, the mechanisms for electron ejection are best described in a velocity space as shown in Fig. 3.

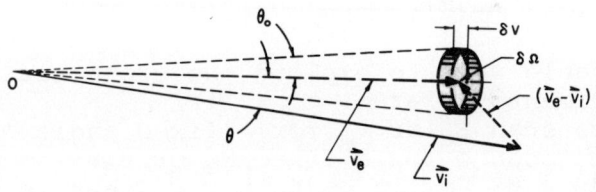

Fig. 3. The $\vec{v}_e$, $\vec{v}_i$ vector diagram in the v_e-space. The shape of the CEC-cross section distribution is described theoretically in terms of $\vec{v}' = \vec{v}_e - \vec{v}_i$. The "resolution volume", made up of the experimental uncertainties in the velocity δv_e and the solid angle $\delta \Omega$, is also shown.

It is therefore convenient to discuss also experimental aspects and results in terms of the ion velocity v_i, the electron velocity v_e and the vectorial electron-ion velocity difference $\vec{v}' = \vec{v}_e - \vec{v}_i$. It is indeed in this velocity space where the theories for ECC (8-13) discuss the cross sections.

According to Dettmann (11) and Dettmann et al. (12) the cross section for emission of electrons with $\vec{v}_e \approx \vec{v}_i$ is, for sufficiently large v_i, given by:

$$\frac{d\sigma}{d\vec{v}_e} = F(\vec{v}_i)\, f(\vec{v}') \qquad (1)$$

Here f describes the shape of the ECC cross section as a function of $\vec{v}'$ and F gives the dependence of the yield of the process on $\vec{v}_i$.

Obviously, in the $\vec{v}_e$ space of Fig. 3 the ion velocity $\vec{v}_i$ determines an axis of symmetry of $d\sigma/d\vec{v}_e$ as given by Eq. 1. Dettmann gives the shape f of the distribution $d\sigma/d\vec{v}_e$ as:

$$f(\vec{v}') = \Pi\alpha \frac{e^{\Pi\alpha}}{\sinh \Pi\alpha}, \text{ with } \alpha = \frac{zv_o}{|\vec{v}'|}, \qquad (2)$$

where v_o is the Bohr velocity and z the charge of the projectile. This expression has been previously obtained by Macek (8). It also results from Salin's theory (9) if one neglects $1/v_e$ when compared with $1/|\vec{v}'|$. Eq. 2 is the natural consequence of describing the final state of the electron with Coulomb wave functions centered at the projectile (18).

According to Eq. 2, the cross section given by Eq. 1 results spherically symmetric in the velocity frame centered at v_i, where it diverges as $|\vec{v}'|^{-1}$ since for large α, $\exp \Pi\alpha \simeq \sinh \Pi\alpha$ and

$$f(\vec{v}') \simeq \Pi z v_o / |\vec{v}'| . \qquad (3)$$

Suitable ways to explore the cross section $d\sigma/d\vec{v}_e$ experimentally are:
  a) to scan E_e or v_e for a fixed angle θ,
  b) to scan θ for a fixed E_e or v_e.
If in a) $\theta = 0°$, or in b) $E_e = E_{eq}$, that is $v_e = v_i$, the cross section is measured "over the peak".

The shape and width of the experimental double differential distributions in energy E_e or velocity v_e and in ejection angle θ are influenced by the resolution of a particular apparatus in energy (δE_e) or velocity (δv_e) and in solid angle ($\delta\Omega$). In the velocity space of Fig. 3 this resolution is represented by a volume element

$$v_e^2 \, \delta v_e \, \delta\Omega . \qquad (4)$$

The apparatus integrates over the cross section $d\sigma/d\vec{v}_e$ contained in this volume.

For a divergent cross section such as given by Eqs. 2 or 3, the measured peak shape would depend drastically on any experimental resolution in v_e and Ω, however narrow. Only in the limiting case $\delta v_e \to 0$ and $\delta\Omega \to 0$, the measured peak would tend towards the diverging theoretical peak of vanishing width at half maximum. On the other hand, if $f(\vec{v}')$ is assumed to be bound, the experimental peak shape will approach the actual one when δv_e and $\delta\Omega$ become sufficiently small compared to the measured peak width.

For the specific case of measuring electron spectra as a function of v_e at $\theta = 0$, Dettmann (11, 12) derived an analytical expression for an "experimental cross section" by integrating Eq. 3 over a solid angle $\delta\Omega$, given by a cone of half angle θ_o as determined by the angular acceptance of a specific electron analyzer. In the velocity space of Fig. 3 this is equivalent to an integration over only a surface element given by $v_e^2 \delta\Omega \approx v_i^2 d\Omega$. The result obtained is:

$$\frac{d\sigma(v_i,\theta_o)}{dv_e} = 4\Pi z v_o\, F(v_i)\, \{(v'+v_i^2\theta_o^2)^{1/2} - |v'|\} \tag{5}$$

This equation results in a cusp, similar in shape to those observed experimentally. The half width at half maximum (HWHM) of this cusp is:

$$\Delta v_e = \frac{3}{4} v_i \theta_o \tag{6}$$

Eqs. 5 and 6 are an example of the above mentioned decisive influence of the experimental resolution, as determined by θ_o, on the measured peak width, if the discussion is based on a divergent cross section, as given by Eq. 3. It is also seen in Eq. 6 that $\Delta v_e \propto v_i$. This results because for constant $\delta\Omega$, the surface over which the integration of Eq. 3 is performed increases as v_i increases and because this equation depends only on $|v'|$, i.e., the velocity distribution of the electrons in the frame moving with the projectile is independent of the velocity of the latter.

We have to bear in mind that the discussion of Dettmann leading to Eqs. 5 and 6 is incomplete in that the finite experimental resolution in electron energy, δE_e, or velocity δv_e, has not been accounted for.

II. EXPERIMENTS

Our present experimental knowledge of ECC at

angles of ejection near 0 arises from a series of investigations carried out at the Universities of Nebraska, Sussex, Georgia and Western Ontario.

A. Measurements at the University of Nebraska

Crooks and Rudd (7) have used beams of protons ranging in energy from 100 to 300 keV and He gas as target. The angle of observation was fixed at 0° with an angular spread of 1.4°. The energy of the electrons was measured with a parallel plate electrostatic analyzer having an energy resolution of 1.6%. A slit in the back plate in the analyzer allowed the proton beam to pass through and be collected by a Farady cup.

B. Experiments at the University of Sussex

Lucas and co-workers (12, 19-21) have used H^+, He^+, H_2^+ and N^+ ion beams in the energy range 60-300 keV with C and Au target foils; 300-1200 keV protons on H_2, He, Ne and Ar gases and 200-300 keV protons in single crystal Au foils, in both <111> channeling and random directions. The $\vec{v}_e \approx \vec{v}_i$ electrons were observed at the fixed angle of 0° with angular dispersions reported between 2.9° and 4.6°. The velocity of the electrons was measured by means of a 180°-deflection magnetic analyzer of 1% velocity resolution, equivalent to an energy resolution of 2% (FWHM).

C. The University of Georgia Experiments

Duncan and Menendez (21-25) have reported measurements using 350-1000 keV H^+ and H_2^+ ions on carbon foils and He and Ar gas targets, and 2 MeV He^+ and He^{++} ions on Ar gas targets. The angle of observation was changed between 0° and 10° with an acceptance half angle of 0.44° and 0.36° in different experiments. The energy analysis were performed by means of a hemispherical electrostatic analyzer with a 2% FWHM resolution. Since the analyzer blocks the ion beam, the measurements were normalized to the signal from a solid-state detector which sampled a fraction of the scattered ion beam. At small θ when the ion beam hits the analyzer aperture determining the angular acceptance of the measured electrons, a large background of secondary electrons produced at this aperture made measurements difficult.

Fig. 4 shows typical spectra obtained by scanning E_e at different angles Θ. It is observed that by increasing Θ from 0.1° to only about 3° the peak height diminishes by almost an order of magnitude, demostrating the sharpness of the angular distribution of the ejected electrons. Such angular distributions were evaluated at Georgia as single differential distributions by integrating over the energy spectra obtained at different angles.

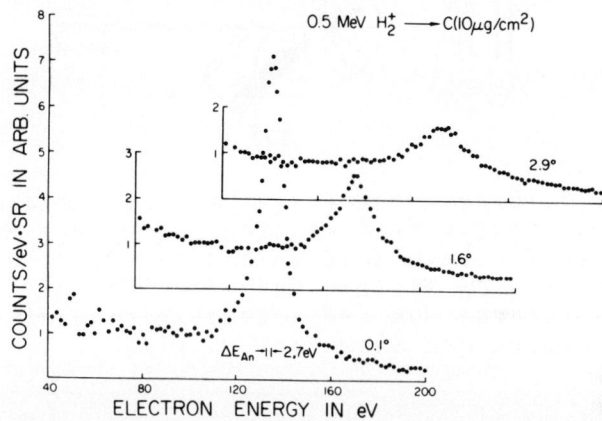

Fig. 4. Energy spectra of secondary electrons emitted at three different angles with respect to the direction of a 0.5 MeV H_2^+-beam passing through a 10 µg/cm² C-foil (from: Duncan and Menendez, ref. 23).

D. Measurements at the University of Western Ontario

Meckbach et al. (26) measured double-differential (in energy and angle) distributions of $\vec{v}_e \approx \vec{v}_i$ electrons for 199 to 489 keV H^+ and 350 to 433 keV H_2^+ on thin carbon foils. The energies of the electrons were measured by means of a sector of a cylindrical electrostatic energy analyzer (see Fig. 5) of 1% resolution (FWHM). The foils were mounted at the entrance focal point of the analyzer, and the angle was scanned by rotating the analyzer around this point, while keeping the foil fixed. The angular acceptance of the system was determined at the exit side of the analyzer and estimated to be less than 1°. Slots in the analyzer plates allowed the beam to pass through without being intercepted, and to be finally collected in a Faraday cage.

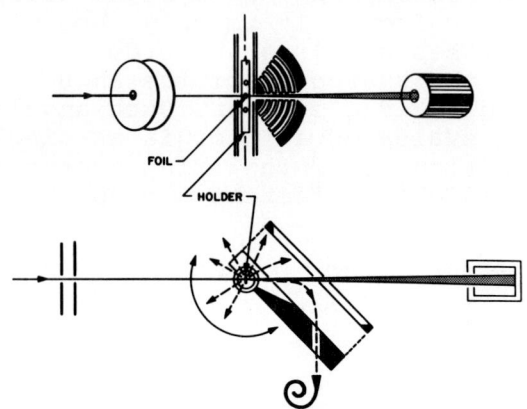

Fig. 5. Experimental equipment for measuring energy and angular distributions of $\vec{v}_e \approx \vec{v}_i$ -electrons produced in ion beam-foil interactions, as used at the University of Western Ontario. The foil holder, ion beam, Faraday cup and rotatable coaxial cylindrical analyzer are shown in two planes. (from ref. 26).

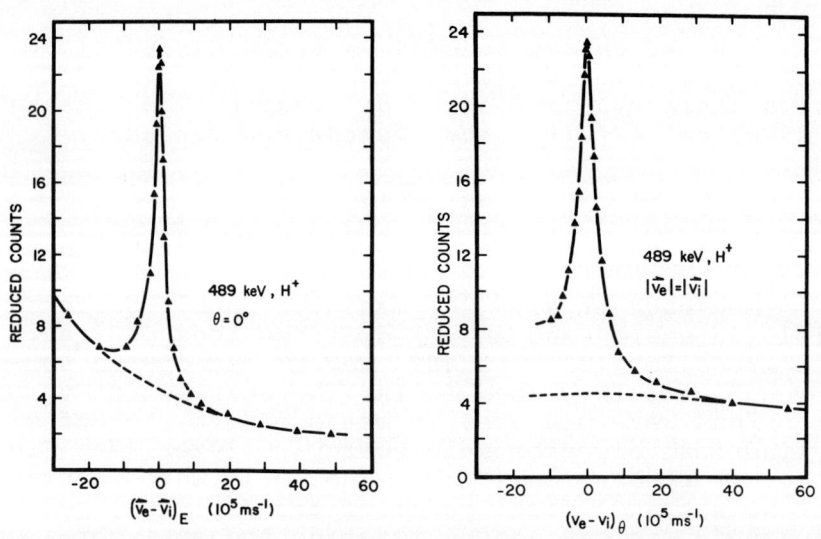

Figs. 6a,b. Reduced electron spectra, measured a) as a function of E_e at $\theta=0°$ and b) as a function of θ at $E_e = E_{eq}$, but represented in terms of the associated velocity differences $\vec{v}'_E = (\vec{v}_e - \vec{v}_i)_E$ and $\vec{v}'_\theta = (\vec{v}_e - \vec{v}_i)_\theta$ (ref. 26).

Figs. 6a and 6b show: a) a typical energy spectrum obtained at $\Theta=0$ represented in terms of $\vec{v}_E'$, the electron-ion velocity difference associated with the measurement of energy and b) a typical angular spectrum obtained at $E_e=E_{eq}$ $v_e = v_i$ and represented in terms of $\vec{v}_e'=v_i\Theta$, the electron-ion velocity difference associated with the measured angle. This method of representation obviously eases comparison with theory and sets the discussion of the double-differential electron distributions obtained experimentally by varying energy and angle, on an equal footing.

III. THE SHAPE OF THE CHARGE EXCHANGE TO THE CONTINUUM PEAKS

Let us first review the existing experimental information concerning the shape and width of the peak in the electron emission, observed when $\vec{v}_e \rightarrow \vec{v}_i$. Most of the measurements have been performed by taking electron energy spectra as a function of the electron energy E_e or velocity v_e at an angle $\Theta=0°$. In Fig. 2, the experimental cusp-shaped peak obtained by the Nebraska group (7) measuring electrons emitted within a cone of $\Theta_0=1.4°$ half angle with respect to the beam direction is seen to lie between cusps calculated using Macek's theory for $\Theta=0$ and $\Theta=1.4°$. A qualitative agreement of experiment and theory is apparent.

A more quantitative accordance of experimental peak shape with Eq. 5 of Dettmann has been shown by Cranage and Lucas (20) for 709 keV H^+ colliding with Ne (Fig. 4 of their paper). It is interesting to note that given the relative resolution in velocity $\delta v_e/v_e=1\%$ and the solid angle of acceptance $\delta\Omega$ (determined by $\Theta_0=2.9°$) of their electron spectrometer, their resolution volume in $\vec{v}_e$ -space is a flat disc of only 1.2×10^5 m/s thickness, but 12×10^5 m/s diameter. Consequently, in this specific case the integration leading to Eq. 5 could be performed over $\delta\Omega$ only and not considering the finite resolution δv_e.

However, careful inspection of experimental results, particularly those obtained recently at Georgia and Western Ontario, which include angular distributions at low Θ, show significant discrepancies with the results of theories of ECC.

We have seen that, according to Eqs. 5 and 6, the half widths Δv_e at half maximum of the electron spectra is proportional to v_i, the ion velocity. As an immediate consequence it follows that the corresponding half widths in electron energy ΔE_e should be proportional to E_{eq}, the energy of an electron with $v_e=v_i$. According to measurements performed at Georgia (21-25), the FWHM of ECC peaks measured as a function of electron energy at a given deflection angle (including $\Theta=0$) increase approximately linear with ion energy such that, in agreement with Eq. 6, $\Delta E/E_{eq}$ is approximately independent of E_{eq}. However, it is observed from Fig. 3(B) of ref. (22) and also stated in ref. 23, that this relationship is not exact, since at larger E_{eq} the ratio $\Delta E_e/E_{eq}$ gives results consistently smaller. This is furthermore in accordance with Fig. 2 of Crooks and Rudd (7) which shows, for protons interacting with He, an increase in peak width which is not proportional to E_{eq}, but rather such that the corresponding widths calculated in terms of electron velocity tend to level off to a constant value above 200 keV ion energy.

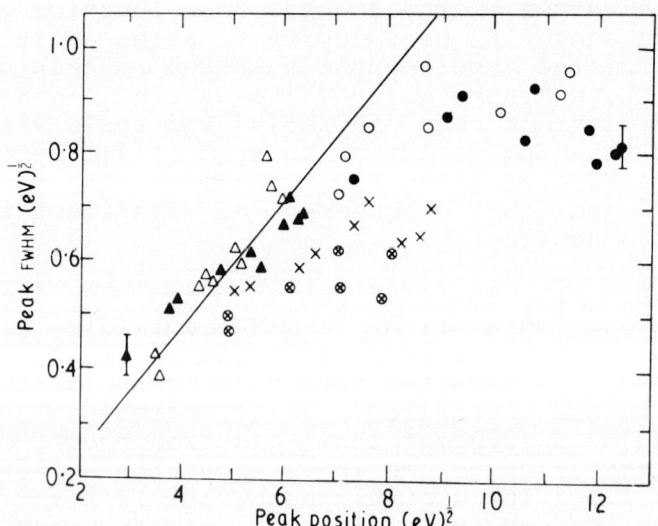

Fig. 7. *The variation of the FWHM of electron velocity spectra at $\Theta=0°$ with 60-300 keV H^+, He^+ and H_2^+ on C and Au foils as a function of the peak position (ion velocity).* ● $H^+\rightarrow C$; ○ $H^+\rightarrow Au$; ▲ $He^+\rightarrow C$; △ $He^+\rightarrow Au$; ✕ $H_2^+\rightarrow C$; ⊗ $H_2^+\rightarrow Au$ *(from: Dettmann et al. ref. 12).*

A similar dependence is seen in Fig. 7 which shows FWHM of velocity spectra taken at the University of Sussex (12) for atomic (H^+, He^+) as well as molecular (H_2^+) ions passing through carbon and gold foils. Also shown is the linear dependence of the FWHM with v_i as predicted by Eq. 6 for $\Theta_o=4.6°$. For H^+ and He^+ at the lower velocities, an increase of the widths is observed which, within the spread of the experimental data, could suggest an agreement with Eq. 6. However, at peak electron velocities above $7(eV)^{1/2}=4.2 \times 10^6$ m/s, the FWHM begin to level off into the approximate constant value of $\approx 0.9(eV)^{1/2}$. On the other hand, the widths obtained with H_2^+ seem to level off already above $5(eV)^{1/2}= 3 \times 10^6$ m/s to a lower constant value of approximately $0.6 (eV)^{1/2}$.

The measurements performed at Western Ontario (25) confirm the independence of the cusp widths (when represented in terms of $\vec{v}'$) on ion velocity.

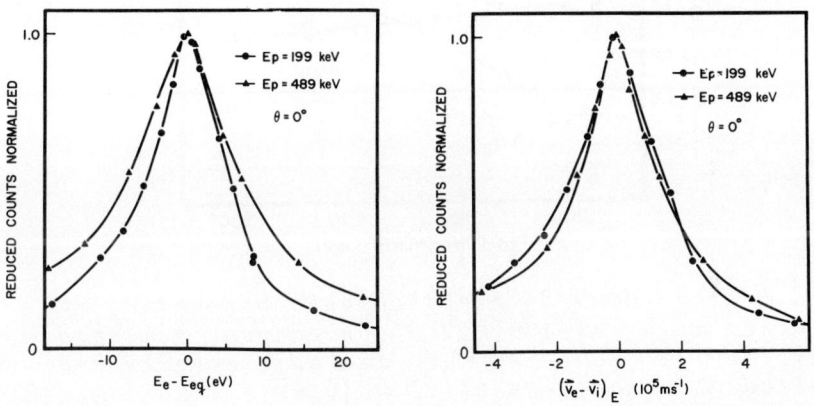

Figs. 8a,b. Normalized electron energy spectra produced at $\Theta=0°$ by 199 and 489 keV H^+ on $10 \mu g/cm^2$ carbon. These spectra are plotted a) as a function of the energy difference (E_e-E_{eq}) and b) as a function of the associated velocity difference $\vec{v}'_E = (\vec{v}_e-\vec{v}_i)_E$ (from: Meckbach et al. ref. 26).

Fig. 8a shows electron energy spectra obtained at $\Theta=0$ with 199 and 489 keV protons, after having substracted the background due to secondary electron emission not contributing to the observed peak (see Fig. 6a) and having normalized the peak heights to 1. These spectra are represented in terms of the energy difference E_e-E_{eq}. Here again we observe a broader distribution at the higher ion energy. However, as shown in Fig. 8b, when these spectra are represented in terms of the electron-ion velocity difference $\vec{v}_E^!$ associated with (E_e-E_{eq}) they become the same within the 10% experimental uncertainties. Accordingly, in Fig. 9, the HWHM of the peaks (ΔE_e) when represented as a function of (E_e-E_{eq}) are seen to increase with increasing proton energy whereas the half widths $(\Delta v)_E$ obtained from the same distributions represented in terms of the associated velocity difference $\vec{v}_E^!$ are seen to be independent of the proton energy, and not to increase proportionally to $E_i^{1/2}$ as would follow from Eqs. 5 and 6.

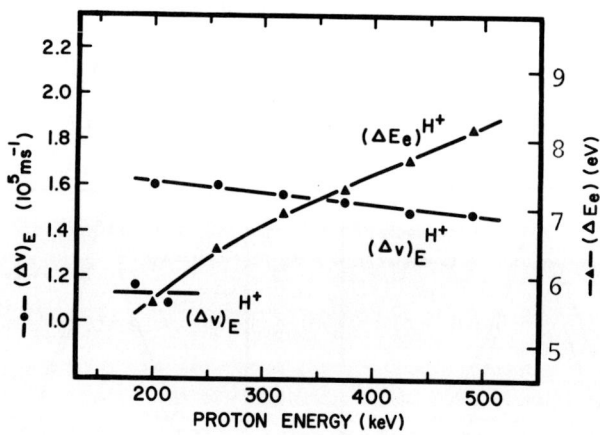

Fig. 9. HWHM of electron energy spectra, ΔE_e, obtained with 199-489 keV H^+ on carbon, compared with HWHM, $(\Delta v)_E$, resulting when the same spectra are plotted as a function of $\vec{v}_E^! = (\vec{v}_e - \vec{v}_i)_E$. Half widths $(\Delta v)_E$, resulting with incident H_2^+, are also shown. They are smaller by a factor of 0.7 (ref. 26).

We will now look at the shape and width of the peaks resulting from measurements of angular distributions. Fig. 10 shows normalized single differential angular distributions as obtained in Georgia (22) at $E_i = 0.35$ and 1.0 MeV. It is seen that the distribution obtained at the lower ion energy is narrower by a factor of about $0.6 \simeq (0.35/1)^{1/2}$ (the ratio of the ion velocities) than for the higher energy. This is confirmed in Table I of ref.26 where it is observed that the half widths $\Delta \theta$ reported by Duncan and Menendez (22) decrease when E_i increases from 0.35 to 1.0 Mev, whereas the half widths $(\Delta v)_\theta = v_i \theta$ stay constant within 3%. This result calls our attention if we remember that the shape of the cross section $d\sigma/d\vec{v}_e$ (Eq. 1) is expressed in terms of $\vec{v}'$ and that the electron-ion velocity difference produced by changes in θ is $\vec{v}'_\theta \simeq \vec{v}_i \theta$.

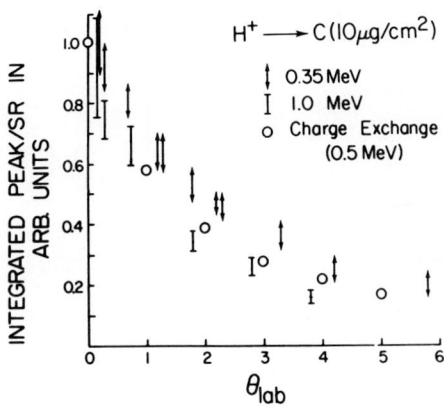

Fig. 10. Normalized single differential angular distributions obtained with H^+ on carbon. The width of the distribution obtained with 1.0 MeV H^+, as compared to the width resulting at 0.35 MeV is smaller by a factor of 0.6 (from: Duncan and Menendez, ref. 23).

At Western Ontario the measured doubly differential angular distributions were interpreted in terms of this variable. Let us then have a look at Fig. 11 a,b where these distributions, again obtained with 199 and 489 keV protons traversing carbon foils, are represented as a function of θ and $\vec{v}'_\theta$ respectively. In Fig. 11a it is confirmed that the distribution in θ obtained at the higher proton energy is narrower. In Fig. 11b it is strikingly shown that the two measured angular distributions become identical when represented in terms of $\vec{v}'_\theta$. Correspondingly, in Fig. 12, the half widths $\Delta\theta$ of the angular distributions measured between 199 and 489 keV are seen to decrease with increasing E_i whereas the half widths $(\Delta v)_\theta \simeq v_i \Delta\theta$ are independent of the proton energy.

The conclusion one can draw from these results is that the shape of the cross section is described by a function $f(\vec{v}')$ which depends only on the vectorial electron-ion velocity difference and that the experimentally measured peak shapes truly reproduce this function. This implies that the results shown in Table I (ref.26) and Figs.8 through 12 have not been

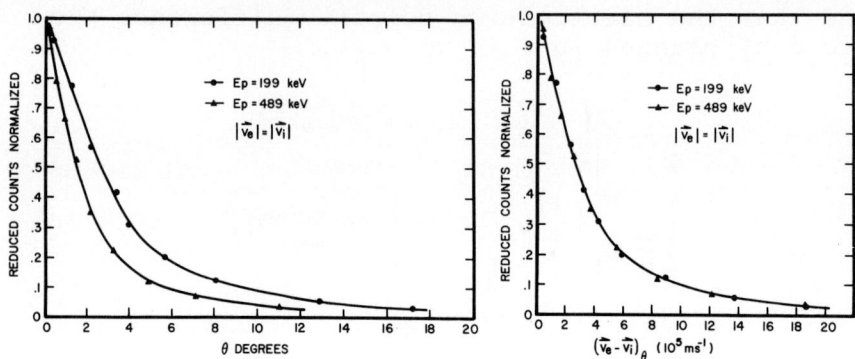

Figs. 11a, b. Normalized angular spectra taken at $v_e = v_i$ for 199 and 489 keV H^+ on $10 \mu g/cm^2$ carbon. These spectra are plotted a) as a function of θ and b) as a function of the associated velocity difference $\vec{v}'_\theta = (\vec{v}_e - \vec{v}_i)_\theta \simeq v_i \theta$ (ref. 26).

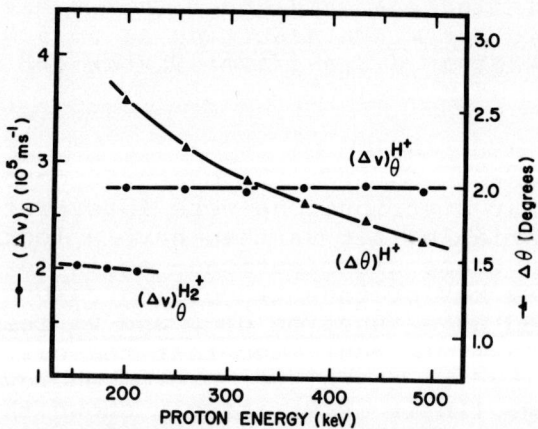

Fig. 12. HWHM of angular spectra, $\Delta\theta$, obtained with 199-489 keV H^+ on carbon, compared with HWHM $(\Delta v)_\theta$ resulting when the same spectra are plotted as a function of $\vec{v}'_\theta = (\vec{v}_e - \vec{v}_i)_\theta$. Half widths $(\Delta v)_\theta$ resulting from incident H_2^+ are also shown. They are smaller by a factor of 0.7 (ref. 26).

sensibly influenced by the experimental resolutions in energy and angle. Then, and according to what was explained in the introduction, we conclude that $f(\vec{v}')$ cannot be an unbound function of $\vec{v}'$ as it results from existing theories of ECC.

A further consequence of these theories was that $f(\vec{v}') = f(|\vec{v}'|)$; i.e. the cross section resulted spherically symmetric in the space of Fig. 3. However, if we compare Figs. 6a and 6b, we find that the peak in $\vec{v}'_\Theta$ is broader than the peak in $\vec{v}'_E$, as can also be seen in that the widths $(\Delta v)_\Theta$ in Fig. 12 are larger by a factor of about 1.8 than the widths $(\Delta v)_E$ in Fig. 9. Careful inspection of the results obtained at the University of Georgia for solids as well as for gases also confirm this asymmetry, i.e. $(\Delta v)_\Theta$ substantially larger than $(\Delta v)_E$, as can be derived from their Figs. 2 and 4 (ref. 23) and Figs. 1 and 2 (ref. 24). We must point out, however, that in this case the widths $\Delta\Theta$ are referred to the single differential distributions obtained by integrating the energy distributions at different angles.

If we accept that the measured distributions are not sensibly influenced by the experimental resolution in energy and angle, the observed asymmetry cannot be attributed to an asymmetry in the shape of the resolution volume of Fig. 3 and must be accepted as real. The conclusion is then that the distribution of the electron yield, given by $d\sigma/d\vec{v}_e \propto f(\vec{v}')$ is an angular dependent vectorial function. We find here again a discordance with the results of the ECC theories.

In the case of single collisions, the asymmetry may result from the distortion of the coulomb wave of the ejected electron set up by the field of the residual target ion. The influence of the field of the target ion has so far not been taken into account in the theory.

The described asymmetry for solid targets can be analyzed on the grounds of the theory of electron capture into the potential throughs occuring in the polarization wake of fast ions in solids (16,17).

The emerging electron wave packets have been calculated (16,17) to be wider in the direction of motion of the ions than in a direction perpendicular to it; therefore the inverse will be true for the velocity distribution, i.e. $(\Delta v)_E$ will be smaller than $(\Delta v)_\Theta$, as observed experimentally.[1]

---

[1] We would like to call the attention of the reader to the fact that Eqs. 4 and 6 of ref. 17 are in error. Eq. 4 should read $Wbe^{-a}|1-e^b E_1(b)| = 2$, and in Eq. 6, α and β should be interchanged. We thank N.R. Arista for recalculating the results of this paper.

On the other hand, contrary to experimental evidence, the peaks expected from the "wake-riding" theory are Gaussian in shape and have widths about three times as large as those observed experimentally in refs. (23) and (26). Furthermore, these widths result independent on the angle of observation, again in discordance with experiments (23). Finally, according to results obtained at Georgia (24), the energy spectra obtained in beam-foil experiments look identical to those measured under otherwise the same experimental conditions but with a crossed molecular beam target.

Does this evidence lead to a conclusion against "wake-riding" of electrons behind an ion moving inside a solid? When these electrons emerge through the surface they are close to the ion, in physical and velocity space? This is a favourable condition for them to be captured by the Coulomb field of the moving ion into a continuum state, their initial distribution being changed into the observed cusp. This picture might furnish an alternative model to the last-layer hypotesis of Dettmann.

Let us now look for evidence for "charge transfer into the continuum" of electrons originating from projectile ionization, a process called "electron loss into the continuum" in section I. In Georgia (25), electron spectra from He^{++} interacting with Ar were found to be indistinguishable to those obtained with He^+ for which, as discussed later in section IV, the main contribution to the electron ejection is from electron loss. The energy distributions measured at 2° and single differential angular distributions were compared with the electron scattering model for projectile ionization of Burch et al. (14). In this model, the projectile electrons are considered as scattered in the field of the target atom. The initial momentum distribution in bound states of the projectile is taken into account, but the Coulomb interaction of the scattered electrons with the moving ion is not considered. Large discrepancies were found between the measured cusps and the much broader and flatter distributions resulting from the scattering model. We may conclude that here again the electrons emitted from the projectile are found in a favourable initial condition to be "carried along" by the Coulomb field of the moving ion in a continuum state and that therefore at small ejection angles with respect to the ion beam, the electron distribution is shaped into the observed cusps, characteristic of the "charge exchange to the continuum" mechanism.

We finally compare the widths of electron energy and angular distributions obtained with protons to those resulting from the interaction of molecular H_2^+ ions at equal velocities, with gaseous and solid foil targets. In Fig. 7 we already observed that, according to measurements performed at Sussex with carbon foil targets, the half widths of the electron velocity spectra taken at $\Theta=0$ are smaller for incident H_2^+ than for incident H^+. The ratio of these half widths is approximately 0.65.

This behaviour is also seen in results of angular distribution measurements. Single differential angular distributions measured at Georgia (23) behind carbon foils and with 1.0 MeV H_2^+ were seen to be narrower by a factor of $\simeq 0.6$ compared with those obtained with 0.5 MeV H^+. This has been confirmed by measurements performed at Western Ontario. In Figs. 9 and 12 half widths corresponding to the double-differential energy and angular distributions, represented in terms of $\vec{v}_E'$ and $\vec{v}_\Theta'$, respectively, are narrower for H_2 than for H by the same factor of about 0.7.

The observation of narrower cusps resulting from molecular-ion collisions seems to be typical for ion-solid interactions. There is evidence from measurements performed at Georgia that under single collision conditions the energy spectra obtained with equal-velocity H^+ and H_2^+ projectiles show similar widths. The difference in behaviour of H_2^+ and H^+ in foils can arise from the difference in the potential troughs of the polarization wake wherein the electrons are captured. The spatial correlation of the two protons resulting from the dissociative ionization of H_2^+ (28, 29) would cause wider troughs and correspondingly narrower electron velocity distributions than in the case of equal-velocity H^+ ions.

IV. YIELDS

In the theory of Dettmann (11, 12) for H impact, the yield for ECC at an angle $\Theta=0$ within a cone of semiangle Θ_o is obtained by integrating the differential cross section $d\sigma(v_i,\Theta_o)/d\vec{v}_e$ (Eq. 5) over the velocity interval $v_i-v_w < v_e < v_i+v_w$, assuming $v_i\Theta_o \ll v_w$. The result is: (7)

$$\sigma(v_i,\Theta_o,v_w) \simeq 4\pi^2 z'v_o(v_i\Theta_o)^2 F(v_i) \ln(2v_w/v_i\Theta_o)$$

where $F(v_i)$ has the same dependence on v_i as the

cross section for electron capture into the ground state. Dettmann defines the cross section as $\sigma(v_i, \theta_o, z'v_o)$, i.e. by setting $v_w = z'v_o$, which is in contradiction with the approximation $|v'| \ll z'v_o$ used to derive Eq. 5. However it must be borne in mind that $\sigma(v_i, \theta_o, v_w)$ depends on v_w logarithmically and therefore is not too sensitive to v_w. If the condition $v_i\theta \ll v_w$ is removed, then:

$$\sigma(v_i, \theta_o, v_w) = 4\pi^2 \ z'v_o F(v_i) \left[v_w(V-v_w) + v_i^2\theta_o^2 \ln(V+v_w/V-v_w)\right]$$

with $V = (v_w^2 + v_i^2\theta_o^2)^{1/2}$

(8)

The height of the peak measured with a velocity resolution $\delta v_e \ll v_i\theta_o$ can be derived from Eq. 8 with $v_w = \delta v_e$, yielding:

$$\sigma(v_i, \theta_o, \delta v_e) = 12\pi^2 z'v_o F(v_i) v_i \theta_o \delta v_e \qquad (9)$$

Experimentally, $\delta v_e/v_e$ is constant, therefore, with $v_e \simeq v_i$:

$$\sigma(v_i, \theta_o, \delta v_e) = 12\pi^2 z'v_o(v_i)^2 F(v_i) \frac{\delta v_e}{v_e} \theta \qquad (10)$$

which has the same velocity dependence as Eq. 7, if the slowly varying dependence of the logarithmic term with v_i is neglected.

For evaluating the yield, the use of a constant v_w in Eqs. 7 and 8 is not justified on the grounds of the theory of Dettmann, since it predicts that the width of $d\sigma(v_i,\theta_o)/dv_e$ is $3v_i\theta_o/4$ (Eq. 6) and therefore, proportional to the ion velocity. The experimental results of Meckbach et al. (26) have shown however, that the widths of the distributions expressed in terms of $\vec{v}'$ are nearly independent of velocity, in contradiction with theory. We could therefore expect on the basis of experimental evidence, that the velocity dependence of σ for C targets is contained only in the function $F(v_i)$; i.e. the same velocity dependence as that of the electron capture cross section σ_c. The second Born approximation prediction for σ_c, if the initial state of the electron in the target is described by hydrogenic wave functions, as done by Dettmann, gives a v^{-12} asymptotic behaviour Here, we prefer to use experimental values of the electron capture cross section.

Table I shows a compilation of k values ($\sigma \propto v^{-k}$) for the cross sections for electron capture to bound states (σ_c) and to continuum states (σ_{cc}). A very

TABLE I

Velocity dependence of the cross sections ($\sigma \propto v_i^{-k}$)

Energy range (keV)	Target	k_c	k_{cc} d)	$k' = k_c - k_{cc}$
310-609	H_2	10.6 (a)	9.0 (d)	1.6
298-509	He	9.7 (a)	7.0 (d)	2.7
200-1000	C	6.4 (b)	6.0 (d)	0.4
			6.8 (d)	-0.4
308-1200	Ne	6.6 (c)	6.4 (e)	0.2
497-1207	Ar	4.3 (a)	4.2 (d)	0.1

a) $\sigma_c \propto v_i^{-k_c}$. Values of k_c from Toburen et al. (30) and Tawara and Russek (31).
b) Derived from cross sections on carbon-containing compounds (29).
c) Interpolated from compilation of Tawara and Russek (30).
d) $\sigma_{cc} \propto v_i^{-k_{cc}}$. Values of k_{cc} for gases from Cranage and Lucas (20). Values for C (6.0) from Dettmann et al. (12) and Duncan and Menendez (23), and (6.8) from Meckbach et al. (26).
e) Mean value; k_{cc} varies slowly with velocity.

close agreement is seen between k_c and k_{cc} except for the lighter targets H_2 and He. This shows that $\sigma_{cc}(v_i)/F(v_i)$ does not depend on velocity as $v_i^{-k'}$ with k'=2 as predicted by the theory of Dettmann; rather, k'=k_c-k_{cc} is seen to be nearly zero for the targets C, Ne and Ar. The theoretical dependence $\sigma_{cc}(v_i)/F(v_i) \propto v_i^2$ is a consequence of the dependence of Eq. 5 on $v_i \theta_o$, where θ_o was the experimental acceptance angle. The similarity of k_{cc} and k_c for the heavier targets gives us an independent support to the conclusion we have reached above in that the experimental angular acceptance and energy resolution does not affect much the measured peak shapes.

According to measurements of Dettmann, Harrison and Lucas (DHL) (12), at velocities $v \simeq 2.8 v_o$, σ_{cc} decreases with v more slowly, reaching a maximum at v near $2v_o$ and then decreasing at lower velocities. These authors normalize the yield to the current of the ion beam which has emerged from the foil. If electron capture processes occur at the surface (i. e. the projectile is ionized inside the solid (12, 31)), it would be better, to get consistency with the results at higher energies, to normalize to the total ion beam flux which goes through the exit surface. If we divide the results of DHL for H by $F_1 - F_{-1}$ (Where F_i is the fraction of the beam with charge i times the proton charge) and those for He by $F_1+2F_2-F_{-1}$ using measured values of these quantities (32, 33 and unpublished data from our laboratory), it is found that the maxima and decay of the yields for decreasing velocities is no longer apparent. The large spread in the data points preclude us however from making any definitive statement about the behaviour of the yield at low velocities.

Let us now look at the difference in yields for different projectiles. The ratio of the yields per proton for equal-velocity H_2^+ and H^+ ions on carbon foils is reported by DHL to be about 2 at $v_i \approx 2v_o$. DHL state that the difference in yields is almost certainly due to the spatial correlation of the H^+ molecular fragments. It is difficult, however, to reconcile this with the measurements by Duncan and Menendez (23) where the yields per proton were found to be the same at $v_i \approx 4v_o$, especially if one takes into account that in the latter experiment, foils with thicknesses about twice of those used by DHL were tried. The average distance between protons at the exit surface of the foil is in first order determined by the transit time of the projectiles in the foil (34) which, in the latter experiments, was about the same as in the experiments by DHL.

In the case of interactions of H^+ and H_2^+ on He, Duncan and Menendez (24) found the electron yield per proton to be approximately 17 times greater for H_2^+ than for H^+; and attributed the excess electrons to come from the H_2^+ projectiles. Similarly, Menendez et al. (25) have found that the yield in 2 MeV He^+ Ar collisions was 3 times larger than that for He^{++} Ar collisions at the same energy. The difference can be attributed to electrons from He^+. It must also be taken into account that in the ECC process, He^{++} should give higher yields than He^+ due to its higher effective charge.

The electron loss peak has been previously observed by different workers at electron ejection angles larger than 20°. Burch et al. (24) have given a simple theory for this process in which the final projectile-electron interaction is not taken into account. As mentioned above, for angles near 0° this interaction, which is also responsible for the shape of the ECC peak, should be included and it should also be included in the simple model of Neelavathi et al.(16,17) for capture of "wake-riding" electrons from solids. This final Coulomb interaction "dragging" the electrons into the path of the projectile is then responsible for the similar shape of the ECC peak in gaseous and solid targets and of the electron loss peak. Single electron detachment from negative ions would, in turn, give a broader peak due to the absence of this final Coulomb interaction. This is supported by yet unpublished results obtained at Georgia with H^- projectiles.

V. CONCLUSIONS

The recent availability of high resolution double differential cross section measurements, in energy and angle of ejection of electrons with velocities $\vec{v}_e$ close to the projectile velocity $\vec{v}_i$, in ion-atom and in ion-solid collisions, have made stringent tests of theories possible. While the available theories succeed in describing the processes qualitatively, they are seen to fail in detail. In particular the experimental evidence is against a divergent behaviour of the differential cross sections. The reason for this may lie in the competing character of the process of radiative capture into bound states of electrons ejected with $\vec{v}_e$ very close to $\vec{v}_i$.

Furthermore it has been shown that the widths of the measured electron distributions, evaluated in terms of $\vec{v}' = \vec{v}_e - \vec{v}_i$, the electron-ion velocity difference, are narrower in the direction of the beam than in a transverse direction. This is against the predictions of electron capture into the continuum (ECC) theories and indicates a possible influence of the field of the residual ion in distorting the Coulomb wave of the ejected electrons.

The theory of electron capture into the polarization wake of ions in solids accounts for bound double-differential distributions and for their asymmetry in a qualitative way. The fact that the shapes of the electron distributions resulting from "wake-riding" are different and that their widths are too large compared with experiments, suggest that a revision of this theory is needed with the inclusion of the Coulomb electron-ion interaction after their emergence into vacuum. This theory can also account for the smaller widths of the electron distributions obtained with H_2^+ on solids as compared with equal-velocity protons, if one takes into account that the effect of the spatial correlation of the two protons, resulting from the dissociative ionization of H_2^+, is to produce wider potential troughs in the wake where electrons are captured. More work is needed, however, to ascertain the relative importance of capture into "wake-riding" states as compared to ECC from collisions with surface atoms.

In the case of $\vec{v}_e \simeq \vec{v}_i$ electrons coming from ionization of the projectile there is few data available but they suggest nevertheless that for H_2^+ and He^+ ions, this process is more probable than ECC. Here, a revision of theories for electron loss seems necessary, taking into account the final electron-ion Coulomb interaction.

Finally, it follows from above that more work is needed to test the suggestions, to perfect the theories and to explore effects due to higher z-ions, molecular ions and different gaseous and solid targets over a wide projectile energy-range. Work along these lines is in progress at our institution.

VI. AKNOWLEDGEMENTS

The authors acknowledge helpful and stimulating discussins with N. Arista and V.H. Ponce. One of the authors (W.M.) wants to express his thanks to T.M. Buck for the hospitality received during the International Workshop on Inelastic Ion-Surface Collisions at Bell Laboratories.

VII. REFERENCES

* Work supported in part by the Multinational Project in Physics of the Organization of American States (OAS).
1. G.N. Ogurtsov, Rev. Mod. Phys. 44, 1 (1972).
2. Y.K. Kim, ANL-Report 7960 (1972).
3. S.T. Manson, L.H. Toburen, D.H. Madison, N. Stolterfoht, Phys. Rev. A12, 60 (1975).
4. M.E. Rudd, Radiat. Res. 64, 153 (1975).
5. M.E. Rudd and J. Macek, "Mechanisms of Electron Production in Ion-Atom Collisions", Case. Stud. At. Phys. 3, 47 (1972).
6. M.E. Rudd, D. Gregoire and J.B. Crooks, Phys. Rev. A.3, 1635 (1971).
7. G.B. Crooks and M.E. Rudd, Phys. Rev. Lett. 25, 1599 (1970).

8. J. Macek, Phys. Rev. A1 235 (1970).
9. A. Salin, J. Phys. B, 2, 631 and 1255 (1969).
   J. Phys. B, 5, 979 (1972).
10. J.B. Band, J. Phys. B, 7, 2557 (1974).
11. K. Dettmann, Proc. Conf. on Interaction of Energetic Charged Particles with Solids, Istanbul, BNL-Report 50336 (1972).
12. K. Dettmann, K.G. Harrison and M.W. Lucas, J. Phys. B, 7, 269 (1974).
13. K. Dettmann, M.N. Khan and M.W. Lucas, J. Phys. C, 9, 1879 (1976).
14. D. Burch, H. Wieman and W.B. Ingalls, Phys. Rev. Lett. 30, 823 (1973).
15. N. Stolterfoht, D. Schneider, D. Burch, H. Wieman and J.S. Risley, Phys. Rev. 33, 59 (1974).
16. V.N. Neelavathi, R.H. Richie and W. Brandt, Phys. Rev. Lett. 33, 302 (1974).
17. V.N. Neelavathi and R.H. Richie, in "Atomic Collisions in Solids" (S. Datz, B.R. Appleton and C.D. Moak, Eds.) Vol. 1, p. 289. Plenum Press, New York and London, 1975.
18. L.D. Landau and M.E. Lifschitz, "Quantum Mechanics, Nonrelativistic Theory", p. 579 ff. Pergamon Press, Inc. London 1958.
19. K.G. Harrison and M.W. Lucas, Phys. Lett. 33A 142 (1970); 35A, 402 (1971).
20. R.W. Cranage and M.W. Lucas, J. Phys. B, 9, 445 (1976).
21. M.G. Menendez and M.M. Duncan, in "Beam-Foil Spectroscopy" (J.A. Sellin and D.J. Pegg, Eds.) Vol 2, p. 623. Plenum Press New York and London 1976.
22. M.G. Menendez and M.M. Duncan, Phys. Lett. 54A, 409 (1975).
23. M.M. Duncan and M.G. Menendez, Phys. Rev. A13 566 (1976).
24. M.M. Duncan and M.G. Menendez, Phys. Lett. 56A, 177 (1976).
25. M.G. Menendez, M.M. Duncan, F.L. Eisele and B.R. Junker, Submitted to Phys. Rev. July 1976.
26. W. Meckbach, K.C.R. Chiu, H. Brongersma and J. Wm. McGowan, Submitted to J. Phys. B. Sept. 1976.
27. D.S. Gemmel, J. Remillieux, J.C. Poinzat, R.E. Holland and Z. Vager, Phys. Rev. Lett. 34, 1420 (1975).
28. Z. Vager, D.S. Gemmel and B.J. Zabransky, Submitted to Phys. Rev. May 1976.
29. W. Brandt, in "Atomic Collisions in Solids", (S. Datz, B.R. Appleton and C.D. Moak, Eds.)

Vol. 1, p. 261. Plenum Press, New York and London, 1975.
30. L.H. Toburen, M.Y. Nakai and R.A. Langley, Phys. Rev. 171, 114 (1968).
31. H. Tawara and A. Russek, Rev. Mod. Phys. 45, 178 (1973).
32. K.N. Berkner, J. Bronstein, R.V. Pyle and J.W. Stearns, Phys. Rev. A6, 278 (1972).
33. B.T. Meggitt, K.G. Harrison and M.W. Lucas, J. Phys. B. 6, L362 (1973).
34. A.D. Bacher, E.A. Mc Clatchie, M.S. Zisman, T.A. Weaver and T.A. Tombrello, Nucl. Phys. A181, 453 (1972).
35. D.R. Bates and G. Griffing, Proc. Phys. Soc. (London) A66, 961 (1953).

orientation and alignment in beam tilted-foil spectroscopy

H. G. Berry
Argonne National Lab, Argonne, 60439,
and
Department of Physics, University of Chicago,
Chicago, Illinois 60637

ABSTRACT

The production of atomic orientation and alignment by anisotropic excitation is analyzed. The Stokes parameters of the light emitted from tilted-foil excited ions provide measurements of orientation and alignment and some examples are given. The variations of the Stokes parameters with foil tilt angle, excited state, ion velocity and foil material are compared with existing theories.

I. INTRODUCTION

The anisotropy of an ensemble of excited atoms can be measured in terms of the polarization and spatial distribution of the light emitted during decay to the ground state or other excited state. It is well known that cylindrical symmetry can yield linearly polarized light: Skinner[1] and Skinner and Appleyard[2] observed excited atoms emitting linearly polarized light after electron beam excitation. Similarly, Rupp[3] and Hirsch and Döpel[4] observed linearly polarized light emitted from canal rays.

Percival and Seaton[5] derived the relationships between the linear polarization of the light emitted and the process of excitation by electron beams. They described the linear polarization fraction in terms of the excitation cross-sections $\sigma(|m_L|)$ to different magnetic substates m_L of the excited upper level of transition being observed.

More recently, in beam-foil excitation and in electron-photon coincidence measurements systems of reduced symmetry have been investigated. These changes of symmetry also affect the polarization and spatial distribution of the emitted light. We shall discuss briefly some of these systems before analyzing in detail the beam tilted-foil experiments.

We shall show that a study first of the symmetry properties of the beam-foil source in terms of the polarization of the light emitted by the moving ions gives considerable insight into the properties of the beam-foil interaction. Detailed measurements of the variations of the linear and circular polarizations observed can then be used to analyze the general problems of fast ions interacting with solids and surfaces. By optimizing the production of aligned and oriented excited states, atomic structures of the fast ions can also be measured using various resonance techniques.

In Fig. 1 we show schematically sources of various symmetries. Fig. 1a represents a spherically symmetric source which has statistically populated magnetic substates and gives isotropic and unpolarized radiation. Each excited state can be defined by its population $N(t)$ at a time t. Fig. 1b is cylindrically symmetric about the z-axis. We can resolve two directions parallel and perpendicular to this axis - e.g. two radiating electric dipoles of independent magnitude - A single additional parameter is required to define the excited state alignment relative to this z-axis. Percival and Seaton[5] thus defined the linear polarization fraction, which for for a $P \to S$ transition observed at $90°$ to the beam axis is

$$\frac{M}{I} = \frac{I^{\parallel} - I^{\perp}}{I^{\parallel} + I^{\perp}} = \frac{\sigma(m_L = 0) - \sigma(|m_L| = 1)}{\sigma(m_L = 0) + \sigma(|m_L| = 1)} \tag{1}$$

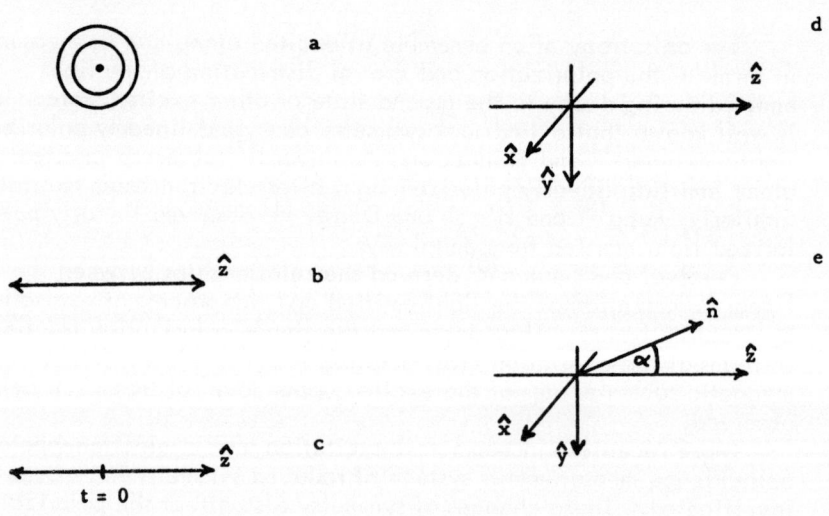

Fig. 1. Source symmetries: (a) spherical symmetry, (b) cylindrical symmetry, (c) with time $t = 0$, (d) reflection asymmetry in x-y plane, (e) orientation axis $\hat{n} \times \hat{z}$.

In beam-foil spectroscopy excitation occurs at a definite time t = 0 (Fig. 1c). Fine and hyperfine states are uncoupled in the short excitation time of $\sim 10^{-15}$s. The subsequent time development of these states, coupled through the fine and hyperfine interactions or by external electric and magnetic fields, gives rise to interference modulations or quantum beats. The frequencies provide measurements of fine and hyperfine coupling constants and g-values of the upper states.

It was realized by Eck[6] that as the foil excited atom leaves the foil surface it sees a potential which is not reflection symmetric in the x-y plane of the foil shown in Fig. 1d. Thus, the excited state need not be parity invariant. A superposition of 2s and 2p excited states of hydrogen form such a non-definite parity state (Eck[6]).

Sellin et al.[7] and Gaupp et al.[8] observed that the 2s, 2p states produce an excited atomic state with an electric dipole moment directed out of the foil as the hydrogen atom leaves the foil.

If the final surface interaction is important in beam-foil excitation, a tilt of the foil (represented by the surface vector $\hat{n}$ at angle α in Fig. 1e) will break the cylindrical symmetry of the preceding cases. An orientation axis $\hat{n} \times \hat{z}$ is introduced and the emitted light may be circularly polarized. This was first observed by Berry et al.[9] in tilted foil excitation of the ^{4}He I transition 2s ^{1}S - 3p ^{1}P at 5016 A.

II. THEORY OF LIGHT EMISSION

The intensity of electric dipole radiation emitted at time t, in the direction $\underline{p}$, of polarization $\underline{\ell}_\lambda$ is[10]

$$I(\underline{\ell}_\lambda, \underline{p}, t) = I_o \Sigma (\underline{\ell}_\lambda \cdot \underline{d})(\underline{\ell}_\lambda^* \cdot \underline{d}) \sigma(t) \quad (2)$$

where I_o is a constant including frequency, solid angle and efficiency factors, the summation is over all spectrally unresolved substates of the initial and final states and $\sigma(t)$ is the density matrix of the excited state.

Using standard angular momentum algebra and introducing the irreducible spherical tensor components

$$<FM|\sigma|F'M'> = (-)^{F+M} \sum_{k,q} \sqrt{2k+1} \begin{pmatrix} F & F' & k \\ M & -M' & q \end{pmatrix} \sigma_q^{FF'k} \quad (3)$$

and the polarization factor[11]

$$\varphi_q^k(\underline{\ell}_\lambda) = \sum_{q_1 q_2} (-)^q \sqrt{2k+1} \, \ell_{q_1} \cdot \ell_{q_2}^* \begin{pmatrix} 1 & 1 & k \\ q_1 & -q_2 & -q \end{pmatrix} \quad (4)$$

we obtain for the transition from states of angular momenta FF´ to state G

$$I(\underline{\ell}_\lambda, \underline{p}, t) = I_o \sum_{FF'G} (-) <F' ||d||> <G||d||F>$$

$$\times \sum_{kq} \begin{Bmatrix} l\,l\,k \\ FF'G \end{Bmatrix} (-)^q \, \omega_{-q}^k (\underline{\ell}_\lambda) \cdot {}^{FF'}\sigma_q^k (t) \quad (5)$$

Ellis[12] has discussed the symmetry properties of equation 5. The Percival-Seaton hypothesis[5] of spin independence of the interaction and the spin independence of the electric dipole operator d allows a direct reduction to orbital angular momentum factors $L_{\sigma q}^k(t)$ and $|<L||d||L_o>|^2$. The 6-j symbol { } limits the values of k to 0, 1 and 2. The k = 0, q = 0 term is proportional to the total population of the excited state; the k = 1, q = 0, ± 1 terms define the orientation or dipole moments of the state, while the k = 2, q = ±2, ±1, 0 terms define the alignment or quadrupole moments of the state. For the case of cylindrical symmetry along the z-axis a spin-independent interaction and reflection symmetry in any plane containing the z-axis, only the terms k = 0, q = 0 and k = 2, q = 0 can be non-zero. The former defines the total population and the latter the alignment, which is proportional to the linear polarization fraction M/I defined in equation (1).

In most beam-foil measurements, a multiplet of states decays with a characteristic lifetime $\tau = 1/\Gamma$ in a vacuum, and neglecting cascades, the time evolution of the density matrix $\sigma(t)$ can be written ($\hbar = 1$)

$$\frac{d\sigma}{dt} = i[H, \sigma] - \Gamma\sigma \quad (6)$$

where H is the atomic hamiltonian giving rise to structure $\Omega_{FF'}$ between the states (FF'). Equation (6) can be integrated to give

$${}^{FF'}\sigma_q^k(t) = {}^{FF'}\sigma_q^k(0) \cdot \exp[(-i\,\Omega_{FF'} - \Gamma)t] \quad (7)$$

Substitution of (7) into (5) clearly can give rise to oscillatory terms of frequencies $\Omega_{FF'}$ in the light emission as has been discussed in detail by numerous authors.[13]

For the decay of a single state without structure, we obtain a single exponential decay and

$$I(\underline{\ell}, \underline{p}, t) = I_o (-)^{F+G} |<F||d||G>|^2 \cdot \sum_{kq} \begin{Bmatrix} l\,l\,k \\ FFG \end{Bmatrix} \cdot$$

$$(-)^q \cdot \omega_{-q}^k \cdot {}^F\sigma_q^k(0) \cdot e^{-\Gamma\tau} \quad (8)$$

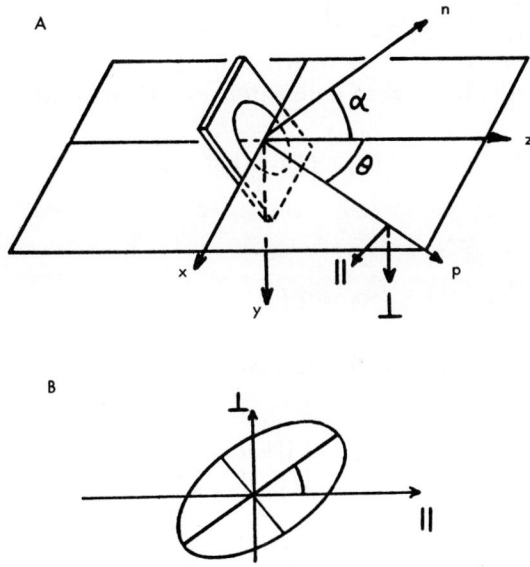

Fig. 2. *A-Tilted foil geometry, B-Polarization ellipse.*

The light is observed in the direction $\underset{\sim}{p}$ and we describe its state of polarization by four stokes parameters I, M, C, S. Fig. 2a shows a general detection geometry. Then if $I(\beta)$ is the intensity of light linearly polarized at an angle β to the direction $I^{\parallel} \equiv I(\beta=0)$ in the plane perpendicular to the propagation direction $\underset{\sim}{p}$,

$$\begin{aligned} I &= I(0) + I(90) \\ M &= I(0) - I(90) \\ C &= I(45) - I(135) \\ S &= I_{RH} - I_{LH} \end{aligned} \quad (9)$$

where S is the net right hand circularly polarized light intensity. I is the total intensity and M and C are the net linear polarizations for two sets of perpendicular axes which then completely define the polarization ellipse of Fig. 2b. Generally, the relative stokes parameters M/I, C/I and S/I are measured.

For an excitation geometry which retains reflection symmetry in the y-z plane (but does not retain cylindrical geometry) it is easily shown[12] that

$$L_{\sigma}{}^{k}_{-q} = (-)^{k} L_{\sigma}{}^{k}_{q} \qquad (10)$$

Thus, $\sigma_0^1 = 0$, $\sigma_1^1 = -\sigma_{-1}^1$, $\sigma_1^2 = \sigma_{-1}^2$, σ_{-2}^2, etc. and only five independent non-zero density matrix elements are needed to describe the excitation. The stokes parameters can then be written

$$\begin{aligned}
I &= \tfrac{2}{\sqrt{3}}\sigma_0^0 + H_2(FG)\left\{(2-3\sin^2\theta)\,\sigma_0^2/\sqrt{6} + \sigma_2^2 \cdot \sin\theta\right\} \\
M &= H_2(FG)\left\{-3\sigma_0^2 \cdot \sin^2\theta/\sqrt{6} - \sigma_2^2(1+\cos^2\theta)\right\} \\
C &= H_2(FG) \cdot 2i\,\sigma_1^2\,\sin\theta \\
S &= H_1(FG) \cdot 2\,\sigma_1^1\,\sin\theta
\end{aligned} \qquad (11)$$

where we have neglected the exponential term $I_o\,e^{-\tau\tau}\,|<F||d||G>|^2$

$$H_k(FG) = \begin{Bmatrix} l\,l\,k \\ F\,F\,G \end{Bmatrix} \Big/ \begin{Bmatrix} l\,l\,0 \\ F\,F\,G \end{Bmatrix} \qquad (12)$$

For convenience we note that Ellis[12] uses $\rho_q^k = \sigma_q^k\,\sqrt{2k+1}$. The orientation parameter O_{1-}^{col} and alignment parameters A_0^{col}, A_{1+}^{col} and A_{2+}^{col} of Fano and Macek[13] are directly proportional to density matrix ratios σ_q^k/σ_0^0. Taking $\sigma_0^0 = 1$ gives

$$\begin{aligned}
A_0^{col} &= -\sqrt{2}\,R_2(F)\,\sigma_0^2, & A_{1+}^{col} &= \tfrac{2i}{\sqrt{3}}R_2(F) \cdot \sigma_1^2 \\
A_{2+}^{col} &= \tfrac{2}{\sqrt{3}}R_2(F)\,\sigma_2^2, & O_{1-}^{col} &= \tfrac{-2}{\sqrt{3}}R_1(F)\,\sigma_1^1
\end{aligned} \qquad (13)$$

where

$$R_k(F) = H_k(FG)/h^{(k)}(FG) = \begin{Bmatrix} l\,l\,k \\ F\,F\,F \end{Bmatrix} \cdot \sqrt{3(2F+1)} \qquad (14)$$

and $h^{(k)}(FG)$ is defined by Fano and Macek[13].

For a cylindrically symmetric source, $\sigma_2^2 = \sigma_1^2 = 0$ $[A_{1+}^{col} = A_{2+}^{col} = O_{1-}^{col} = 0]$ and $C = S = 0$, and the alignment σ_0^2 (or A_0^{col}) is proportional to M/I as discussed above.

For an LS coupled state (LSJ) decaying to the state $(L_o S_o J_o)$, $H_k(FG)$ must be multiplied by the recoupling factor $K_k(SLJ)$ where

$$K_k(SLJ) = \begin{Bmatrix} L\,L\,k \\ J\,J\,S \end{Bmatrix} \Big/ \begin{Bmatrix} L\,L\,0 \\ J\,J\,S \end{Bmatrix} \qquad (15)$$

A second recoupling multiplier $K_k(IJF)$ is needed for a transition with resolved hyperfine structure.

For unresolved fine and hyperfine structures the polarizations are reduced by the factors Q_k:

$$Q_k = \frac{1}{(2I+1)(2S+1)} \sum_{JF} [F]^2 [J]^2 \begin{Bmatrix} F & F & k \\ J & J & I \end{Bmatrix}^2 \begin{Bmatrix} J & J & k \\ L & L & S \end{Bmatrix}^2 \quad (16)$$

where $[F] = 2F+1$, etc. and I and S are the total nuclear spin and electronic spin angular momenta for LS coupled states.

III. INITIAL TILTED FOIL EXPERIMENTS

Berry et al. (9) used the source geometry shown in Fig. 2a. A thin carbon foil was tilted an angle α_o to the beam axis ($\hat{n} \cdot \hat{z} = \cos \alpha$, $\hat{n} \times \hat{z} = \hat{x} \sin \alpha$) and light from the 5016Å transition $2s\,^1S - 3p\,^1P$ in ^{4}He I was observed in the x - z plane at two angles $\theta = 90°$ and $\theta = 53°$. The stokes parameters of the detected light were measured by a polarization sensitive system consisting of a quarter-wave plate followed by a rotating polarizer and Hanle depolarizer. The depolarizer cancelled any instrumental polarization in the monochromator and photomultiplier detection system.

The first results showed the emitted light to be partially elliptically polarized after tilted foil excitation. Thus, both alignment and orientation have been produced. The latter is necessarily produced by the final surface interaction. The alignment, which occurs also for perpendicular foil excitation also changed with tilt angle, and we reproduce these results in Table I.

Table I

Density matrix components in tilted foil excitation of ^{4}He I $2s\,^1S - 3p\,^1P$ at 130 keV beam energy

Foil Tilt angle (deg)	$\frac{\sigma_0^2}{\sqrt{6}}$	σ_2^2	$\frac{i\sigma_1^2}{\sqrt{2}}$	$\frac{\sigma_1^1}{\sqrt{2}}$
0	-0.030(12)[a]	-0.016(9)	-	-
20	-0.027	-0.012	-0.017(5)	0.009(8)
30	-0.024	-0.008	-0.015	0.027
45	-0.018	+0.0002	-0.028	0.029

a. Numbers in parentheses give accuracy of last quoted figure.

The stokes parameters measure the expectation values of the angular momenta of the excited state, and for observation in the x-direction of Fig. 2a, they are

$$M/I = <L_y^2 - L_z^2>/<L_x^2>$$
$$C/I = 2\,\text{Re}<L_y L_z>/<L_x^2> \qquad (16)$$
$$S/I = -\hbar<L_x>/<L_x^2>$$

Clearly the circular polarization is proportional to the angular momentum along the x-direction which in this case is the orientation axis $\hat{n} \times \hat{z}$. S/I will be a maximum in the x-direction for any system with reflection symmetry in the y-z plane. In such a case $<L_y L_z> = <L_z L_x> = 0$. However, the experimental results showed that $<L_x^2> \approx <L_y^2>$, an equality which is not derived from any symmetry principle.

The orientation and alignment components σ_q^k can also be measured by observations of magnetic field induced quantum beats (14). The light emitted at a distance d downstream from the exciter foil is observed with a polarization analysis detection system. As the magnetic field H, usually perpendicular or parallel to the beam direction, is varied the light intensity is modulated at the Larmor frequency $\omega = \gamma H$ so that

$$I(H) = A + \sum_{kq} \beta_k\, \sigma_q^k\, \cos\left(\frac{k\omega d}{v} + \Phi_k\right) \qquad (17)$$

where the parameters A, β_k, Φ_k depend upon the observation direction and the state of polarization observed, and k takes the values 1 and 2. (v is the beam velocity.)

With a perpendicular foil, no modulations appear for a magnetic field parallel to the beam axis (parallel to the alignment vector, and only $k = 2$ alignment beats can be observed using other field directions. Using a tilted foil, Church et al. (15) and Liu et al. (16) showed that $k = 2$ and $k = 1$ beats would be observed in linearly and circularly polarized light. Thus, they found oriented states in He I (^{1,3}D states $3p\ ^1P$), O II ($3p\ ^2F_{7/2}$) and Ar II ($4p\ ^2F_{7/2}$, $4p\ ^2P_{3/2}$). In Fig. 3 we show $k = 2$ and $k = 1$ frequencies for the O II term.

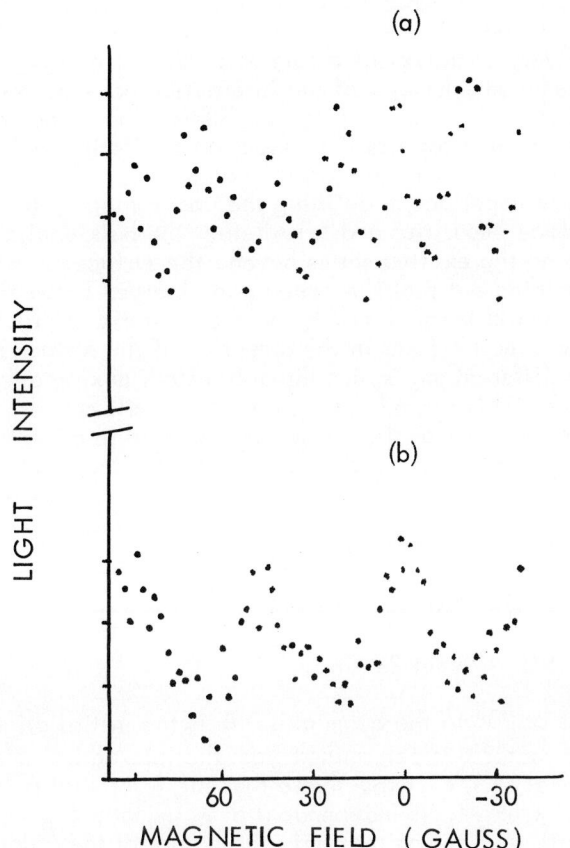

Fig. 3. Magnetic field quantum beats for O II 3p $^2F_{7/2}$ (a) linear polarization, (b) circular polarization, with B perpendicular to v (adapted from ref. 16).

IV. THEORY OF TILTED-FOIL EXCITATION

The interaction of the fast moving ion in the solid bulk of the foil (e.g. carbon) polarizes the material producing a wake of induced electric dipoles (17). The ion reaches an equilibrium charge state although many of the electrons associated with the ion may be merely travelling at approximately the same velocity. Thus, as the ion exits the foil many electrons are liberated, a forward peak occurring at the ion velocity (18). The state of excitation of the valence electrons of the ion while within the solid may be a function of the solid traversed and, complementarily, the particular excited states available to the electrons in the solid may affect this excitation distribution. The interactions occurring with this excited ion state as it passes the final surface may be very complex but a number of simplified models based mostly on the reduced symmetry of the interaction have been proposed. Some questions to be answered are (a) the interaction ranges which are appropriate for electron transfer, electron excitation and electron arrangement (alignment and orientation changes) (b) which parameters of the surface are important in defining the above ranges, (c) the importance of surface impurities and (d) whether the bulk excitation state helps to determine the excited states beyond the surface.

Eck (19) pointed out that the coherence observed between different m_L states could be produced by an electrostatic interaction at the surface. An electric field in the direction of the surface normal can couple the different m_L states through a stark mixing of other electronic states. This second order stark effect will convert an alignment of the excited state at the surface into a mixture of orientation and alignment. Thus, for a ^1P state he obtains for the relative stokes parameters

$$M/I = R[1 - 2 \sin^2 2\alpha \sin^2 \varphi/2]$$

$$C/I = R \sin 4\alpha \sin^2 \varphi/2 \qquad (18)$$

$$S/I = R \sin 2\alpha \sin \varphi$$

for observations at 90° to the beam axis. R is the initial alignment while the phase factor φ is determined from the interaction strength and is proportional to $1/v \cos \alpha$. Note that the fractional polarization $F_p = (M^2 + C^2 + S^2)^{\frac{1}{2}}/I$ is independent of α (in fact, $F_p = R$ the initial alignment), indicating that only the nature of the polarization is changed.

Further experiments (20) showed that the degree of coherence in general increases with foil tilt angle rather than remaining constant. Extensions of Eck's initial model can clearly be made to account for this increase by assuming the surface interaction does more than just redistribute the alignment from the bulk interaction. The work functions of most materials are large compared with the atomic level separations, and image charge potentials are also strong enough to produce excitation and ionization of the outgoing beam ion.

Lombardi (21) included first order stark mixing of the type already observed in lyman α production (6-8). Agreement between experiment and Lombardi's theory for the angular dependence of the stokes parameters appears good as shown by Berry et al. (22) and in Fig. 4. However, the fitting parameters have been varied independently to obtain agreement. Physical interpretation of the parameters is difficult both because they may themselves vary with foil tilt angle (they are assumed constant in the fit of Fig. 4) and also because mixing can occur with many different close electronic states.

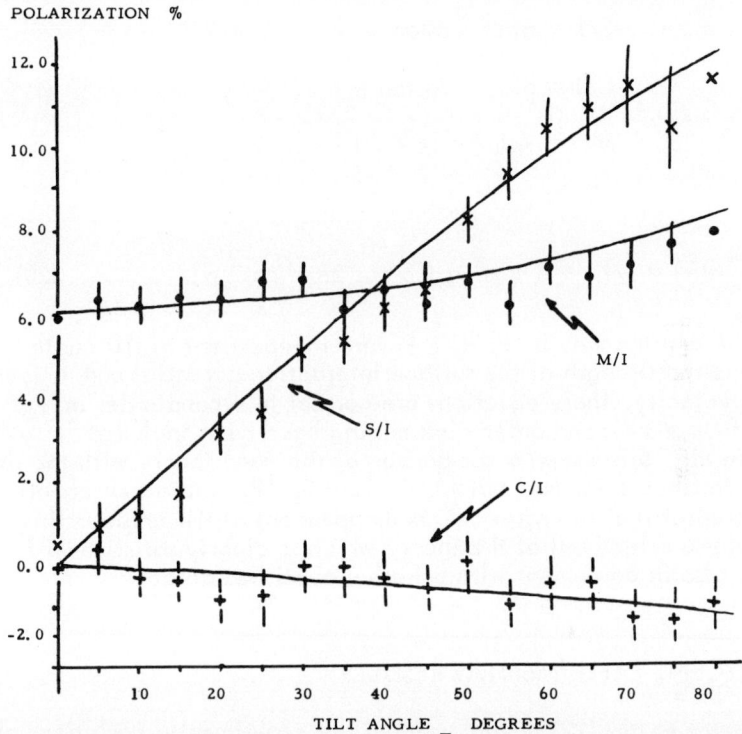

Fig. 4. Stokes parameters for Ne III, 2866Å, $3s'$ $^1D - 3p'$ ^1F ($\theta = 90°$, $E = 1$ MeV), fitted to the Lombardi theory (21) giving $I = 1+0.200 \cos\alpha$, $M = 0.090 - 0.012 \cos\alpha$, $C = -0.012 \sin\alpha$ and $S = 0.130 \sin\alpha$ (ref. 22).

Herman (23) used symmetry considerations to obtain agreement for the orientation and alignment parameters at small tilt angles (24). Lewis and Silver (25) and Band (26) have provided general formulations of the symmetry properties of the bulk and surface interactions with results that compare well with experimental variations of the stokes parameters with foil tilt angle. Refs. 25 and 26 generalize Eck's basic electric field model to include excitation at the surface and the effects of the image charge potential. Band finally obtains the general angular variations of the stokes parameters with foil tilt angle for all electric dipole transitions:

$$S/I = p(\alpha) \cdot c \tan \alpha$$
$$C/I = p(\alpha) \cdot 2d \tan \alpha \qquad (19)$$
$$M/I = -p(\alpha) \cdot [e + (f + d \cos 2\alpha)/\cos^2 \alpha]$$

where

$$p(\alpha) = 1/[1 + a/\cos^2 \alpha + b \cos 2\alpha / \cos^2 \alpha] \qquad (20)$$

and the constants a, b, c, d, e f are independent of tilt angle. If V_o is the strength of the surface interaction potential and v is the beam velocity, these equations are correct to second order in V_o/v - i.e. first and second order stark mixing have been included.

In Fig. 5 is shown a comparison of the Band theory with the experimental data for He I 5015Å, 2s ^{1}S - 3p ^{1}P. Note that equations (19) predict that the ratio S/C is independent of tilt angle. This provides a direct test of the theory which is clearly satisfied in Fig. 5, but is also in agreement with all other published data.

V. RECENT EXPERIMENTAL RESULTS

Berry et al. (22) have compared the experimental variations of S/I with foil tilt angle for several transitions in He I and ionized neon. They find satisfactory fits to the theories of both Lombardi (21) and Band (26). Hence it would appear that the general character of the surface interaction is understood. However, only the necessary symmetry characteristics of the interaction need be invoked to obtain agreement.

Variations of alignment, and orientation production with ion velocity, excited state and foil material have not been studied thoroughly. We indicate some of the preliminary results below.

Beam Velocity Dependence

The orientation of the 3p ^{1}P state of He I increases with beam velocity over the range 0.5 v_o to 2 v_o, where v_o is the bohr velocity (20), for a foil tilt angle of 30°. However, the Ne III $^1F_3$ orientation and alignment decrease over the velocity range v_o to 1.6 v_o as shown in Fig. 6 for all foil tilt angles.

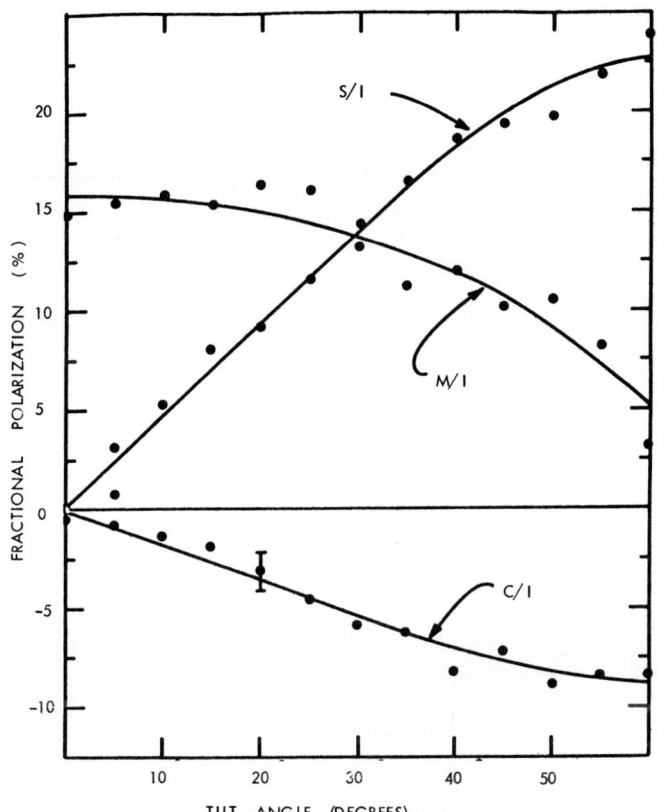

Fig. 5b   Stokes parameters for He I 2s ^1S - 3p ^1P, 5016Å (θ = 90°, E = 246 keV) fitted to the Band theory (26).

The monotonic increase in orientation of 3p ^1P of He I is in contrast to the modulations observed in the alignment (M/I) for the same transition (ref. 27 and Fig. 10 below). This may be an indication that two processes contribute differently to the orientation and alignment of the excited atomic states.

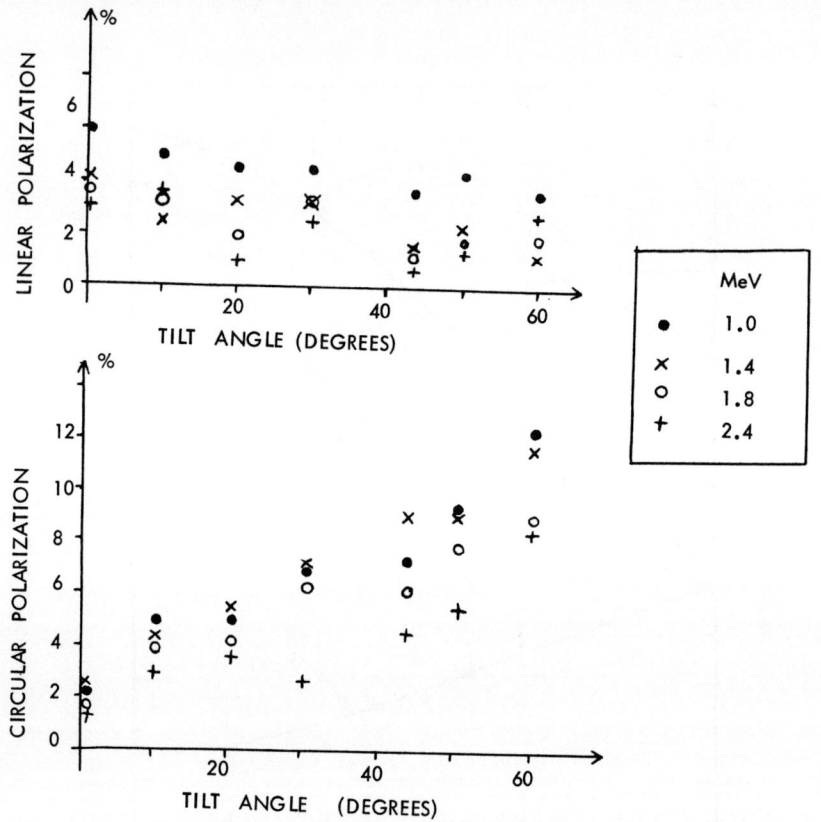

Fig. 6. Angular dependence of (a) M/I and (b) S/I for the Ne III 2866Å transition, for different incident beam energies.

Excited State Dependence

Alignment and orientation of unresolved multiplets can be measured by observing the time-resolved quantum beats. Such modulations of the 3889Å He I transition 2s ^{3}S – 3p ^{3}P in a magnetic field have been measured by Silver and McIntyre (28). The ^{14}N IV multiplet 3s ^{3}S – 3p ^{3}P at 3480Å has hyperfine structure separations shown in Fig. 7c producing possible quantum beat frequencies ω, 2ω, 3ω and 5ω where ω is proportional to the hyperfine splitting constant. The frequency ω, corresponding to an F = 0 – 1 transition can only be observed in circular polarized light and when the atomic orientation is non-zero. It is, therefore, an excellent test of the presence of orientation. The frequency is observed in these conditions as shown in Fig. 7a and b. The structure of the linear polarization quantum

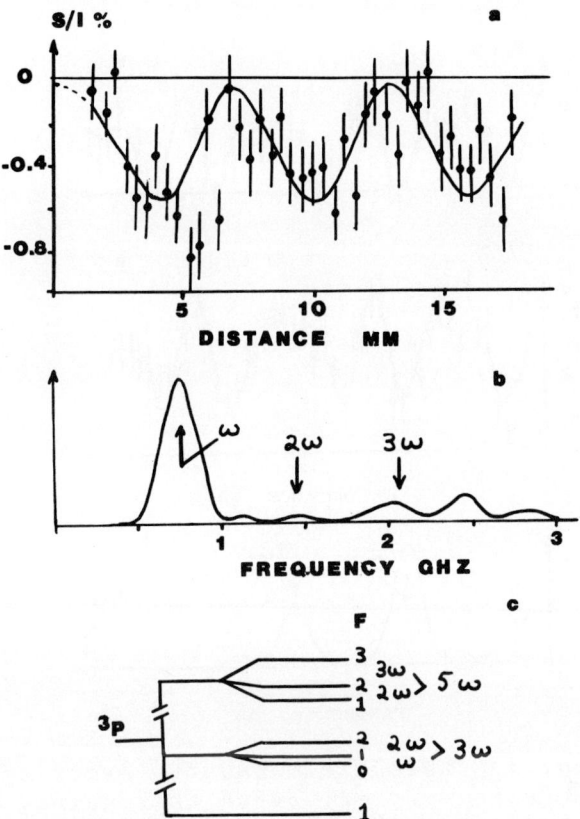

Fig. 7. (a) Zero field quantum beats of ^{14}N IV $3s\ ^3S - 3p\ ^3P$, 3480Å, in circularly polarized light, foil tilted at 45°, (b) Fourier transform, (c) energy levels.

beats is unchanged with foil tilt angle as shown in Fig. 8, and no ω component appears in the Fourier transform of Fig. 8c.

The dependence of orientation on excited state energy, principal quantum number and orbital angular momentum is unexplored. However, orientation appears to be a quite general phenomenon and occurs even when the alignment is close to zero for a perpendicular foil (20). In Fig. 9 we show the stokes parameters of the transition $4s'\ ^2D_{5/2} - 4p'\ ^2F_{7/2}$ in Ar II.

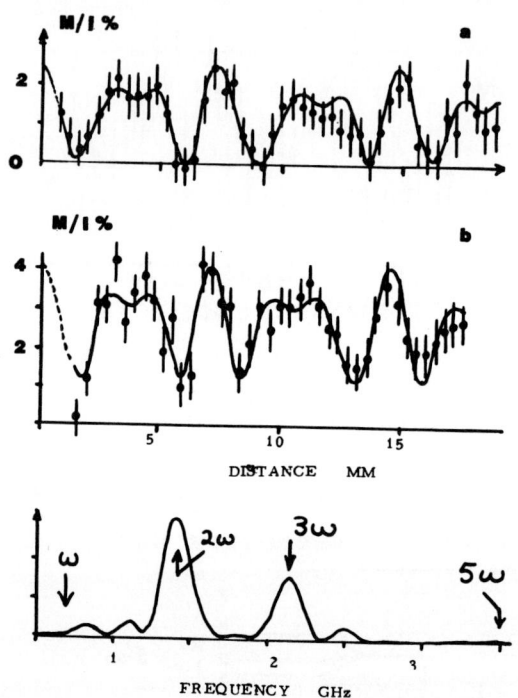

Fig. 8. Zero field quantum beats of ^{14}N IV $3s\ ^3S - 3p\ ^3P$, 3480Å in linearly polarized light. (a) -for 0° foil, (b) for 45° tilted foil, (c) Fourier transform of b.

Foil Material Dependence

Almost all beam-foil experiments have used thin carbon foils in vacua of $10^{-5} - 10^{-6}$ torr. At such vacua, impurities coat the surface layer. Carbon deposits, presumably from cracked hydrocarbons, build on the final foil surface unless the foil is surrounded in a liquid nitrogen temperature enclosure (29).

However, we find that with a vacuum of about 10^{-6} torr, metallic foils such as gold and silver take from 1000 to 4000 seconds to degenerate to carbon-like surfaces. Thus, the alignment M/I for the 5016Å line in He I is shown in Fig. 10 for a set of beam energies, as a function foil material (all foils perpendicular to the beam). M/I for Au and Ag foils approaches the carbon value in roughly 2000 seconds.

Church (30) has found that both the orientation and alignment parameters change when gold is deposited on the down beam side of thin carbon foils for the $2p\ ^1P - 4d\ ^1D$ transition of He I at 4922Å.

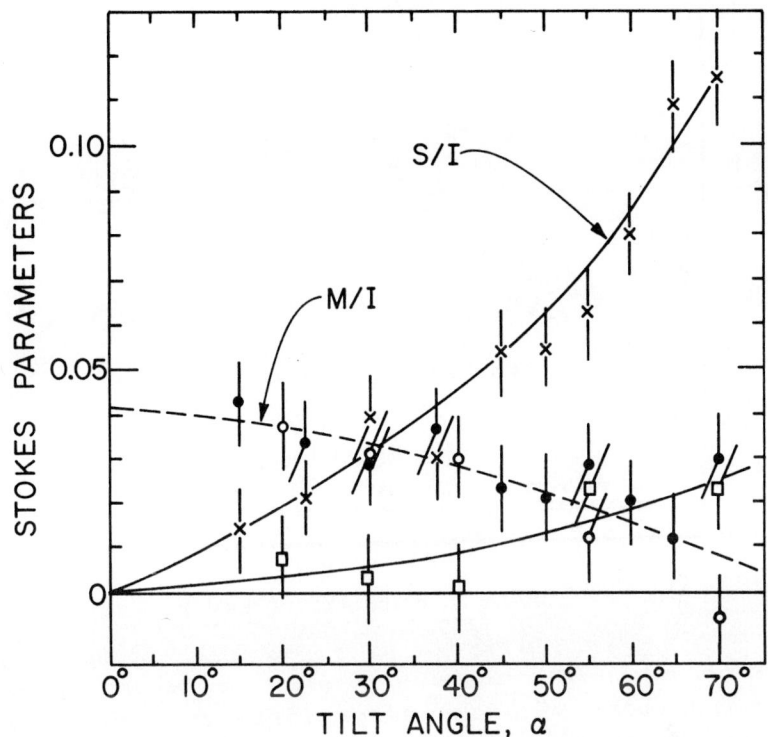

Fig. 9. Stokes parameters for the Ar II $4s\ ^2D_{5/2} - 4p\ ^2F_{7/2}$ 4610Å transition excited by a thin carbon foil ($\theta = 90°$, $E = 1$ MeV).

VI. CONCLUSIONS

The production of atomic orientation by tilted foil excitation has shown the importance of the final surface interaction in beam-foil excitation. The mostly preliminary results discussed here show that the technique may become useful as a probe of the surface interaction and for atomic structure work through level crossing and magnetic field beat and zero field quantum beat measurements.

The theories for the production of the orientation cannot yet predict the magnitude of the effect nor its dependence on excited state (both signs of S/I have been observed in neon ions excited by carbon foils), foil material and ion beam velocity.

Experimentally, very few investigations have been made of the variation of orientation and alignment with excited state, foil material and beam velocity. It is expected that at improved vacua of 10^{-9} to 10^{-11} torr in which clean, better known surfaces can be prepared, that more resolvable information will be obtained.

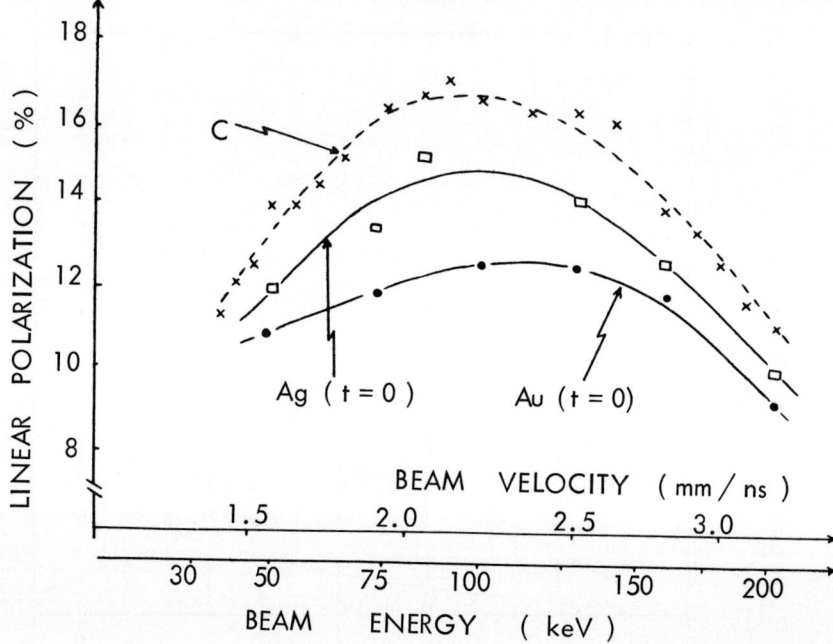

Fig. 10. M/I for He I $2s\ ^1S - 3p\ ^1P$, 5016Å, excited by thin foils of carbon, gold and silver as functions of beam energy (Berry, Gabrielse and Gay, to be published).

VII. ACKNOWLEDGEMENTS

The paper describes measurements made at the University of Toledo in collaboration with R. M. Schectman, L. J. Curtis and D. G. Ellis and at the University of Chicago and Argonne National Laboratory with G. Gabrielse, T. Gay and J. Desesquelles. I thank them and other authors who gave permission to use their published figures.

Work supported in part by ERDA and NSF.

VII. REFERENCES

1. Skinner, H. W. B., Proc. Roy. Soc. A112, 642 (1926).
2. Skinner, H. W. B. and Appleyard, E. I. S., Proc. Roy. Soc. A117, 224 (1927).
3. Rupp, E., Ann. d. Phys. 79, 1 (1926); 85, 515 (1928).
4. Hirsch, R. V. and Döpel, R., Ann. d. Phys. 82, 16 (1927).

5. Percival and Seaton, M., Phil. Trans. Roy. Soc. (London) A25, 113 (1958).
6. Eck, T. G., Phys. Rev. Lett. 31, 270 (1973).
7. Sellin, I. A., Mowat, J. R., Peterson, R. S., Griffin, P. M., Lambert, R. and Hazelton, H. H., Phys. Rev. Lett. 31, 1335 (1973).
8. Gaupp, A., André, H. J. and Macek, J., Phys. Rev. Lett. 32, 268 (1974).
9. Berry, H. G., Curtis, L. J., Ellis, D. G. and Schectman, R. M. Phys. Rev. Lett. 32, 751 (1974).
10. Cohen-Tanoudji, C., Ann. Phys. 7, 423 (1962).
11. Dyakonov, Y., Opt. and Spectr. 19, 372 (1965).
12. Ellis, D. G., J. Opt. Soc. Am. 63, 1232 (1973).
13. Ref. 12, Macek, J., Phys. Rev. A1, 618 (1970); Macek, J. and Jaecks, O. H., Phys. Rev. A5, 2288 (1971); Berry, H. G., Subtil, J. L. and Carré, M., J. Phys. (Paris) 33, 947 (1972); Bosse, J. and Gabriel, H., Z. Phys. (1974); Fano, U. and Macek, J., Rev. Mod. Phys. 45, 553 (1973).
14. Gaillard, M., Carré, M., Berry, H. G. and Lombardi, M., Nucl. Inst. Meths. 110, 273 (1973).
15. Church, D. A., Kolbe, W., Michel, M. C. and Hadeishi, T., Phys. Rev. Lett. 33, 565 (1974).
16. Liu, C. H., Bashkin, S. and Church, D. A., Phys. Rev. Lett. 34, 933 (1974).
17. Gemmell, D. S., et al., Phys. Rev. Lett. 34, 1420 (1975).
18. Harrison, K. G. and Lucas, M. W., Phys. Lett. 33A, 142 (1970).
19. Eck, T. G., Phys. Rev. Lett. 33, 1055 (1974).
20. Berry, H. G., Bhardwaj, S. N., Curtis, L. J. and Schectman, R. M., Phys. Lett. 50A, 59 (1974).
21. Lombardi, M., Phys. Rev. Lett. 35, 1172 (1975).
22. Berry, H. G., Curtis, L. J., Ellis, D. G. and Schectman, R.M., in Beam Foil Spectroscopy (I. A. Sellin and D. J. Pegg, Eds.), Vol. 2, p. 755. Plenum, New York, 1976.
23. Herman, R. H., Phys. Rev. Lett. 35, 1626 (1975).
24. Church, D. A., Michel, M. C. and Kolbe, W., Phys. Rev. Lett. 34, 1140 (1975).
25. Lewis, E. L. and Silver, J. D., J. Phys. B8, 2697 (1975)
26. Band, Y., Phys. Rev. Lett. 35, 1272 (1975); Phys. Rev. 13, 2061 (1976).
27. Berry, H. G. and Subtil, J. L., Nucl. Inst. Meths. 110, 321 (1973).
28. Silver, J. D. and McIntyre, Jr., L. C., Beam Foil Spectroscopy, Vol. 2, p. 773 (1976).
29. Dumont, P. D., Livingston, A. E., Baudinet-Robinet, Y., Weber, G. and Quaglia, L., Physica Scripta 13, 122 (1976).
30. Church, D. A. and Michel, M. C., Phys. Lett. 55A, 167 (1975).

optical polarization in high energy ion-surface scattering at grazing incidence

H. J. Andrä, R. Fröhling, and H. J. Plöhn
Fachbereich Physik der Freien Universität Berlin

ABSTRACT

We report large circularly polarized light fractions emitted from surface scattered ions when energetic ion beams are impinging on mechanically polished solid surfaces at grazing incidence. First results on the dependence of the observed circular polarization on the surface quality, on the incident grazing angle, on the incident beam energy and on the incident ions are presented. From the analysis of various multiplets of NeII, ArII, KrII we interpret the excitation mechanism as an orbital angular momentum effect which is governed by the symmetry of the process. Measurements with 10 different incident ions prove this effect to be a general phenomenon. As an application in atomic physics, a zero-field level-crossing experiment is presented.

Our studies of the optical polarization of light emitted from surface scattered particles was first initiated by pure atomic physics interests (1). As soon as we successfully observed circular polarizations larger than 50%, however, we got of course also concerned about the study of the interaction mechanism itself (2).

In atomic physics we are interested in the preparation of nonisotropically excited, i.e. aligned or oriented states, in order to use this anisotropy for the application of high resolution techniques like quantum beats or levelcrossing (3). In Fig. 1 we shortly sketch what is meant by oriented or aligned sublevel population distributions, where the circles represent relative sublevel populations. In this chapter we shall concentrate on the oriented distribution (dipole-polarization) which is responsible for the circularly polarized light emission of the atomic ensemble and which is equivalent to $<L_x> \neq 0$ in the coordinate systems as depicted in all figures. The isotropic population (which also represents the

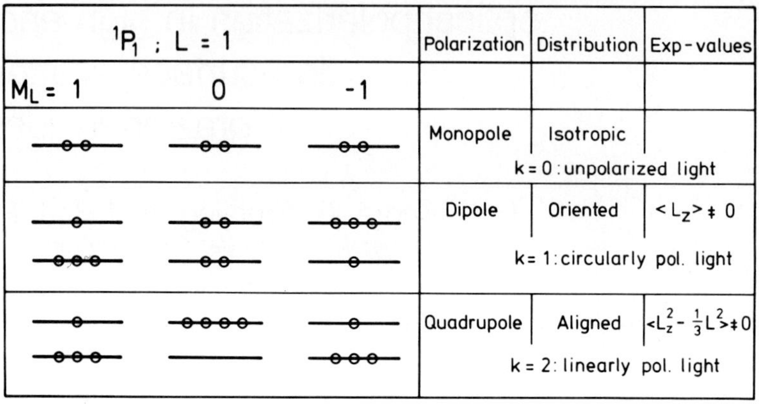

$^1P_1$; L = 1			Polarization	Distribution	Exp-values
M_L = 1	0	-1			
──●●──	──●●──	──●●──	Monopole	Isotropic	
				k = 0 : unpolarized light	
──●──	──●●──	──●●●──	Dipole	Oriented	$<L_z> \neq 0$
──●●●──	──●●──	──●──		k = 1 : circularly pol. light	
──●──	──●●●●──	──●──	Quadrupole	Aligned	$<L_z^2 - \frac{1}{3}L^2> \neq 0$
──●●●──		──●●●──		k = 2 : linearly pol. light	

Fig. 1. Multipole polarizations and sublevel population distributions of a L = 1 - $^1P_1$ excited state.

total population) and the aligned one will be responsible for noncircularly polarized light emission in our investigation. If now an excitation process leads to such nonstatistical population of sublevels then one cannot only perform the above atomic physics experiments, but one also can expect to receive some information on the understanding of the ion beam surface interaction at grazing incidence (IBSIGI) by the measure of the polarization state of the emitted light. These are the two aspects we want to discuss in a situation where neither one is anywhere close to complete understanding or application.

The underlying idea for the choice of this particular geometry stems initially of course from tilted foil experiments (4). According to the observation that the circular polarization in tilted foil experiments (Fig. 2a) increases with tilt angle β roughly proportional to sinβ the best situation for maximum polarization should be obtained at grazing incidence (Fig. 2b). This is of course an impractical region for the work with foils. Therefore, with the experiments of Carl Rau in mind (5), we thought, why not use surface scattered ions since we believed that the outgoing particle direction z' in Fig. 2d with respect to the surface must be responsible for the circular polarization observed.

This rather crude picture leads to the first experiments made with arbitrarily chosen surfaces and beams with a very simple set-up schematically shown in Fig. 2c. The measurable quantity for all our experiments is the normalized stokes parameter $S/I = [I(\sigma^-)-I(\sigma^+)]/[I(\sigma^-)+I(\sigma^+)]$ observed along the x-axis, where $I(\sigma^-)$ = right hand and $I(\sigma^+)$ = left hand circularly polarized light intensity in the optical convention

Optical Polarization in High Energy   331

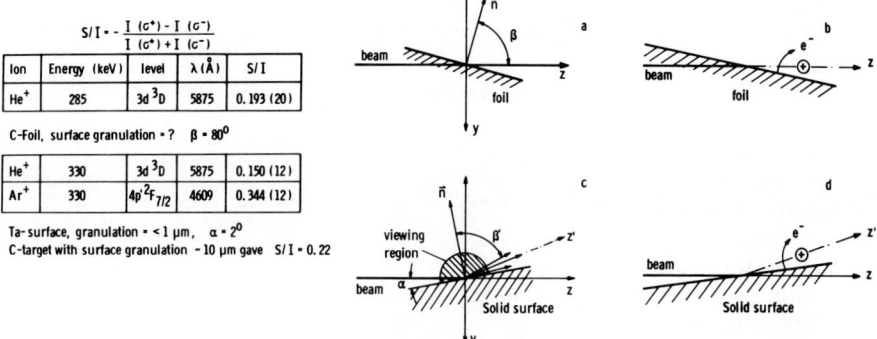

Fig. 2. Comparison of the tilted foil excitation geometry with the excitation regime of an ion beam impinging on a solid surface at grazing incidence.

(σ^- corresponds to negative and σ^+ to positive helicity of the photons) (6). In order to compare our surface scattering results with a tilted foil experiment we made a He-tilted foil experiment at β=80° and observed the 5875 Å-transition starting from the $3d^3D$-level. For S/I we obtained 0.19 which is in good agreement with similar results by Berry and co-workers. The positive sign of S/I indicates that the electrons rotate clockwise around the positive ion-core as shown in Fig. 2b. In our first surface scattering experiment (1) we repeated this He-experiment and found a value S/I = 0.15 in astonishingly close agreement with the above tilted foil result. As surface we used mechanically polished Ta with a surface granulation <1μm at a grazing angle of 2 degrees. The great surprise came when we used Ar^+-ions and observed the 4609 Å-transition starting from the $4p'^2F_{7/2}$-level which gave S/I = 0.344, a surprisingly high value compared to all earlier tilted foil results. A repetition of this experiment with a graphite target with a surface granulation of the order of 10μm gave S/I = 0.22 for this same transition. In all these first surface scattering experiments the S/I was positive so that again the electrons seem to move clockwise around the positive ion-core as indicated in Fig. 2d. The obvious problem, however, in these first surface scattering experiments was the bad vacuum pressure condition of typically 10^{-5} Torr which lead to a black dirt built up on the surfaces under ion bombardment which seemed to reduce the S/I-values with the time of beam exposure.

In spite of this problem we felt encouraged enough to pursue this new excitation regime with an improved experimental set-up shown in Fig. 3 with which we hoped to obtain some

answers on the following questions:
1) Do the scattered ions have a predominant average forward direction thus defining reasonably well a scattering angle θ?
2) How does the surface quality influence the observed polarizations?
3) How does the degree of circular polarization depend on the grazing anlge α?
4) Does the polarization effect depend on the incoming beam energy?
5) How general is the polarization effect?
6) Can we use this excitation regime for atomic level-crossing experiments?

With the axes indicated in Fig. 3 the direction of observation was along Ox. The incident beam direction is along Oz, the target surface normal $\vec{n}$ makes an angle α with the (-y)-axis in the y-z plane, and we anticipate the results given below by ascribing an angle θ to the average forward scattered particle direction. The target was amorphous or microcrystalline copper, of dimensions 30mm x 15mm x 1mm, and its surface was mechanically polished flat to 2 µm. It was mounted on a shaft which was along Ox, and the angle α could be set to a precision of $\pm 0.17°$. A 1 mm diameter pick-up wire was mounted parallel to and off center from the x-axis in such a way as to describe a circle of radius 7 cm about that axis. The scattered beam current could be measured with this wire as a function of γ in $1.4°$-steps. The pick-up wire and

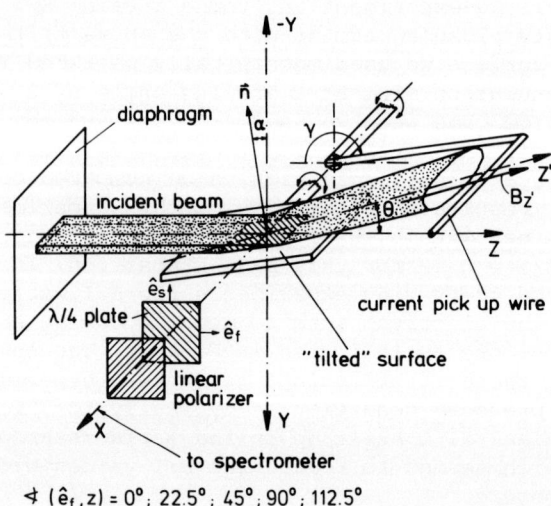

Fig. 3. Improved experimental set-up. The incoming beam cross section is 0.5×5 mm^2.

the target could be held independently at variable potentials
to suppress secondary electron emission. The optical detection
system consisted of an achromatic λ/4-plate and a linear
polarizer, followed by a single fused quartz imaging lens, a
0.3 m grating spectrometer and a photomultiplier tube in
photon counting mode. The viewing region was approximately
7 mm in diameter in the y-z plane and was centered on the x-
axis. The linear polarizer's E-vector made a fixed angle of
45° with respect to the z-axis. For the measurement of all
Stokes parameters it was then sufficient to step the λ/4-
plate's fast axis to 5 different angular positions relative
to the z-axis as given in Fig. 3. From these 5 independent
measurements it is then easy to calculate all 4 Stokes parameters. Spectra were built up by repetitive sweeping of the
wavelength proportional to collected charge of the unbiased
target (our results are independent of secondary electron
emission). At each setting of the wavelength the λ/4-plate was
stepped to the 5 indicated angles and the photon counts were
recorded in 5 corresponding memory blocks of a multiscaler
synchronized parallely to the wavelength. In particular for
S/I measurements only the angles 0° and 90° are necessary such
that 2 spectra are simultaneously recorded in two different
memory blocks, one storing the $I(\sigma^-,\lambda)$ and the other storing

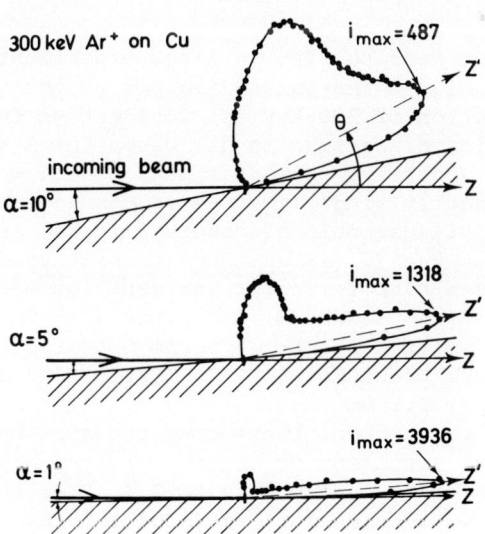

Fig. 4. *Polar diagrams of the scattered ion current at
three grazing angles α. Relative values of the maximum
scattered current are given (i_{max}, in arbitrary units), and
each curve is normalized such that $i_{max}=1$ for that curve.*

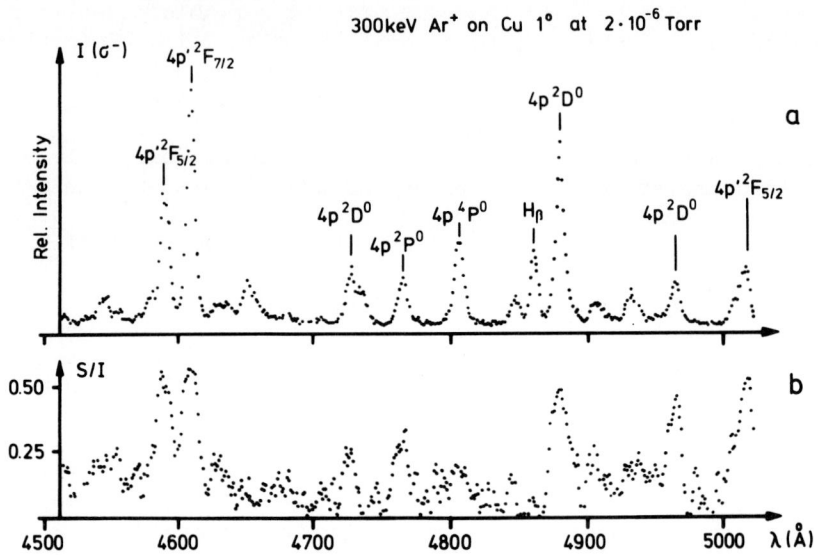

Fig. 5. $I(\sigma^-)$-spectrum (a) and $S/I(\lambda)$ (b) obtained with 300 keV Ar^+ on Cu at a grazing angle of $\alpha=1°$ at $2 \cdot 10^{-6}$ Torr vacuum pressure and an incident ion current density of $1\mu A/2.5mm^2$.

$I(\sigma^+, \lambda)$.

As our first results with this new experimental arrangement we show in Fig. 4 the polar diagrams of the scattered ion current distribution of 300 keV Ar^+ incident on Cu-surfaces at 3 different angles α. We find in all cases for $\alpha < 5°$ a well defined forward scattered peak at around $\theta = 2\alpha$ which we attribute to predominantly singly charged scattered beam ions (7) because no ArIII-lines could be observed in the spectra. The further peak near $\theta = 70°$ is assigned to sputtered Cu-ions since it increases with different incident ion masses in agreement with sputtering yields (8) and since weak CuII-lines were observed in the emitted spectrum. These measurements were performed at a vacuum pressure of $2 \cdot 10^{-6}$ Torr and at a current density of $1\mu A$ per 2.5 mm^2.

A spectrum taken under these conditions is shown in Fig. 5a for 300 keV Ar^+ incident on Cu at a grazing angle of $1°$. Only $I(\sigma^-)$ is shown from 4500 Å to 5000 Å. From the simultaneously measured $I(\sigma^+)$-intensity S/I is calculated channel by channel in Fig. 5b and shows significant polarizations $S/I>0.5$ for several lines. The appearance of the Hydrogen H_β-line in Fig. 5a which does not show any significant degree of polarization is a clear indication for still not clean enough surfaces under these conditions. One can easily imagine that it primarily stems from deposited H_2O or hydrocarbons on the surface.

In this respect it was very interesting to note that the finally obtained S/I was strongly dependent on the incident ion current density. Although no direct study of this phenomenon was performed we could obtain from our notebooks the dependence in Fig. 6. At the same time it was observed that the surfaces appeared as rough and dirty after long exposure to ion current densities below $1 \mu A/2.5$ mm^2 whereas a bombardment with ion current densities above $2 \mu A/2.5$ mm^2 yielded clean and polished surfaces according to inspection by eye. We interpret this behaviour by two competing effects, the deposition of impurities under ion impact which dominates at low current densities, and the cleaning by sputtering which dominates at high current densities. In particular one can expect from the latter one that it polishes and flattens the surface according to the steep decrease of the sputtering yield when grazing angles are approached. This behaviour has been experimentally observed for lower energies (9) (10) and is expected also for higher energy incident ions. Whether this behaviour indeed leads to flat and polished surfaces is still a matter of debate in the literature (11) but if it does, then it may become of great importance for obtaining clean and flat surfaces in future applications of IBSIGI. The competing impurity deposition can be of course reduced by better vacuum conditions

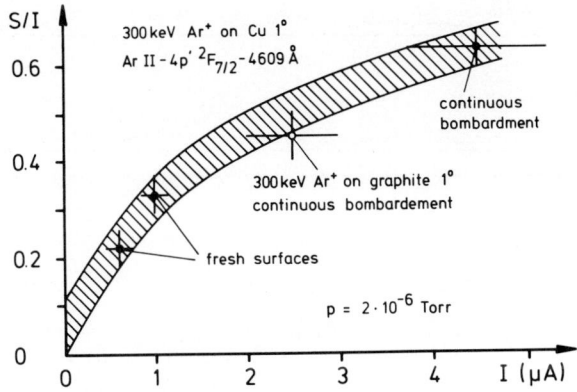

Fig. 6. Dependence of S/I of the ArII-4609 Å-transition on the incident ion current at a beam cross section of 2.5 mm^2. Fresh surfaces are mechanically polished to a surface granulation of about 2μm prior to installation.

or by enclosing the target by a cold trap. It is the latter approach which we have used quite successfully for all the following experiments: No impurity deposition was anymore observed and the additional polishing by sputtering at grazing incidence could be clearly seen under an optical microscope.

The influence of sputtering may become on the other hand very disturbing for the analyis of spectra. This problem is instructively demonstrated in Fig. 7 where the upper half shows a spectrum from 4300 to 4800 Å of light emitted during 300 keV Ar^+ bombardment of a stainless steel target at 1.5° grazing angle. The lower half represents the repetition of the upper experiment with a Cu-target. With some of the most prominent FeI-lines indicated it is obvious that sputtered ion disturbs considerably the spectrum of the Ar-beam. The choice of a Cu-target is therefore quite advantageous in this spectral region since CuI and CuII have a much lower line density in this region compared to FeI and FeII. For the understanding of the ion beam surface interaction mechanism it is important to note that the light emitted from sputtered iron is unpolarized and that the degree of polarization of the various ArII-lines is quite different with a stainless steel target from the polarization obtained with a Cu target! This is an

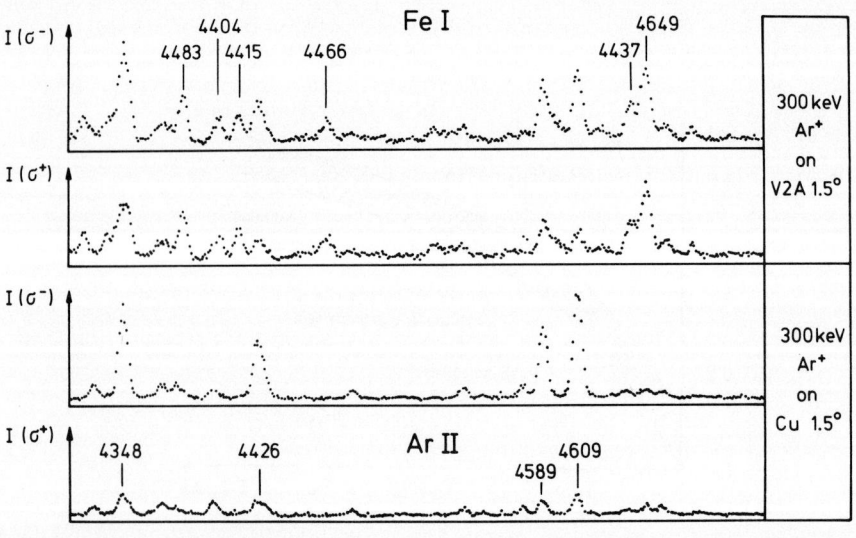

Fig. 7. Comparison of the spectra obtained with a stainless steel (V2A)- and a Cu-target with 300 keV Ar^+ ions impinging at a grazing angle of $\alpha=1.5°$.

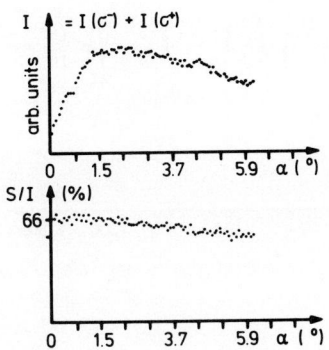

Fig. 8. Angular dependence of the total Intensity I and the normalized Stokes parameter S/I of the ArII-4609 Å-transition on the grazing angle α.

interesting hint for a possible material dependence of the polarization and therefore also a hint for a possible future use of polarization measurements in this geometry as an analytical tool in surface physics.

In Fig. 8 we show the angular dependence on the grazing angle α for the total intensity I and the normalized Stokes parameter S/I of the 4609 Å line of ArII when 300 keV Ar$^+$-ions are incident on a copper surface. S/I starts out at rather high values of 0.67 and decreases only slowly with increasing angle up to 6°. This could very crudely be interpreted as a rough agreement with the $\sin(90-\alpha)$-dependence with $\theta \simeq 2\alpha$ as observed by Berry and coworkers in tilted foil experiments (4). However, the present experiment is not as clear cut as it should be since an averaging over all scattering angle has to be taken into account in this geometry. Future experiments should therefore measure S/I as a function of $\theta-\alpha$ and could then better reveal information on the interaction process.

Further information on the collisional dynamics of the ion beam surface interaction process can be obtained from the variation of S/I with the incoming beam energy. Only one such experiment has been performed in our laboratory on H_α which has unfortunately only little to do with the rest of the data presented here. However, it reveals an interesting behaviour with a maximum around 100-120 keV which make it worth presenting here in Fig. 9. This energy dependence was the same for 3 different grazing angular settings of 1, 1.5, and 2°. Whether it is typical for all incoming ions is an open question at the moment, especially since the hydrogen fine structure levels

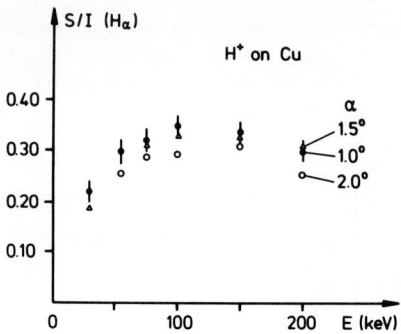

Fig. 9. Incoming H^+-beam energy dependence of S/I of the Hydrogen H_α-line.

can be easily mixed by surface electric fields via Stark effect such that the energy dependence for hydrogen maybe not at all typical for other elements. It would be of great importance therefore to measure more energy dependencies of S/I for various elements.

In order to demonstrate that the occurence of circular polarized light emission after ion beam surface interaction at grazing incidence is a general phenomenon we show a complete spectrum from 3000 to 5500 Å in Fig. 10, recorded with a line width of $\Delta\lambda = 6$ Å when 300 keV Ar^+- ions were impinging on a copper surface at 1.5° grazing angle. The experimental conditions where good with a vacuum pressure of $1.5 \cdot 10^{-6}$ Torr and a liquid nitrogen cooled shielding of the target, so that the H_β-emission was completely absent. Besides a few prominent CuI lines and a few weak CuII lines from sputtered copper atoms or ions a rather clean ArII spectrum is obtained with most lines showing a significant fraction of positive circular polarization. This general result is disturbed however, by 3 lines indicated by arrows which clearly yield a negative S/I. They therefore form a stringent test for any theoretical explanation for the various relative circular polarizations observed in the whole spectrum. Besides such low intensity lines with negative circular polarization the general finding is however a strong averaged positive circular polarization. This is exactly what was also found in all elements listed in Table 1. Thus we deal here with a very general phenomenon of circular polarization accompanying IBSIGI.

For an understanding of the anisotropic excitation process one can draw conclusions from the magnitude and signs of the measured S/I values of the spectrum in Fig. 10. For an elegant description of the polarized light emission from an atomic ensemble we choose the density matrix formalism with the density matrix expanded with respect to irreducible tensor

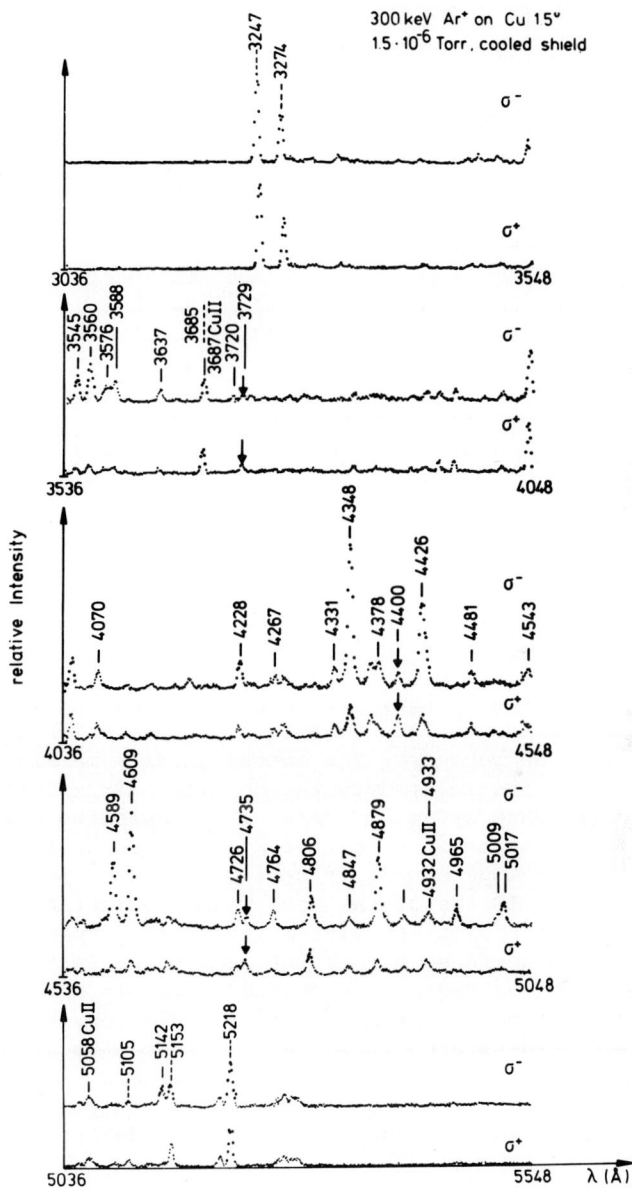

Fig. 10. Complete spectrum from 3036 to 5548 Å of the σ^- and σ^+ light emission when 300 keV Ar^+ ions are impinging on a Cu target at 1.5° grazing angle. The wavelengths together with straight lines indicate ArII-lines and together with broken lines indicate CuI-lines. CuII-lines are assigned separately. ArII-lines which exhibit negative S/I are marked by arrows.

Effect observed in:	surface quality + = "clean" 0 = rough and dirty	S/I averaged observed effect
$H^+ \longrightarrow Cu$	+	35%
$He^+ \longrightarrow C, Ta$	0	20%
$Li^+ \longrightarrow Cu$	0	20%
$O^+ \longrightarrow Cu$	0	17%
$Ne^+ \longrightarrow Cu$	+	50%
$P^+ \longrightarrow Cu$	0	10%
$Ar^+ \longrightarrow Cu$	+	50%
$Kr^+ \longrightarrow Cu$	+	50%
$Xe^+ \longrightarrow Cu$	+	40%
$Sn^+ \longrightarrow Cu$	0	20%

Table 1. Compilation of elements tested under various target and surface quality conditions. The averaged S/I values where observed on various lines in the region from 3500 to 5500 Å.

operators (12-15). Some background for the understanding of this formalism was presented by Gordon Berry during this workshop. The emitted intensity for a single multiplet component $J \rightarrow J_o$ is then given by Equ. 1 in Fig. 11. Here the $\phi_q^{(k)}$'s are the polarization density tensor components which are constants for a given geometry. The actual density matrix tensor components $\rho_q^{(k)}$ are related to the density matrix components in the standard base by Equ. 2 of Fig. 11. Assuming now LS-coupling for the beam and target atoms the spin orbit interaction is essentially turned off during the short interaction time of less than 10^{-13}s of the ions with the surface. Thus, an oriented total orbital angular momentum $\vec{<L>}$ can be generated by pure ion surface Coulomb interaction. When the ion is leaving the surface, the corresponding isotropic spin S has to be coupled to the oriented $\vec{<L>}$ in order to form all allowed fine structure eigenstates (LS)J which then can decay radiatively within a multiplet to lower states (L S)J_o. Under such circumstances the emitted light intensity is only determined by the initial irreducible spherical orbital density matrix components $^{LL}\rho_q^{(k)}$. These $^{LL}\rho_q^{(k)}$ are constrained by the reflection symmetry of the excitation process with respect to the y-z plane according to Equ. 4 of Fig. 12 when the z-axis is chosen as quantization axis (16). With Eqs. 3 and 4 it is now a straightforward procedure to calculate with the proper $\phi_q^{(k)}$'s for the given experimental geometry a closed expression for S/I in Equ. 5 of Fig. 12 which only depends on 4 unknown $^{LL}\rho_q^{(k)}$ components.

Of these four components $^{LL}\rho_1^{(1)}$ and $^{LL}\rho_o^{(o)}$ are defining

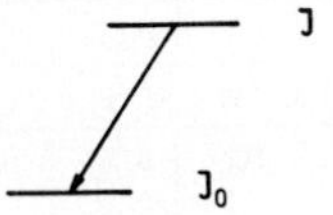

$$I \propto \sum_{kq} \begin{Bmatrix} J & J & k \\ J & J & J_0 \end{Bmatrix} \times \phi_q^{(k)} \times {}^{JJ}\rho_q^{(k)} \qquad (1)$$

$$^{JJ}\rho_q^{(k)} = \sum_{MM'} \rho_{MM'} (-1)^{J-M} \begin{pmatrix} J & k & J \\ -M & q & M' \end{pmatrix} \sqrt{2k+1} \qquad (2)$$

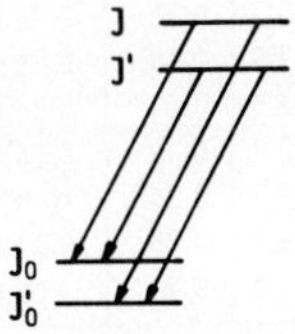

$$I \propto \sum_{kq} (-1)^k \times \begin{Bmatrix} J & J & k \\ J & J & J_0 \end{Bmatrix} \begin{Bmatrix} L & J & S \\ J & L & k \end{Bmatrix} \times \phi_q^{(k)} \times {}^{LL}\rho_q^{(k)} \qquad (3)$$

$$\rho_q^{(k)} = (-1)^{(k)} \rho_{-q}^{(k)} \; ; \; \rho_0^{(0)} \; ; \; \rho_1^{(1)} = -\rho_{-1}^{(1)} \; ; \; \rho_0^{(2)} \; ; \; \rho_1^{(2)} = -\rho_{-1}^{(2)} \; ; \; \rho_2^{(2)} = \rho_{-2}^{(2)} \qquad (4)$$

$$S/I = -\frac{\begin{Bmatrix} 1 & 1 & 1 \\ J & J & J_0 \end{Bmatrix} \begin{Bmatrix} L & J & S \\ J & L & 1 \end{Bmatrix} {}^{LL}\rho_1^{(1)}}{\sqrt{1/3} \begin{Bmatrix} 1 & 1 & 0 \\ J & J & J_0 \end{Bmatrix} \begin{Bmatrix} L & J & S \\ J & L & 0 \end{Bmatrix} {}^{LL}\rho_0^{(0)} - \begin{Bmatrix} 1 & 1 & 2 \\ J & J & J_0 \end{Bmatrix} \begin{Bmatrix} L & J & S \\ J & L & 2 \end{Bmatrix} (\sqrt{1/24}\, {}^{LL}\rho_0^{(2)} - 1/2\, {}^{LL}\rho_2^{(2)})} \qquad (5)$$

Set of parameters: $^{LL}\rho_0^{(0)} = 1$ $\quad$ $^{LL}\rho_0^{(2)} = -0.5$

$\quad\quad\quad\quad\quad\quad\quad\quad$ $^{LL}\rho_1^{(1)} = -0.5$ $\quad$ $^{LL}\rho_2^{(2)} = 0.3$ $\qquad (6)$

Fig. 11. Compilation of relevant formulae for the interpretation of the S/I-values obtained for various multiplet components.

the gross value of S/I whereas $^{LL}\rho_0^{(2)}$ and $^{LL}\rho_2^{(2)}$ only add small corrections to it.

For homologous $np^4 n'p$-configurations of three different ions Ne^+, Ar^+, Kr^+ an attempt was made to describe all observed S/I_{exp} by a single crudely fitted set of $^{LL}\rho_q^{(k)}$-components. This set is given in Equ. 6 of Fig. 11 and leads to the theoretical S/I_{th} in Table 2 for various multiplets of these ions. The experimental S/I_{exp}-values are compared with these S/I_{th} in the neighbouring column of Table 2. Besides such cases where blending uncertainties add to the normally given counting statistical errors an excellent agreement is found between our simplified theory and experiment including the three mentioned negative S/I_{exp}-values. This allows us to conclude that
  a) our description must be correct at this level of accuracy,
  b) $^{LL}\rho_1^{(1)} = -0.5$ implies for all excited states of Table 2 the generation of a unique orbital angular momentum $<-L_x>$ exactly in accord with the picture in Fig. 2b, d and
  c) that therefore the IBSIGI can be understood as a pure orbital angular momentum effect as long as LS-coupling is a good approximation.

J_0	J	λ [Å]	I_{th}	S/I_{th}	S/I_{exp}
ArII [^{2}D] 4s' ^{2}D - 4p' ^{2}F					
5/2	7/2	4609.56	100	0.57	0.76 (4)
5/2	5/2	4637.23	5	0.27	0.17 (7)
3/2	5/2	4589.90	70	0.58	0.72 (4)
ArII [^{3}P] 4s ^{2}P - 4p ^{2}D					
3/2	5/2	4879.86	100	0.58	0.65 (4)
3/2	3/2	4726.86	11.1	0.38	0.33 (5)
1/2	3/2	4965.07	55.6	0.60	0.64 (4)
ArII [^{3}P] 4p ^{2}D - 4d ^{2}F					
5/2	7/2	3559.51	100	0.57	0.58 (5)
5/2	5/2	3656.05	5	0.26	0.26 (10)
3/2	5/2	3545.60	70	0.58	0.60 (5)
ArII [^{3}P] 3d ^{4}D - 4p ^{4}D					
7/2	5/2	4400.99	100	-0.38	-0.27 (10)
ArII [^{3}P] 4s ^{4}P - 4p ^{4}P^0					
5/2	5/2	4806.02	100	0.20	0.15 (3)
5/2	3/2	4735.91	42.9	-0.18	-0.14 (4)
3/2	5/2	5009.33	42.9	0.54	0.50 (10)
3/2	3/2	4933.21	12.7	0.10	-
3/2	1/2	4847.82	39.7	0.14	0.10 (3)
1/2	3/2	5062.04	39.7	0.35	blended
1/2	1/2	4972.16	7.9	-0.29	-

J_0	J	λ [Å]	I_{th}	S/I_{th}	S/I_{exp}
NeII [^{1}D] 3s' ^{2}D - 3p' ^{2}F					
5/2	7/2	3568.53	100	0.57	0.58 (4)
5/2	5/2	3574.23	5	0.28	-
3/2	5/2	3574.64	70	0.58	0.58 (4)
NeII [^{3}P] 3s ^{4}P - 3p ^{4}P^0					
5/2	5/2	3694.20	100	0.20	0.11 (3)
5/2	3/2	3664.11	42.9	-0.18	-0.11 (3)
3/2	5/2	3766.29	42.9	0.54	0.36 (5)
3/2	3/2	3734.94	12.7	0.10	0.08 (4)
3/2	1/2	3709.64	39.7	0.14	blended
1/2	3/2	3777.16	39.7	0.35	0.17 (5)
1/2	1/2	3751.26	7.9	-0.29	-0.08 (4)
KrII [^{1}D] 5s' ^{2}D - 5p' ^{2}F					
5/2	7/2	4577.20	100	0.57	0.56 (5)
5/2	5/2	4691.28	5	0.27	-
3/2	5/2	4633.88	70	0.58	0.38 (15)
KrII [^{3}P] 5s ^{4}P - 5p ^{4}D^0					
5/2	7/2	4355.47	100	0.56	0.41 (5)
5/2	5/2	4301.53	22.5	0.17	0.09 (4)
5/2	3/2	3912.59	2.5	-0.30	-
3/2	5/2	4765.74	52.5	0.51	0.49 (4)
3/2	3/2	4292.92	26.7	0.20	0.15 (4)
3/2	1/2	3987.78	4.2	-0.25	-
1/2	3/2	4811.76	20.8	0.50	0.18 (10)
1/2	1/2	4431.67	20.8	0.50	0.28 (10)

Table 2. Measured normalized Stokes parameters S/I_{exp} of ArII, NeII, and KrII multiplets with 300 keV ions impinging on a Cu-target at 1.5° grazing angle and comparison with theoretical values S/I_{th}. J represents the total electronic angular momentum of the upper and J_0 of the lower state of a multiplet.

This latter assumption is, however, already disturbed by the results shown in Table 3. According to our theoretical description ^{4}S-levels should not be oriented at all in LS-coupling. The experimental values including the largest negative S/I_{exp} found in the spectrum of Fig. 10 obviously contradict this statement. The only way out of this puzzle, we can offer at the moment, is a description of the ^{4}S-levels by jl-coupling (17). We assume that only the outermost p-electron is oriented by IBSIGI whereas the ^{3}P-core stays isotropic. One obtains then a set of 11 $\rho^{(k)}$-components for this particular electron with which we calculate S/I in jl-coupling. Only with a truncated and crudely adjusted set 11 $\rho_0^{(o)}$ = 1, 11 $\rho_0^{(1)}$ = -.29, we obtain the theoretical gross values S/I_{th} in the last column which favorably compare with the measured values. One should, however, mention that such a strong jl-coupling contribution is very surprising for these ^{4}S-levels (18). Therefore

J_0	J	λ [Å]	I_{th}	S/I_{th}	S/I_{exp}
Ar II [^{3}P] 4s ^{4}P - 4p ^{4}S^0					
5/2	3/2	3729.31	100	-0.13	-0.16 (6)
3/2	3/2	3850.58	66.6	0.09	0.08 (4)
1/2	3/2	3928.63	33.3	0.21	0.20 (5)
Kr II [^{3}P] 5s ^{4}P - 5p ^{4}S^0					
5/2	3/2	3460.09	100	-0.13	not measured
3/2	3/2	3754.24	66.6	0.09	0.06 (4)
1/2	1/2	4145.12	33.3	0.21	0.16 (8)

Table 3. Measured S/I_{exp} of $^4P-^4S$ multiplets, which should exhibit $S/I_{exp} = 0$ in pure LS-coupling, and comparison with S/I_{th} obtained with a calculation in jl-coupling.

other mechanisms such as the partial breakdown of LS-coupling during the collision or a Fano-Lichten type collision with a single surface atom (19) cannot be fully excluded and future studies concerning this problem may be very fruitful for a deeper understanding of the interaction mechanism.

The large orientations obtained with IBSIGI suggest its application in level-crossing and magnetic resonance studies of excited atomic or ionic levels. In order to demonstrate such an application we show in Fig. 12 the zero-field level-crossing (Hanle-effect) signals obtained for the 4p $^2F_{7/2}$-state in the 4609 Å-emission of ArII. These curves were obtained by applying a magnetic flux $\vec{B}_z$, along the average scattered beam direction z' in Fig. 3 with Helmholtz-coils and sweeping it proportional to time at less than ± 5% beam fluctuations with simultaneous recording of photons of a particular polarization. It is easy to show that the intensities of B = 0 and B → ∞ again depend on a set of four $_{LL}\rho_q^{(k)}$-components as derived before. With the same set as deduced from the measured S/I-values of ArII we can therefore account for the Hanle-effect intensities reasonably well with the expressions given in Fig. 12. In principle, the width of such Hanle signals can also yield a value to the lifetime τ of the excited state. However, the finite observation time due to the unknown scattered ion velocity and the observation window leads to uncontrollable broadening (factor ~2 in Fig. 12) of the unperturbed line shape, so that in the present geometry lifetime determinations are not possible.

The positions of high field level-crossings on the other hand are not affected by such broadenings. Therefore this new technique can easily be extended to high field level-crossing measurements of fine and hyperfine structures of ions and atoms and offers a number of advantages owing to the indestructible nature of the targets, the large orientation,

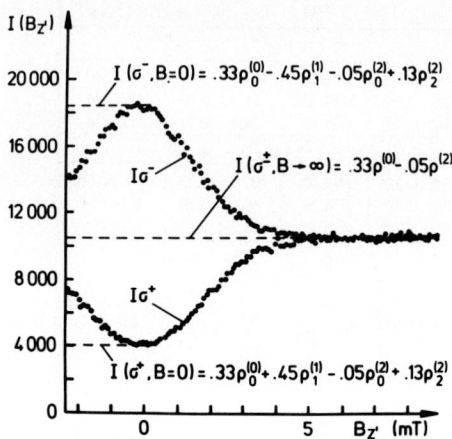

Fig. 12. Zero-field level-crossing signals for $I(\sigma^-)$ and $I(\sigma^+)$ observed in the ArII-4609 Å emission.

and the large magnetic fields which can be used along z' with minimal beam bending.

After this presentation of experimental data and their limited interpretations we should add some remarks on our present understanding of the IBSIGI. Since we deal with a general phenomenon one can expect that basic symmetry principles are governing the process. We therefore start with a discussion of the symmetries of the various excitation geometries with the z-axis as quantization axis.

As an excitation with the highest symmetry one can consider a gas discharge. In such a situation electrons are impinging from all directions isotropically on the atom to be excited (Fig. 13a). This we call an "isotropic excitation" and consequently the emitted light is completely unpolarized. In terms of the $\rho_q^{(k)}$-components this means that only $\rho_0^{(0)} \neq 0$.

If one moves now to the excitation of the atomic ensemble by a directed beam of particles, i.e. electrons, ions, or equivalently a beam of ions which is excited by traversing a foil, then the isotropy of space is disturbed for the atoms to be excited. The symmetry is therefore lowered to rotational symmetry around the incoming beam axis and to reflection symmetry with respect to any plane containing this axis (16). We call this an "aligned excitation" which can lead to linearly polarized or unpolarized light emission. In terms of the $\rho_q^{(k)}$-components this means that only components with k = even can be unequal from zero. In Fig. 13b we give a simplified and figurative explanation for the elementary excitation of a

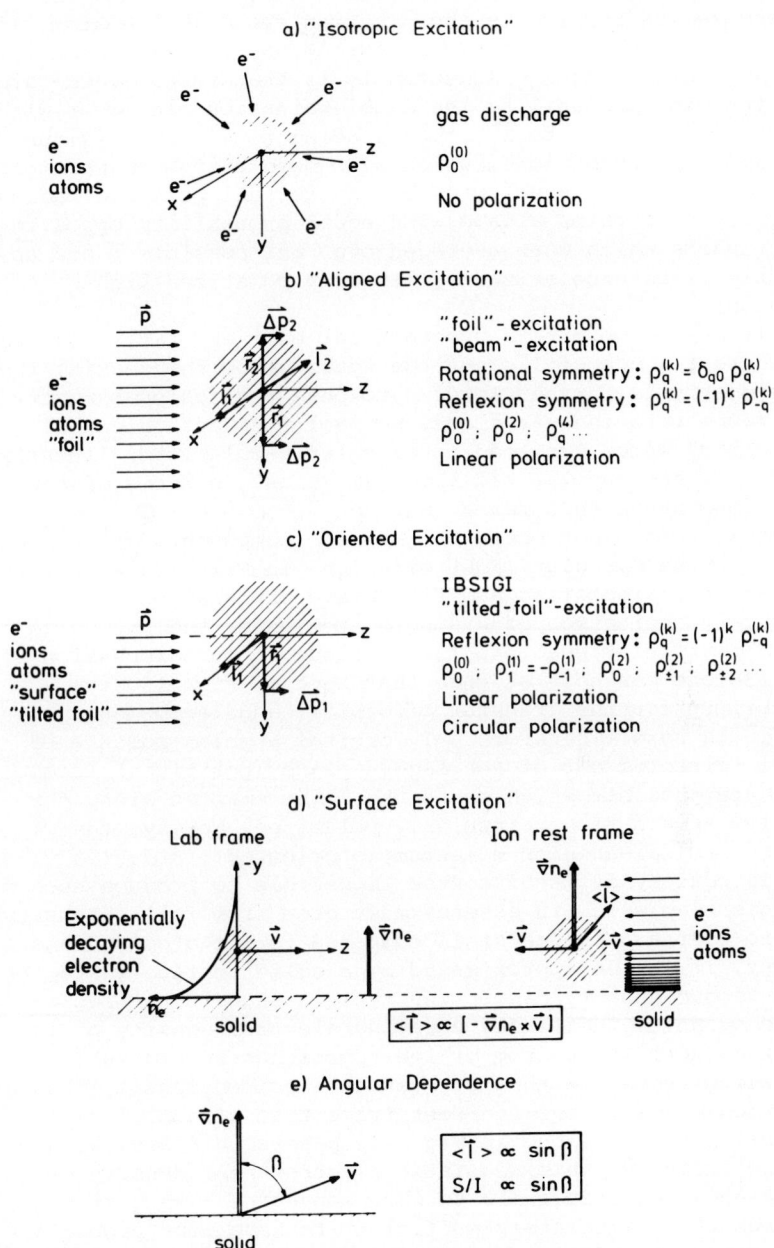

Fig. 13. Figurative explanation of IBSIGI via symmetry considerations and approximation of the surface by an exponentially decaying electron density.

single atom. The nucleus is located at the origin and the hatched region represents the electron cloud of the atom. If we now consider a momentum transfer $\vec{\Delta p}_1$ at radius vector $\vec{r}_1$, then an orbital angular momentum $\vec{l}_1$ is transfered to the atom. With the same probability the same excitation can occur with $\vec{\Delta p}_2$ and $\vec{r}_2$ opposite to $\vec{r}_1$, thus leading to $\vec{l}_2$. That is, under such symmetry conditions, which also mean that beam particles pass the atom with equal probability for all possible radius vectors, one obtains always with equal probability opposite $\vec{l}_1$, $\vec{l}_2$ pairs which when averaged over all possible $\vec{r}$ and $\vec{\Delta p}$ can only yield unpolarized or linearly polarized light emission.

If we now move to a hypothetical beam excitation in Fig. 13c where the beam fills out the space below the z-axis, then the symmetry is further lowered to pure reflection symmetry with respect to the y-z plane. We call this an "oriented excitation" which can lead to circularly-polarized, linearly-polarized, and unpolarized light emission. In terms of the $\rho_q^{(k)}$-components this means that $\rho_o^{(o)}$, $\rho_1^{(1)} = -\rho_{-1}^{(1)}$, $\rho_q^{(2)}$.. components can occur where the k = 1 - components are responsible for the circular light emission. In Fig. 13c we try to give again a simplified and figurative explanation for the elementary excitation of a single atom under the influence of such a particular beam. In full analogy to our discussion of Fig. 13b one can note at once that here in Fig. 13c the probability for the transfer of orbital angular momenta of type $\vec{l}_2$ is zero. Therefore only orbital angular momenta of type $\vec{l}_1$ directed out of the figure plane (positive l_x-component) are possible which cause the light emitted along the positive x-axis to be circularly polarized. We say, the average orbital angular momentum is oriented.

In reality it is of course impossible to produce such a particle beam which in essence must drop from full intensity to zero over a distance small compared to the atomic diameter. However, it can be approximated by a solid surface where the solid-vacuum density discontinuity is assumed to be represented by an exponentially decaying electron density n_e in Fig. 13d (20). If an atom or ion approaches such a surface at grazing incidence then it sees an electron density gradient $\vec{\nabla} n_e$ which means in the ion rest frame that a beam of electrons very similar to the one in Fig. 13c passes this atom with beam velocity $-\vec{v}$. This realistic electron beam density of course does not drop fully to zero above the z-axis in Fig. 13c, but it is absolutely sufficient to have less probability for passage of particles above the z-axis compared to passage below the z-axis in Fig. 13c in order to obtain an oriented averaged orbital angular momentum. In Fig. 13d we show this type of realistic beam seen by the ion in its rest frame with the surface represented by a mixture of electrons, ions and atoms. The surface density gradient $\vec{\nabla} n_e$ thus determines a unique direction (parallel to the surface normal) for the interaction process which determines via the vector product $-[\vec{\nabla} n_e \times \vec{v}] = \langle \vec{L} \rangle$ directly the direction of orientation of the orbital angular momentum in the ion rest frame which corresponds to $[\vec{\nabla} n_e \times \vec{v}] = \langle \vec{L} \rangle$ in the lab-frame.

This formula now dominates any detailed interaction mechanism one may think of and therefore directly explains the sinβ-dependence of S/I observed in tilted-foil experiments where the outgoing velocity vector $\vec{v}$ in Fig. 13e is assumed to be responsible for the interaction geometry. These interaction mechanisms include direct electron collisional excitation, direct electron capture from the fermi sea, direct inelastic electron capture of electrons extracted from the surface by the image force, resonant or nonresonant charge transfer, collisional interaction with surface atoms, or Fano-Lichten type collisions with surface atoms (19). In all cases the dynamic description will lead to an expression proportional to the above vector product and thus readily explains our large orientations observed. At present it is of course a completely open question which of the above mechanisms is the most important one under the various experimental conditions. Therefore a large number of systematic measurements under controllable and reproducible surface conditions is urgently needed for the more detailed understanding of the elementary processes taking place during the interaction. Such measurements may also give an answer as to whether IBSIGI in combination with S/I-observation can be developed to a new analytical tool in surface physics or not.

We wish to thank R. Burghardt and J. Hensel for their skilled technical assistance, and E. Kupfer and H. Schröder for several clarifying discussions. We also enjoyed the collaboration with J.D. Silver during some of the experiments presented here. This work is supported by the Sonderforschungsbereich 161 der Deutschen Forschungsgemeinschaft.

REFERENCES

1. Andrä, H.J., Phys. Lett. 54A, 315 (1975)
2. Andrä, H.J., et al., Phys. Rev. Lett. 37, 1212 (1976)
3. Andrä, H.J., Physica Scripta 9, 257 (1974)
4. Berry, H.G., in "Beam-Foil Spectroscopy" (I.A. Sellin and D.J. Pegg, Eds.), Vol. 2, p. 755. Plenum Press, New York 1976.
5. Rau, C., and Sizmann, R., in "Atomic Collisions in Solids" (S. Datz, B.R. Appleton, and C.D. Moach, Eds.), p. 295. Plenum Press, New York, 1975.
6. Born, N., and Wolf, E., in "Principles of Optics", 4th ed.,

p. 30, p. 554. Pergamon Press, Oxford, 1970.
7. Behrisch, R., et al., in "Atomic Collisions in Solids" (S. Datz, B.R. Appleton, and C.D. Moack, Eds.), p. 315. Plenum Press, New York, 1975.
8. Sigmund, P., Phys. Rev. 184, 383 (1969)
9. Chenney, K.B., and Pitkin, E.T., J. Appl. Phys. 36, 3542 (1965)
10. Evdokimov, I.N., and Molchanov, V.A., Can. J. Phys. 46, 779 (1968)
11. Townsend, P.D., Kelly, J.C., and Hartley, N.E.W., "Ion Implantation, Sputtering and their Applications", chapter 6. Academic Press, London, 1976.
12. Happer, W., Rev. Mod. Phys. 44, 169 (1972)
13. Gabriel, H., and Bosse, J., in "Proc. Int. Conf. on Angular Correlation in Nuclear Disintegration", Delft 1970 (H. v. Krugten, B. v. Nooijen, Eds.) p. 394. Rotterdam Univ. Press, Groningen, 1971.
14. Bosse, J., and Gabriel, H., Z. Phys. 266, 283 (1974)
15. See appendix of ref. 3.
16. Ellis, D.G., J. Opt. Soc. Am. 63, 1232 (1973)
17. Sobel'man, I.I., "Introduction to the Theory of Atomic Spectra", p. 185. Pergamon Press, Oxford, 1972.
18. Minnhagen, L., Ark. Fys. 25, 203 (1963)
19. Fano, U., and Lichten, W., Phys. Rev. Lett. 14, 627 (1965)
20. Schröder, H., and Kupfer, E., Z. Phys. 279, 13 (1976)

index

A

A/a, 14, 43, 108, 116, 189, 190, 210, 217–225
Adsorbed species, 73, 105, 153, 154, 171, *see also* Impurities
Alignment, 310–347, *see also* Optical polarization
Angular dependence
  of backscattering, 31, 36, 44, 66, 69, 73, 74, 84, 90, 105, 110–112
  of sputtered particles, 122, 140, 149, 157, 169, 170
  neutral fraction, 270
Auger processes, 1–25, 273–277, *see also* Electron emission, Neutralization
Autoionization, 22–24, 38, 39, 100, 153, 186

B

Backscattering, 28, 33, 36, 44, 47–119
Backscatter radiation, 203, 214–225
  effect of oxygen, 224
  from H^-, 225
  from He^+, 214–225
  from H^+, H_2^+, and H_3^+, 214–225
Beam foil excitation, 310–326
Beam foil experiments, 300, 310–326
Binary collisions, 27, 28, 74, 105, 106, 111, 113
Binding energy, 261
Bond breaking, 208, 214
Bound states, 260–270
Bremsstrahlung, 233, 235, 236

C

C/I
  definition, 314, *see also* Stokes parameters, Alignment, and Optical polarization
  dependence on ion current density, 335
Cathodoluminescence, 235

Cascade (collision), 121, 129–146, 150, 155, 191
Capture cross sections, 255, 272
Channeling, 66, 68, 84, 154, 169–170, 209, 243, 276
Charge density, 260
Charge equilibrium length, 266
Charge fractions, *see* Ion fractions
Charge state, 29, 48–66, 68, *see also* Ion fraction
Charge states, 253–308
  passing through gases, 265–268
  passing through solids, 258–264, 269–280
Clusters, 146–148, 153, 158–164, 167, 174–181, 192
Collisional excitation, 201–252, 310–347
  aligned excitation, 344
  isotropic excitation, 344
  oriented excitation, 345
Computer simulation, 31, 33–36, 62, 63, 107, 132, 148
Continuum radiation, 182, 202, 203, 237–248
  from Mo, 237–243
  from Nb, 237–239
  from Ta, 237–238
  from W, 237–238
Cusp in energy distribution of ejected electrons, 284, 289, 295

D

De-excitation, 1–25, 30, 39, 41, 99, 118, 189,
  Auger, 1–25, 99, 192, 214–225
  resonance, 1–25, 212, 214–225
Density of electron states, 10–13, 16, 22
Density matrix, 338
Depth
  distribution of energy, 134–136, 138
  particle penetration, 48, 51, 53, 142, 143, 154, 171–173
Dissociative excitation, 213
Dissociative ionization, 306
Doppler broadening, 14, 182, 203, 205, 208, 215, 217

E

Electric dipole moment, 310
Electric dipole radiation, 312–215
Electron capture, 50, 64, 253–308, 272, 318
Electron ejection, 283–308
　into continuum states, 285
　with H^+, H_2^+, 290–298, 301–308
　with He^+, 290–291, 300, 304
　with He^{2+}, 290
　with N^+, 290
Electron emission
　Auger, 1–25, 44, 153, 155, 174, 182, 184, 191, 210
　kinetic, 12, 44, 210
Electron excitation, 310
Electron-hole recombination, 28, 225, 233
Electron pickup, *see* Electron capture
Electron promotion, 208
Electron recombination, 272
Electron screening, 47, 50, 51, 108
Electronic transition, 1, 2, 20
Emission functions, 207
Energy distribution
　electron, 5, 8–24, 40–42, 153, 191
　ion, 30–38, 52–53, 58–66, 73–77, 105, 156–166, 179, 190, 191
　neutral, 52, 53, 58–66, 122, 125, 126, 138, 140–142, 146–148, 156, 179, 190
Energy loss processes, 261
Equilibrium charge states, 269, 276
Excitation, 28, 30, 35, 37, 38
　outer-shell, 35, 201–252, 253
　inner-shell, 165, 183
Excitation recombination, 204

F

Fractional polarization, 318
Franck–Condon principle, 9

G

Glancing angle reflection, 273, *see also* Grazing incidence collisions
Grazing incidence collisions, 329–347
Grazing incidence excitation, 329–347
　angular dependence, 337
　with Ar, 331–344
　with H, 338
　with He, 331, 340

H

Hanle effect, 343
Hyperfine structure, 315, 343

I

Image force, 9, 10, 13, 24
Implantation, 153, 155, 164, 170–173
Impurities, 153, 158, 170–181
Interference, quantum mechanical, 110
Ion fraction, 48, 51–66, 69, 254, *see also* Charge states
Ionization, 4, 5, 17, 19, 41, 68, 69, 127, 131, 142, 155–164, 169, 170, 186–191, 210
　inner shell, 64
　kinetic, 153, 187, 188
　resonance, 4, 5, 17, 19
Ion neutralization spectroscopy (INS), 1–25, 273
Ion-surface scattering (ISS), 48, 73, 105
Ionization energy, effective, 8, 14
Isotope effect, 88–90

L

Larmor frequency, 316
Level crossing, 329, 332, 343
Line intensity, 219–221
Linear response theory, 256
Loss cross sections, 255
Luminescence, 203, 225–235
　from alkali–halide crystals, 226–233
　from CaF_2, 225
　from MgF_2, 226
　from metals, 233–235

M

M/I, definition of, 314, *see also* Stokes parameters, Alignment, and Optical polarization
Molecular bombardment, 222–225
Molecular excitation, 203, 204, 237–248
　in front of surface, 237–245
　on surface, 204, 245–248
Metastable species, 15, 18–24

Index   351

N

Near-resonant charge exchange, 73, 86–90, 98, 109, 113, 116
Negative ion, 30, 62, 153, 156, 158
Neutral fraction, 258, 261
Neutralization
  Auger, 1–25, 29, 38–44, 64, 68, 69, 80, 105–108, 113, 116
  nonadiabatic, 105, 106, 113–118
  radiative, 2
  resonance, 1–25, 41, 43, 64, 80, 105–108, 210

O

Optical polarization, 233, 236, 309–348
  circular, 310–347
  elliptical, 310–347
  foil material dependence, 324
  linear polarization, 310–347
  measurement, 315
Orientation, 310–326, 329–347, *see also* Optical polarization
Oscillatory scattering yields, 67, 73–119
  with He^+–Bi, 81, 82, 86, 116–118
  with He^+–Ga, 79–84, 96–99, 111–113, 116, 117
  with He^+–Ge, 81, 82, 99, 116–118
  with He^+–In, 81, 82, 85–95
  with He^+–Sb, 81, 82, 85, 99
  with He^+–Sn, 81, 82, 85, 99
  with He^+–Pb, 81, 82, 86, 116, 117

P

Phase interference effects, 264
Photon emission, 28, 101, 153, 175, 181–184, 190, 192, 201–252
Plasma–wall interaction, 48, 51
Plasmons, 233, 235–237, 258
Polarization, *see* Optical polarization
Probability
  survival, 6, 14, 19, 40, 43, 76, 106, 107, 110, 114
  transition, 5, 6, 10, 11, 106, 107

Q

Quantum beats, 311, 316, 322–324
Quasimolecular states, 208

R

Radiation damage, 121, 127
Radiationless de-excitation, *see* De-excitation
Resonant tunneling, 272, 274–277

S

S/I, definition of, 314, *see also* Stokes parameters, Orientation, and Optical polarization
Screening, dielectric, 256, 260, 274, *see also* Electron screening
Secondary ion mass spectrometry (SIMS), 154, 155, 158, 177, 214
Secondary ion production, 142, 153–199
Spike (thermal), 121, 128–146, 148
Spin-orbit coupling, 116
Sputter radiation, 203–214
  from Al, 213
  from Cu, 204, 209, 213
  effect of oxygen, 212, 214
  from Mg and MgO, 212
  from molecules, 237–248
  from Ni, 204
  from Si and SiO_2, 206
Sputtering, 121–152, 153–199
Sputtering yield, 123, 124, 126, 130, 131, 137, 139, 145, 149, 155, 156, 167–169
Stark mixing, 318–319
Stokes parameters, definition of, 314
Stopping, 27, 28, 31, 32, 61
  electronic, 27, 28, 31, 32, 61, 124, 127, 128, 221
  nuclear, 28, 32, 61, 124, 125, 127, 128, 136, 137, 143
Surface composition analysis, 203–214
  by ion emission, *see* Secondary ion mass spectrometry (SIMS)
  by light emission (SCANIIR), 204, 214, 337
Survival probability, 211

T

Temperature, effective collision, 208
Thermodynamic equilibrium, 208
Thin foils, 32, 47–49
Thomas–Fermi screening, 257
Tilted foil excitation, 311–326, 330–331
  with Ar, 317, 320–325, 331
  with He, 311, 315, 320–326, 331

Tilted foil excitation (*cont.*)
  with Ne, 319–322
  with O, 317
Transition radiation, 233, 235–236
Tunneling, 2, 17, 22, 29, 43, 44, 50, 80

V

Velocity distribution, 208
  of sputtered particles, 208

W

Wake riding states, 287, 305
Work function, 213, 218

X

X-Ray emission, 183, 184